Study and Problem Solving Guide
to accompany

PRINCIPLES OF MODERN CHEMISTRY

Oxtoby / Nachtrieb

WADE A. FREEMAN
University of Illinois at Chicago

Saunders Golden Sunburst Series
SAUNDERS COLLEGE PUBLISHING
Philadelphia New York Chicago
San Francisco Montreal Toronto
London Sydney Tokyo Mexico City
Rio de Janeiro Madrid

Printed in the United States of America

Study and Problem Solving Guide to accompany PRINCIPLES OF MODERN CHEMISTRY

ISBN # 0-03-070654-8

78 066 987654

FOREWORD

The textbook *Principles of Modern Chemistry*, by David Oxtoby and Norman Nachtrieb, presents a thorough introduction to University chemistry in three major parts:

- The first nine chapters cover the macroscopic principles;

- The next six chapters cover modern theories of atomic structure and chemical bonding;

- The final seven chapters apply the principles of the first fifteen chapters to an organized survey of descriptive chemistry with emphasis on important chemical problems and trends.

Each chapter ends with a list of concepts and skills and an extensive problem set. The problems sets are organized as a series of *paired problems*. Problems in pairs treat the same material. If you can solve one of them you should be able to solve the other. They are followed by individual or *unpaired* problems which draw on the concepts and skills covered in the chapter both singly and in combinations. **You cannot expect to succeed in your chemistry course without working the problems.**

Your degree of success in learning chemistry depends on how effectively you comprehend the text and work the problems. This Guide has been written to help you do both more effectively.

Each numbered chapter in this book corresponds to the same numbered chapter in the text. The first part of each chapter is a review of concepts and skills with particular emphasis on the *type* and *point* of the problems that arise from them. Concepts are referenced to the problems that use them.

In the second portion of each chapter, this book gives detailed solutions not to just the odd-numbered paired problems but to *all* the odd-numbered problems. The solutions mention alternative avenues of attack on the problems and point out common pitfalls.

The serious student will try to solve the odd-numbered paired problems on his or her own, reviewing concepts and checking results against the answers in this Guide. The even-numbered paired problems can then be used as a test. Finally, the unpaired problems will allow a final honing of skills. The answers and details of solution of about half of these problems, as given here, will provide a valuable reference and check.

I am grateful to Norman Nachtrieb, David Oxtoby and Sally Freeman for reviewing and editing this manuscript. My thanks also go to Ritu Bala and Virginia Yip for checking the problem solutions and to Victoria Rias for her invaluable help in preparing the manuscript.

Wade A. Freeman

CONTENTS

Chapter 1

CHEMICAL STOICHIOMETRY AND THE ATOMIC THEORY OF MATTER

Chapter 1 presents the fundamental concepts of stoichiometry (chemical arithmetic) as they developed historically.

1.1 *CHEMISTRY AND THE NATURAL SCIENCES*

Chemistry occupies a central role among the natural sciences. Its many areas of overlap with other fields provide fertile sources of new discoveries.

Progress in chemistry and other sciences involves a blend of theory and experiment. The formulation of theories is the construction of models to explain experimental results. Such models are always subject to revision in the light of newly discovered facts. Theories are always held provisionally. Experimentation amounts to asking nature a carefully-crafted question. A good experiment gets an unambiguous reply. Experimental facts, once verified by repetition, endure.

1.2 *THE COMPOSITION OF MATTER*

Substances and Mixtures

Mixtures consist of two or more pure substances that can in principle be separated from each other by ordinary physical means. Mixtures are *homogeneous* (having properties that are uniform from region to region throughout the sample) or *heterogeneous* (having identifiable regions with different properties). If all efforts by physical means to separate a material into portions having different properties fail, then the material is a pure *substance.*

Analysis and synthesis are the poles of a major dichotomy in chemistry. Analysis means taking things apart; synthesis means putting them together. Chemists do both. An analytical chemist separates the component substances in mixtures and then subjects these substances to elemental analysis. A synthetic chemist constructs new substances that have theoretical or practical interest and subjects them to analysis to verify their composition.

Elements

Elements are substances that cannot be decomposed into two or more simpler substances by ordinary physical or by ordinary chemical means. Compounds are substances that contain two or more different elements.

Every element has a name and a symbol. The symbol consists either of a capital letter or a capital letter followed by a lower case letter.

¤ *It is essential to learn the correctly spelled names of the elements and their symbols in proper correspondence.*

This chore of memorization cannot in general be completed overnight. It is less tedious if it is done in conjunction with a study of the structure of the periodic table of the elements (Chapter 2). It is wise to learn the names and symbols of the *common* elements first.

1.3 *ATOMIC THEORY OF MATTER*

The Law of Conservation of Mass

Matter is neither created nor destroyed in chemical reactions. Reactions in which the products appear to weigh more or less than the reactants typically have picked up a reactant from the surroundings (like oxygen from the air) or evolved a gas into the surroundings. When proper experimental design eliminates these confounding factors the law of conservation of mass always holds for chemical processes. The law of conservation of matter can be of real help in practical problem-solving. **See problems 23 and 33.**

The Law of Definite Proportions

The law of definite proportions states that in a pure chemical compound the proportions by weight of the constituent elements are fixed and definite, independent of the history of the sample.

The composition of some solids, *non-stoichiometric* compounds, violates the law of definite proportions. See Chapter 15. Thus the solid compound of nominal formula $K_2Pt(CN)_4$ can, depending on its method of preparation, be isolated as $K_{1.3}Pt(CN)_4$. The law of definite proportions, however, applies rigorously to all gaseous and liquid compounds.

A *chemical formula* specifies unequivocally the definite composition of a compound. It tells the number and kind of atoms in a compound. Common problems are:

- To calculate the weight percent of an element in a compound from the chemical formula. The answer always comes from multiplying the ratio of the atomic weight of the element in question to the molecular weight of the compound by the number of atoms of interest and then converting to percent.

Example: Calculate the weight percent of oxygen in WO_3.

Solution. The molecular weight of WO_3 is 183.85 + 3(15.9994) or 231.85. The weight *fraction* of oxygen is 3(15.9994) / 231.85 or 0.20702. The weight percent of oxygen is 100 times larger: 20.702 percent. It is not necessary to know what the symbols stand for in the formula to do this problem. All that is required is finding the symbols on a periodic table and reading off the atomic weights.

- To calculate the weight percent of some *group* of elements in a compound. For example, to calculate the weight percent of water in $CuSO_4 \cdot 5H_2O$. The "·5" in this formula means that it contains 5 molecules of loosely-bound water. This is an elaboration of the first type of problem. **See problem 3b.**

The Atomic Theory of Dalton

Atomic theory postulates a lower limit to the subdivisibility of matter. Dalton called this lower bound of matter's graininess the *atom.* Acceptance of the law of definite proportions obliges acceptance of such a view. **See problem 37** for an application of the law of definite proportions.

The Law of Multiple Proportions

Often two elements can combine in different proportions to give more than one compound. When this occurs, the masses of one element that combine with a fixed mass of the other stand to each other in the ratio of small whole numbers. The four oxides of vanadium (VO, V_2O_3, VO_2, V_2O_5) are an excellent example of the law of multiple proportions. See the solution to **problem 1** for an illustration of the law.

The Law of Combining Volumes

Joseph Gay-Lussac established experimentally that when two gases react the volumes that combine stand in the ratio of simple integers (if the two are at the same temperature and pressure). The theme of small whole-number ratios introduced with the law of definite proportions recurs in Gay-Lussac's work. **See problem 15.**

Avogadro's Hypothesis

Avogadro offered a theoretical explanation of Gay-Lussac's results. He hypothesized that equal volumes of different gases at the same conditions contain equal numbers of particles. Thus, 1 L of oxygen contains the same number of molecules as 1 L of argon provided the volumes are measured under the same conditions.

Avogadro's hypothesis affords a method for the determination of the relative weights of the molecules of gases. The vapor densities of gaseous compounds at any specified conditions of temperature and pressure are directly proportional to their molecular weights. Vapor density is the mass of a quantity of gas divided by its volume. **See problems 3, 15 and 39** for applications of Avogadro's hypothesis.

1.4 *ATOMIC AND MOLECULAR WEIGHTS*

The absolute masses of atoms cannot be measured directly. Fortunately it is sufficient in chemistry to know the relative masses of atoms on a scale in which a particular element is assigned an arbitrary value.

The internationally accepted scale today assigns ^{12}C atoms an arbitrary relative mass of exactly 12. ^{12}C refers to an *isotope* (see below) of carbon. The periodic table of the elements tabulates relative atomic weights on this scale. Relative atomic weights are ratios and therefore do not have units. Once a reference standard is assigned, several methods are available for determining relative atomic weights.

Canizzaro's Method for Relative Atomic Weights

This method is based on the Avogadro's hypothesis. It is limited to gases. An arbitrary molecular weight is assigned to one gas. Then, the densities of that gas and of others at set conditions of temperature and pressure are tabulated. The ratio of the density of any gas on the list to the density of the reference gas is identical to the ratio of the molecular weights of the two. **See problem 3.**

The volume of one mole of an ideal gas at standard conditions of temperature and pressure is 22.414 L. Standard conditions of temperature and pressure are 0° C and 1 atm. This number, the *STP molar volume*, can exert an unhealthy fascination for some who force it into improper applications in solving problems. 22.414 L is a *gas* volume and has no application to liquids or solids. 22.414 L approximates the volume of *one* mole of gas *only* under specific conditions. Even then, 22.414 L is *still* approximate; the STP molar volumes of *real* gases vary slightly. Some results of careful measurements on various gases:

Gas	Volume of 1.000 Mole at STP
Ar	22.401 L
H_2	22.410
He	22.398
N_2	22.413
Ne	22.430
O_2	22.4137

In view of these variations among common gases it is a waste of effort to use the STP molar volume to more than three significant digits, 22.4 L, in applications.

The law of Dulong and Petit allows determination of the relative atomic weights of solids. It states that the *molar heat capacities* of solid elements are all approximately the same: 25 J $mol^{-1}K^{-1}$. This heat capacity figure means that it takes 25 joule of heat to raise the temperature of 1 mol of the element by 1 K. The law holds within about six percent for solid elements beyond the second row of the periodic table. Because the *specific heat capacities* of many solid elements are readily measured, (typically in J $g^{-1}K^{-1}$) the law of Dulong and Petit provides an easy estimate of molar weights. Note that the units of molar heat capacity divided by the units of specific heat capacity are g mol^{-1}, proper units for molar weight.

Careful elemental analysis of pure compounds would by itself give the relative atomic weights of the constituent elements, *if* the number of atoms of each kind of element in the compound were known. This is a big if when there is no easy way to know subscripts in compounds' formulas. Fortunately the law of Dulong and Petit gives estimates of molar weights. These estimates allow computation of approximate formulas. Approximate formulas are as good as exact because atoms combine in small whole-number ratios. Errors vanish in the rounding-to-whole number process. Study the solutions to **problems 5 and 41b** as examples of this logic.

Mass Spectroscopy and Isotopes

In a mass spectrometer positively charged ions are accelerated by an electric field and then passed through a magnetic field. More massive ions curve less in the magnetic field and the ions in a mixture are separated according to mass. Careful control of the strength of the magnetic and electric fields allows accurate determination of the ratio of charge to mass of the various ions. Because the ions' charge is known, the instrument gives accurate values of the masses of the ions.

The advent of the mass spectrometer completely outmoded analytical chemical methods in the determination of atomic weights. No one has weighed anything to determine an atomic weight for 60 years or more.

Mass spectrometry revealed the presence of *isotopes*. *Atoms that have essentially the same chemical properties but different masses* are isotopes. Isotopes are designated by the proper chemical symbol prefixed with a superscript to show the mass number of the isotope. The mass number is the integer nearest to the relative atomic weight of the isotope. Because $^{17}O^{+}$ ions, for example, are heavier than $^{16}O^{+}$ ions the mass spectrometer separates them. In addition the instrument measures the relative abundance of isotopes. The masses of an element's isotopes combined in a weighted average with the isotopes' relative abundance give the effective atomic weights of the elements. See the solution to **problem 7** for help with weighted averages.

The Mole Concept

The essence of the concept of the mole is to substitute the convenience of weighing for the tedium of counting. When factories inventory small parts (like screws and washers), they often determine the mass of a set number (*e.g.* 1000), and then weigh the rest of the supply. The inventory number is the total weight divided by the weight per thousand. The process is identical in chemistry. Items as small as atoms are impossible to count individually so the count-by-weighing approach is particularly important. The set number in chemistry is *Avogadro's number*: 6.022×10^{23}. Avogardo's number of "elemental entities" is a mole of those entities.

The mole is a unit for *amount of substance, or "chemical amount."* Formally it is the number of elementary entities in exactly 12 g of ^{12}C. The entity in the case of ^{12}C is the carbon atom. Moles of molecules, ions and electrons, all different elemental entities, are both possible and common. Other units for amount of substance are the dozen (12), the gross (144), the pair (2). A mole contains 6.022×10^{23} elementary particles, which is a lot more than 12, 144 or 2 but not any different in principle.

The mole is a fundamental SI unit. It is accorded this distinction because chemical interactions proceed in small whole-number ratios of moles of the various substances.

The mass of a mole of an element is its atomic weight expressed in grams. The mass of a mole of a compound is its molecular or formula weight expressed in grams. Both of these are the *molar mass* (*molar weight*) of the element or compound. These definitions allow conversion at will between the mass of a sample of a substance and the chemical amount of the substance. **Problem 47** shows the nature of mole concept fully.

1.5 *CHEMICAL EQUATIONS AND STOICHIOMETRY*

A chemical equation represents in brief form what happens in a chemical transformation. *People* write chemical equations, and such equations obviously are incapable of influencing what goes on in real life in a beaker or flask. Equations are *models* of the reality of chemical reactions.

A chemical equation gives the formulas of the reactants on the left and the formulas of the products on the right, linking them by an arrow to emphasize the change. In a *balanced* chemical equation the number of atoms of each kind of element represented on the left-hand side equals the the number of atoms of the same element shown on the right-hand side. This is *material balance* and is an immediate consequence of the law of conservation of matter. Also, the algebraic sum of all of the electrical charges represented on the left-hand side must equal the sum of the charges on the right-hand side.

• • Mass is conserved, element by element; charge is conserved.

Balance is achieved using *coefficients* in front of the formulas representing the molecules.

Arbitrarily changing the chemical formulas of reactants and products is not allowed in balancing equations but is a common error. Misreading chemical formulas, carelessly altering or omitting subscripts and superscripts during transcription, omitting one or more compounds entirely during transcription, and incorporation into formulas of numbers from coefficients are other common reasons for failure properly to balance chemical equations. For example the formula $[Co(NH_3)_6]^{3+}$ includes $3 \times 6 = 18$ H atoms and has a +3 electrical charge. Taking the formula to represent fewer H atoms or to have an zero electrical charge leads to disaster in balancing equations.

Many chemical equations can be balanced by inspection. Consider the compound with the most atoms or type of atoms first and assign it a coefficient of 1. Assign coefficients to the other species to achieve balance. **See problem 11 and problem 49.** A more abstract problem in equation balancing occurs in **problem 55.**

The formulas in chemical equations may refer to moles or molecules of substances, depending on context. Because specifying a half molecule of a substance is absurd, many chemists eliminate any fractional coefficients from their balanced equations, writing:

$$H_2O + N_2O_5 \rightarrow 2HNO_3$$

instead of:

$$\tfrac{1}{2}H_2O + \tfrac{1}{2}N_2O_5 \rightarrow HNO_3$$

However both versions represent the necessary balance.

Stoichiometric Calculations

Stoichiometry is sometimes called *chemical arithmetic.* Balanced chemical equations provide quantitative weight relationships among the reactants and products in a reaction. Chemical formulas provide similar weight relationships among the elements comprising a compound. Stoichiometry concerns the use of these relationships. Typical practical problems in stoichiometry:

- *Computation of a Molar Weight From a Formula.* To determine the molar weight of any substance from its formula add up the atomic weights of all of the atoms represented in the formula. Express this number in grams.

- *Yield Problems.* The task is, given the weight of a reactant and a balanced chemical equation, to determine the possible weight of one or more products if all of the reactant is consumed by the reaction. First calculate the chemical amount (moles) of the *known* reactant; from this calculate the chemical amounts of all the other substances through simple ratios:

$$n_i = (a_i / a_j) \times (n_j)$$

where n_i and n_j are the chemical amounts of substance i and j and a_i and a_j are the coefficients (in the balanced chemical equation) of these substances. Once the chemical amount (moles) of a substance is determined then multiplication by the molar weight (g mol^{-1}) quickly gives its weight (grams). **See problems 21 and 49.** Variations of this problem recognize the fact that in real procedures some product may be lost to side-reactions or during purification. The yield computed above is the *theoretical yield. Actual yield* refers to the weighed quantity of product finally isolated from a reaction. *Percent yield* is the actual yield of product divided by its theoretical yield then multiplied by 100 percent.

- *Determination of Empirical Formulas.* Elemental analysis often gives the weight percent or weight fraction of all of the various elements in a sample of a compound. Convert weight percents to relative chemical amount by *dividing* each one by the molar weight of the corresponding element. The empirical formula of the compound has as its subscripts the smallest whole numbers that have the same ratios as the results of these divisions. **See problem 41.**

- *Limiting Reagent Problems.* In general a chemical reaction consumes one of the reactants before the others. The first reactant to run out is the *limiting reactant.* A fool-proof method of determining which reactant is limiting is to compute the yield of a product (it does not matter which) assuming that all reactants but one are present in *unlimited supply.* Repeat the computation making the same assumption for each of the different reactants. The reactant that gives the smallest yield of product when the answers are compared is the limiting reactant. All other are *in excess.* **See problems 33 and 35.**

- **See problems 19, 21, 27, 31, 49.**

DETAILED SOLUTIONS TO ODD-NUMBERED PROBLEMS

1. The problem deals with a series of four binary compounds of vanadium and oxygen. Since it asks for the *relative* number of atoms of oxygen for a given mass of vanadium in the four compounds, the answer must involve the formation of a ratio. A good first step is to determine what mass of oxygen is present in each compound in combination with some convenient mass of vanadium, for example 1.000 g. This idea is to "reduce to one" for comparison. From the percentages given in the table in the problem, the first compound contains 23.90 g of O for every 76.10 g of V. Putting this on the basis of 1.000 g of vanadium merely requires dividing 23.90 by 76.10. The quotient is 0.3140 or, with the proper units, 0.3140 g O/g V.

For the second, third and fourth compounds, similarly constructed ratios come out 0.4710 g O/g V, 0.6281 g O/g V, and 0.7851 g O/g V, respectively. The ratios *increase* going down the table, showing the increasing oxygen-richness of the compounds.

The next step is to compare the ratios. To compare the first two divide the second by the first: 0.4710/0.3140 = 1.500. Because the relative mass of oxygen in the second compound is 1.500 times larger than in the first, the second compound has 1.500 times more atoms of oxygen than the first. Compare the third and fourth compounds to the first in exactly the same way, forming the ratios 0.6281/0.3140 and 0.7851/0.3140, which equal 2.000 and 2.500 respectively. The relative numbers of atoms of oxygen for a given mass of vanadium in these four compounds are therefore $1 : 1\frac{1}{2} : 2 : 2\frac{1}{2}$ or 2:3:4:5.

Suppose that 50.942 g of vanadium is chosen as the mass to consider. This choice is only apparently arbitrary because 50.942 is the atomic weight of vanadium. The masses of oxygen that would chemically combine with 50.942 g of V in the four compounds are, going down the table: 16.00 g, 24.00 g, 32.00 g and 39.99 g. Division of these numbers each by the smallest, 16.00 g, gives the ratios $1:1\frac{1}{2}:2:2\frac{1}{2}$, just as before. Because 16.00 g of oxygen is 1.000 mol of oxygen the numbers of moles of oxygen combined with 1.00 mol of V (50.942 g) in the four compounds are 1, $1\frac{1}{2}$, 2, and $2\frac{1}{2}$.

---

3. ***a)*** The unknown binary compound is gaseous. From Avogadro's hypothesis it follows that at any particular set of conditions the vapor densities of gaseous compounds are directly proportional to their molecular weights:

$$\rho_{\text{vapor}} = k\,(\text{Mol. Wt.})$$

The value of k, the constant of proportionality, depends on the exact conditions of temperature and pressure. The density of hydrogen (H_2) is 0.0900 g L^{-1} at STP. This is the molar weight of H_2 (2.016 g mol^{-1}) divided by its molar volume at STP (22.4 L). The density of the unknown is 2.77 g L^{-1}. Because the unknown is denser than H_2 and yet has the same number of molecules per liter under the same conditions, the molecules of the unknown *must* weigh proportionately more than H_2 molecules. Therefore the unknown's molecular weight is greater than that of H_2 by a factor of (2.77 / 0.0900). (This amounts to a summary statment of Canizzaro's method for determining relative atomic weights.) The molecular weight of H_2 is 2.016. The molecular weight of the unknown is $(2.77 / 0.0900) \times 2.016$ which equals 62.0.

b) 1.21 g of water arises from 0.500 L of the gaseous compound and includes *all* of the hydrogen present. Like all molecular compounds, H_2O contains a *definite proportion* of its constituent elements. The definite proportion of hydrogen is 2.016 / 18.0153. The numerator of this fraction is two times the atomic weight of H and the denominator is the atomic weight of O added to two times the atomic weight of H. Therefore the 0.500 L sample of gas contains 0.1354 g of H.

A 22.4 L sample of gas (one mole at STP) contains $(22.4/0.500) \times 0.1354$ g H = 6.07 g of H. The atomic weight of H is 1.008 which means 6.07 g of H amounts to 6.02 mol of H. Because 22.4 L of the unknown gas at STP comprises 1.00 mole, there are 6.02 mol of H for every 1.00 mol of the unknown gas. This means there are six *atoms* of H per *molecule* of unknown. The answer must be a whole number.

c) The same 22.4 L (at STP) of gas which contains 6.02 g of H weighs $22.4\ \text{L} \times 2.77\ \text{g L}^{-1} = 62.0$ g. The gas is binary, it has only two kinds of atoms in it: H and the other, call it Y. The mass of Y in 22.4 L of the unknown gas (at STP) is $62.0 - 6.02 = 55.98$ g. If there is only one mole of Y in the 22.4 L then the atomic weight of Y is 55.98. This is the maximum value of the atomic weight of Y. There cannot be *less* than one mole of Y in one mole of unknown just as there cannot be less than one atom of an element in one molecule of a compound.

d) The 55.98 g of Y could include *more* than one mole of Y. If the 55.98 g contains *two* moles, then the atomic weight of Y is 28.0. If there is three mol of Y in the sample, the atomic weight of Y is 18.67. There would seem to be an infinite range of possible atomic weights for Y. In the table below the subscripts of Y could get larger and larger without bound, corresponding to ever smaller atomic weights of Y.

Formula of Binary Compound	*Atomic Weight of Y*	*Mol. Wt. of Unknown*
YH_6	56.00	62.0
Y_2H_6	28.00	62.0
Y_3H_6	18.66	62.0
Y_4H_6	14.00	62.0
Y_5H_6	12.40	62.0
Y_6H_6	9.33	62.0
Y_9H_6	6.22	62.0
$Y_{14}H_6$	4.00	62.0

e) Compare the atomic weights in the above table with those on the periodic table of the elements to determine the identity of the element Y. If the subscript of Y is 2 or 4, numbers quite close to the atomic weights of Si and N respectively come out. A subscript of 56 (not in the table) would give Y an atomic weight of 1.00, making it H. But then the compound would no longer be binary, having the formula H_{62}. A subscript of 14 gives an atomic weight of 4.00, but He, which has close to that atomic weight, forms no known compounds. All other possible atomic weights of Y fail to occur on the periodic table. Therefore the compound is Si_2H_6 (silane) or N_4H_6 (tetrazane). Both molecules exist, but Si_2H_6 is much more stable.

---

5. The heat capacity of the metallic element is 0.631 J $g^{-1}K^{-1}$ so, by the law of Dulong and Petit, its approximate molar weight is:

$$\text{At. wt.} = 25 \text{ J mol}^{-1}\text{K}^{-1} / 0.631 \text{ J g}^{-1}\text{K}^{-1} = 39 \text{ g mol}^{-1}$$

The ratio of the mass of the element to the mass of fluorine it combines with is 51.33 to 48.67 which is 1.0546 to 1. From the periodic table the atomic weight of fluorine is 18.9984. If the metallic element's fluoride had one M for every F (the formula M_1F_1) then the atomic weight of M would be $1.0546 \times 18.9984 = 20.036$. But 20.036 conflicts with the approximate atomic weight. If the formula of the fluoride instead were M_1F_2, then the atomic weight of M would be $1.0546 \times (2 \times 18.9984) = 40.071$ which is consistent with the approximate value. The metallic element is without doubt calcium. Subscripts of 1 are generally not written in chemical formulas but taken as understood. They appear here for emphasis.

---

7. The atomic weight of naturally-occurring Si is the *weighted* mean (weighted average) of the atomic weights of the three isotopes listed. What does weighting an average imply? The *un*-weighted mean (symbolized $\bar{n}$) of the mass of the three isotopes would be:

$$\bar{n} = {}^1/_3\,(27.97693) + {}^1/_3\,(28.97649) + {}^1/_3\,(29.97376)$$

Weighting corresponds to replacing the ${}^1/_3$'s in this expression with values telling each isotope's *true* contribution to the total. These values are the abundances. Fractional abundances (which add up to 1.00) rather than percent abundances (which add up to 100.0) must be used.

The weighted mean is:

$$\bar{n}_{wght} = 0.9221 \times 27.97693 + 0.0470 \times 28.97649 + 0.0309 \times 29.97376 = 28.086$$

---

9. The correct answer must be on the order of 10^{23} atoms because a volume of 15.0 cm^3 of corundum is on the ordinary human scale. The given volume of Al_2O_3 is multiplied by a series of conversion factors:

$$15.0 \text{ cm}^3 \text{ Al}_2\text{O}_3 \times \frac{3.97 \text{ g}}{\text{cm}^3} \times \frac{1 \text{ mol Al}_2\text{O}_3}{101.96 \text{ g Al}_2\text{O}_3} \times$$

$$\frac{6.022 \times 10^{23} \text{ molecules Al}_2\text{O}_3}{1 \text{ mol Al}_2\text{O}_3} \times \frac{2 \text{ atoms of Al}}{1 \text{ molecule Al}_2\text{O}_3}$$

$$= 7.03 \times 10^{23} \text{ atoms of Al}$$

---

11. Many chemical equations can be quickly balanced by trial and error. Make sure that correct formulas are written for all products and reactants. Start with an element that occurs in only one compound on each side of the equation.

Avoid starting with H and O. Adjust the coefficients of those compounds so that equal numbers of this first element are represented on each side. Extend this procedure to other elements.

a) $2\ Al + 6\ HCl \rightarrow 2\ AlCl_3 + 3\ H_2$

b) $2\ C_6H_6 + 15\ O_2 \rightarrow 12\ CO_2 + 6\ H_2O$

c) $2\ ZnS + 3\ O_2 \rightarrow 2\ ZnO + 2\ SO_2$

d) $2\ Fe + O_2 + 2\ H_2O \rightarrow 2\ Fe(OH)_2$

---

13. The mass of one mole of sodalite is the mass of eight moles of Na plus that of six moles of Al plus that of six moles of Si, 24 moles of O and two moles of Cl. The molar mass (molar weight) of any element is its atomic weight (from the periodic table) expressed in grams. Substituting these numbers, sodalite weighs 969.215, g mol^{-1}. The Na contributes 183.918 g to this total so its weight percent is (183.918 / 969.215) × 100 = 18.976 percent. Similarly Al comprises 16.703 percent, Si 17.387 percent, O 39.618 percent and Cl 7.316 percent.

---

15. Avogadro's hypothesis holds that equal volumes of gases at the same temperature and pressure contain equal numbers of molecules.

In this reaction ethylene, C_2H_4, and nothing else burns. Two molecules of $CO_2\,(g)$ must arise from each molecule of C_2H_4 to carry off the carbon, and two molecules of $H_2O(g)$ must form to take off the hydrogen. Therefore, the volume of $CO_2\,(g)$ that forms is $2 \times 3 = 6$ liters, as is the volume of $H_2O\,(g)$. To supply all the oxygen that appears among the products requires three molecules of $O_2\,(g)$ per molecule of C_2H_4 or 9 L of $O_2\,(g)$ for the three L of $C_2H_4\,(g)$.

The preceding discussion amounts to balancing the equation:

$$C_2H_4\,(g) + 3\ O_2\,(g) \rightarrow 2\ CO_2\,(g) + 2\ H_2O(g)$$

$$3\ L + 9\ L \rightarrow 6\ L + 6\ L$$

with the added line emphasizing the volume relationship.

---

17. *a)* The compound gives 0.692 g of H_2O and 3.381 g of CO_2. Determine the weight of elemental H and elemental C in these weights of compound:

$$0.692\ g\ H_2O \times \frac{2.016\ g\ H}{18.015\ g\ H_2O} = 0.0774\ g\ of\ H$$

$$3.381 \text{ g } CO_2 \times \frac{12.01 \text{ g C}}{44.01 \text{ g } CO_2} = 0.9226 \text{ g of C}$$

b) Because the masses of C and H in the CO_2 and H_2O respectively add up to 1.000 g the compound contains no other elements.

c) The compound is 7.74 percent H and 92.26 percent C by weight.

d) The 1.000 g of compound contains 0.0774 g of H and 0.9226 g of C. To get the compound's empirical formula, convert these quantities to the chemical amounts (in moles) and determine their ratio:

$$0.0774 \text{ g H} \times \frac{1 \text{ mol H}}{1.008 \text{ g H}} = 0.0767 \text{ mol of H}$$

$$0.9226 \text{ g C} \times \frac{1 \text{ mol C}}{12.01115 \text{ g C}} = 0.0768 \text{ mol of C}$$

The C and H are present in a 1:1 molar ratio. CH is the empirical formula.

---

19. ***a)*** According to the equation:

$$Mg + 2\,HCl \rightarrow H_2 + MgCl_2$$

the reaction produces 1 mol of H_2 for every 1 mol of Mg it consumes.

H_2 has a molecular weight of $2 \times 1.00797 = 2.01594$. Mg has an atomic weight of 24.305. Therefore the 1 mol Mg $\rightarrow$ 1 mol H_2 relationship implies that 24.305 g Mg yields 2.01594 g H_2. The problem is to compute the weight of Mg that yields 1.000 g of H_2. Since 1.000 g is about half of 2.01594 g, the answer should be about half of 24.305 g. More exactly:

$$1.000 \text{ g } H_2 \times \frac{24.305 \text{ g Mg}}{2.01594 \text{ g } H_2} = 12.06 \text{ g Mg}$$

b) The equation specifies that 1 mol of I_2 arises from every 2 mol of $CuSO_4$:

$$2\,CuSO_4 + 4\,KI \rightarrow 2\,CuI + I_2 + 2\,K_2SO_4$$

We can write the following train of conversions, starting with the 1.000 g of I_2:

$$1.000 \text{ g } I_2 \times \frac{1 \text{ mol } I_2}{253.809 \text{ g } I_2} \times \frac{2 \text{ mol } CuSO_4}{1 \text{ mol } I_2} \times \frac{159.602 \text{ g } CuSO_4}{1 \text{ mol } CuSO_4}$$

$$= 1.258 \text{ g } CuSO_4$$

Notice that the reaction wastes 2 mol of I (in CuI) for every mole of I_2 it makes, which would be bad if the aim of doing the reaction were to make I_2.

c) According to the balanced equation, 1 mole of $NaBH_4$ yields 4 moles of H_2. Some might dispute that such a reaction is possible, arguing that no reaction can transform the 4 H atoms of $NaBH_4$ into the 8 H atoms of 4 H_2. As the equation however shows, the missing H comes from the *other* reactant, water. Write a series of conversion factors:

$$1.000 \text{ g } H_2 \times \frac{1 \text{ mol } H_2}{2.0159 \text{ g } H_2} \times \frac{1 \text{ mol } NaBH_4}{4 \text{ mol } H_2} \times \frac{37.833 \text{ g } NaBH_4}{1 \text{ mol } NaBH_4} = 4.692 \text{ g } NaBH_4$$

---

21. The logical progression for this problem is: mass KCl to chemical amount of KCl to chemical amount of K_2PtCl_6 to mass of K_2PtCl_6. To accomplish the first and third conversions use the molecular weights of KCl and K_2PtCl_6. The balanced equation provides the information for the second of the conversions.

The molecular weight of KCl, the sum of the atomic weights of K and Cl, is 74.555. The molecular weight of K_2PtCl_6 is 486.01. Then:

$$0.1000 \text{ g KCl} \times \frac{1 \text{ mol KCl}}{74.555 \text{ g KCl}} \times \frac{1 \text{ mol } K_2PtCl_6}{2 \text{ mol KCl}} \times \frac{486.01 \text{ g } K_2PtCl_6}{1 \text{ mol } K_2PtCl_6}$$

$$= 0.3260 \text{ g } K_2PtCl_6$$

---

23. **a)** The small whole-number relationships given by chemical equations and formulas *always* refer to chemical amount, never to mass. The chemical equation in this problem relates the chemical amounts of XCl_2 and XBr_2, stipulating that they are equal to each other.

To use this fact one must convert from the mass of XBr_2 to the chemical amount of XBr_2, an operation requiring the molar weight of XBr_2. Similarly, one needs to go from the mass of XCl_2 to the chemical amount of XCl_2, a conversion which requires the molar weight of XCl_2. Arriving at these molecular weights requires the atomic weight of X. Let x stand for the atomic weight of X. Then the molecular weight of XBr_2 is [2 × 79.909 (for the 2 Br's) + x] or (159.818 + x). The molecular weight of XCl_2 is (70.906 + x). Then:

$$1.500 \text{ g } XBr_2 \times \frac{1 \text{ mol } XBr_2}{159.818 + x \text{ g } XBr_2} = \frac{1.500}{159.818 + x} \text{ mol } XBr_2$$

$$0.890 \text{ g } XCl_2 \times \frac{1 \text{ mol } XCl_2}{70.906 + x \text{ g } XCl_2} = \frac{0.890}{70.906 + x} \text{ mol } XCl_2$$

Since the chemical amounts of the XBr_2 and XCl_2 are the same the above two quantities equal each other:

$$\frac{1.500}{(159.818 + x)} \text{ mol of } XBr_2 = \frac{0.890}{(70.906 + x)} \text{ mol of } XCl_2$$

This equation is easily solved for x, the atomic weight of the unknown element. $x = 58.8$.

b) Find the atomic weight in the periodic table. The unknown element is nickel.

Another approach to this problem is to recognize that the mass of X in the sample *cannot* be changed by the chemical reaction. Let y equal this mass. Then $(1.500 - y)$ g is the mass of Br originally present and $(0.890 - y)$ g is the mass of Cl combined with X after the reaction. The chemical equation shows that the chemical amount of Br in the system before the change equals the chemical amount of Cl after the change:

$$n_{Cl} = n_{Br}$$

The chemical amount of a substance in moles is its mass in grams divided by its molar weight in g mol^{-1}. In this problem:

$$n_{Cl} = (0.890 - y)\text{ g} / 35.453\text{ g mol}^{-1}$$

$$n_{Br} = (1.500 - y)\text{ g} / 79.909\text{ g mol}^{-1}$$

Combining these two expressions and solving for y reveals that 0.403 g of X was present both before and after the reaction.

The chemical formula XCl_2 guarantees that this 0.403 g of X must amount to exactly one-half the chemical amount of Cl present in (0.890 − 0.4035) g of Cl. The latter is:

$$(0.890 - 0.4035)\text{ g} / 35.453\text{ g mol}^{-1} = 0.0137\text{ mol Cl}$$

so the 0.403 g of X is (0.0137 / 2) or 0.00686 mol of X. The sample has 0.4035 g of X per 0.00686 mol of X or 58.8 grams of X per *one* mole, that is, the atomic weight of X is 58.8.

---

25. "Quantitatively" means without any loss of material. It means in this case that all of the uranium in the original sample makes it into the product, and none ends up, for example, on the floor. This quantity of U is computed as follows, where mmol stands for millimole and mg stands for milligram:

$$200.0\text{ mg } UO_2 \times \frac{1\text{ mmol } UO_2}{270.03\text{ mg } UO_2} \times \frac{1\text{ mmol U}}{1\text{ mmol } UO_2} \times \frac{238.03\text{ mg U}}{1\text{ mmol U}} = 176.3\text{ mg of U}$$

This weight of uranium must appear in the new oxide. Meanwhile the total weight of the *new* oxide (U_aO_b) is 1.04 times greater than 200.0 mg or 208.0 mg. Therefore the weight of oxygen (O) in the new oxide is 208.0 mg − 176.3 mg = 31.7 mg:

$$31.7\text{ mg O} \times \frac{1\text{ millimole O}}{16.0\text{ mg O}} = 1.98\text{ mmol O}$$

Returning to the uranium, its chemical amount is:

$$176.3 \text{ mg U} \times \frac{1 \text{ mmol U}}{238.4 \text{ mg U}} = 0.741 \text{ mmol U}$$

The ratio of the chemical amount of O to the chemical amount of U is 1.98 to 0.741 or 2.67 to 1 or, in whole numbers, 8 to 3. The empirical formula is therefore U_3O_8.

---

27. ***a)*** The original sample of substance weighed 1.000 g. Because it contained *only* Mn and O the mass of O present in it was clearly 1.000 − 0.632 = 0.368 g. The chemical amounts of the Mn and O are computed as follows:

$$0.632 \text{ g Mn} \times \frac{1 \text{ mol Mn}}{54.938 \text{ g Mn}} = 0.0115 \text{ mol Mn}$$

$$0.368 \text{ g O} \times \frac{1 \text{ mol O}}{15.99949 \text{ O}} = 0.0230 \text{ mol O}$$

The chemical amount of O in the sample is twice that of Mn so the empirical formula is MnO_2.

b) $MnO_2(s) + 2\,H_2(g) \rightarrow Mn(s) + 2\,H_2O(g)$

c) The balanced equation shows that 2 mol of water forms for every 1 mol of Mn*(s)*. Use this chemical equivalency as a conversion factor:

$$0.0115 \text{ mol Mn} \times \frac{2 \text{ mol } H_2O}{1 \text{ mol Mn}} \times \frac{18.02 \text{ g } H_2O}{1 \text{ mol } H_2O} = 0.414 \text{ g of } H_2O$$

---

29. ***a)*** The reaction produces elemental sulfur by combination of two different sulfur-containing compounds. In the following, the first conversion factor comes from the definition of the mole and the second from the balanced equation. The third represents the STP molar volume of the gas:

$$1000 \text{ g S} \times \frac{1 \text{ mol S}}{32.064 \text{ g S}} \times \frac{2 \text{ mol } H_2S}{3 \text{ mol S}} \times \frac{22.4 \text{ L } H_2S}{1 \text{ mol } H_2S}$$

$$= 466 \text{ L of } H_2S(g)$$

b) The reaction consumes twice the volume of $H_2S(g)$ as it does of $SO_2(g)$. This follows from the balanced equation, which shows two molecules of H_2S reacting with one of SO_2, and Avogadro's hypothesis: as long as they are held under the same conditions equal volumes of two different gases contain equal numbers of molecules. The volume of $SO_2(g)$ is thus 466 L / 2 = 233 L.

To get the weight of $SO_2(g)$:

$$1000 \text{ g S} \times \frac{1 \text{ mol S}}{32.064 \text{ g S}} \times \frac{1 \text{ mol SO}_2}{3 \text{ mol S}} \times \frac{64.06 \text{ g SO}_2}{1 \text{ mol SO}_2}$$

$$= 666.0 \text{ g SO}_2$$

Another way to compute the mass of SO_2 is to reason that 233 L of SO_2 *(g)* at STP is (233 L / 22.4 L mol^{-1}) mol of SO_2 and then employ the fact that 1 mol of SO_2 weighs 64.06 g:

$$(233 \text{ L} / 22.4 \text{ L mol}^{-1}) \times 64.06 \text{ g mol}^{-1} = 666 \text{ g SO}_2$$

The first method is better than the second. The second computation involves the number 22.4, which is precise to only 3 significant digits. The first method avoids the use of 22.4. Interestingly, if the computation of the volume of SO_2 (part a) is combined with the conversion *from* volume of SO_2 *to* weight of SO_2, then the factor 22.4 cancels out. A computation just like the first one given in this part of the problem results.

31.

$$50.0 \text{ kg of NaNO}_2 \times \frac{1000 \text{ g NaNO}_2}{1 \text{ kg NaNO}_2} \times \frac{1 \text{ mol NaNO}_2}{69.985 \text{ g NaNO}_2}$$

$$\times \frac{1 \text{ mol NO}(g)}{1 \text{ mol NaNO}_2} \times \frac{22.4 \text{ L (at STP) NO}(g)}{1 \text{ mol NO}(g)}$$

$$= 1.60 \times 10^4 \text{ L of NO}(g)$$

The key factor in the above set-up is the recognition from the balanced equation that the chemical amount of NO*(g)* consumed is equal to the chemical amount of $NaNO_2$ produced.

---

33. The first step is to write the balanced chemical equation representing the reaction:

$$HCl(g) + NH_3(g) \rightarrow NH_4Cl(s)$$

One approach now is to note that one mole of HCl gas weighs 36.46 g and one mole of NH_3 weighs only 17.03 g. The equation shows the two gases reacting in a one-to-one *molar* ratio. Therefore, given equal masses of the two the *heavier* molecule will be used up first. It takes fewer heavy molecules than light molecules to make up a specific weight. Because HCl is heavier it is the limiting reactant: all of it will be used up to produce NH_4Cl and then the reaction will stop, leaving excess NH_3. The weight of NH_4Cl produced will be:

$$10.0 \text{ g HCl} \times \frac{1 \text{ mol HCl}}{36.46 \text{ g HCl}} \times \frac{1 \text{ mol NH}_4\text{Cl}}{1 \text{ mol HCl}} \times \frac{53.49 \text{ g NH}_4\text{Cl}}{1 \text{ mol NH}_4\text{Cl}} = 14.7 \text{ g of NH}_4\text{Cl}$$

Since 20.0 g of matter was present originally the weight of left-over NH_3 is (20.0 − 14.7) = 5.3 g.

Such reasoning works well in this relatively simple case. However it becomes tedious in applications where reactants are consumed in 2 to 3 or 5 to 4 or even more complex ratios and where the initial amounts of the reactants are unequal. A general approach is to compute the weight of product that would form first based on the given amount of one reactant and assuming an unlimited supply of the second. Then repeat the computation of yield, this time based on the given amount of the second reactant and assuming an unlimited supply of the first. Comparison of the results quickly identifies the limiting reactant: the one giving *less* product.

In this problem, 10.0 g of NH_3 and unlimited HCl would give 31.4 g of NH_4Cl. On the other hand, 10.0 g of HCl and unlimited NH_3 would give only 14.7 g of NH_4Cl. The HCl is the limiting reactant.

---

35. ***a)*** The reaction system initially contains 10.0 g of $BaCl_2$ and 20.0 g of $AgNO_3$. After the reaction there will be 30.0 g of a mixture of the products (AgCl and $Ba(NO_3)_2$) with *either* $BaCl_2$ or $AgNO_3$ left in excess. (This sets aside the remote chance that the two reactants are present exactly in a 1:2 molar ratio and *neither* is left over in excess.)

The reactant *not* in excess is the limiting reactant. Compute the chemical amounts of the two reactants:

$$10.0 \text{ g } BaCl_2 \times \frac{1 \text{ mol } BaCl_2}{208.25 \text{ g } BaCl_2} = 0.0480 \text{ mol } BaCl_2$$

$$20.0 \text{ g } AgNO_3 \times \frac{1 \text{ mol } AgNO_3}{169.87 \text{ g } AgNO_3} = 0.118 \text{ mol } AgNO_3$$

By comparing these chemical amounts one concludes that $AgNO_3$ is present in excess: its 0.118 mol would require 0.118 / 2 = 0.0589 mol of $BaCl_2$ to react fully, and there is only 0.0480 mol of $BaCl_2$ available.

b) The chemical amount of $AgNO_3$ left in excess is (0.118 − 0.0960) mol. The 0.0960 mol arises because each 0.0480 mol of $BaCl_2$ uses up *twice* that chemical amount of $AgNO_3$. The 0.02166 mol of $AgNO_3$ left in excess is 3.68 g of $AgNO_3$. The total products then weigh 30.0 − 3.68 = 26.2 g.

Suppose the problem is interpreted as asking the weights of *each* of the two products, AgCl and $Ba(NO_3)_2$. The weight of AgCl is its molar weight (143.323 g mol^{-1}) multiplied by the chemical amount formed (2 × 0.0480 mol). The result is 13.8 g of AgCl. The difference between this figure and 26.2 g is 12.4 g, the weight of the $Ba(NO_3)_2$ formed.

---

37. Problems of this sort appear not to furnish enough information for solution. What is overlooked is that both NaCl and KCl furnish only a set, definite proportion of their mass to the total chloride. These fractions are readily available from the atomic weights and formulas of KCl and NaCl.

Thus, in addition to the immediately obvious relationship:

$$x + y = 1.0000 \text{ g}$$

where x and y are the masses in grams of the NaCl and KCl respectively, there is the relationship:

mass of Cl from NaCl + mass of Cl from KCl = mass of Cl in AgCl

Computing the molecular weights of NaCl, KCl and AgCl and using the law of definite proportions to get the fraction of each compound that is Cl:

$$\frac{35.453x}{58.4428} + \frac{35.453y}{74.555} = \frac{35.453(2.1476)}{143.323}$$

If both sides of this second equation are divided by 35.453:

$$x / 58.4428 + y / 74.555 = 2.1476 / 143.323$$

Algebraic combination of this equation with $x + y = 1.0000$ gives:

$$x = 0.4250 \text{ g} \quad and \quad y = 0.5750 \text{ g}$$

The weight percentages of NaCl and KCl in the original mixture of NaCl and KCl are then 42.50 and 57.50 percent respectively.

---

39. ***a)*** Assuming ideality, 22.4 L of a gaseous compound at STP is one mole of that compound. Use this fact to compute the molar weights of the three compounds:

	Density				*Molar weight*
Compound 1	1.162 g/L	×	22.4 L/mol	=	26.03 g mol^{-1}
Compound 2	1.969 g/L	×	22.4 L/mol	=	44.11 g mol^{-1}
Compound 3	2.595 g/L	×	22.4 L/mol	=	58.13 g mol^{-1}

b) The weight percentages of H in the three compounds are simply (100 − 92.26), (100 − 81.71), and (100 − 82.66) percent respectively. If we consider one mole of each of the three:

	Molar Weight	*Percent H*	*Mass H per mole compound*	*Moles H per mole compound*
Comp. 1	26.03 g mol^{-1}	7.74	2.02 g	2
Comp. 2	44.11	18.29	8.07	8
Comp. 3	58.13	17.34	10.08	10

Note that the chemical amount (in moles) of H is numerically close to the mass (in grams) of H. This is because the atomic weight of H is close to 1.

c) Get the maximum possible atomic weight of element E in each compound by subtracting from the molecular weight of each compound the portion that comes from the H:

	Total mass per mole of compound	*Mass of H per mole of compound*	*Mass of E per mole of compound*
Comp. 1	26.04 g	2.02 g	24.02 g
2	44.13 g	8.07 g	36.06 g
3	58.16 g	10.08 g	48.08 g

The final column in the table allows conclusions about the atomic weight of E. From the definition of atomic weight, the atomic weight of E equals the number of grams of E in a mole of E. The maximum atomic weight of E can be neither 48.08 nor 36.06 because, from the data on the first compound, E's atomic weight is 24.02 or less. But 24.02 is still too large to be the atomic weight for E. If it were, then compound 2 would have $1\frac{1}{2}$ E's per molecule, impossible under atomic theory. Hence, the maximum atomic weight of E is 12.01. Atomic weights of 6.00, 4.00, 3.00, etc. are not ruled out by the data from the problem. The periodic table however reveals that E can only be carbon. The three compounds are C_2H_2, C_3H_8 and C_4H_{10}.

---

41. **a)** The original sample of the violet oxide weighs 50.0 g. When the pure metal is extracted it weighs 35.2 g. Therefore the mass of O was 50.0 − 35.2 = 14.8 g of O. Because 1 mole of O weighs 16.0 g we have:

$$14.8 \text{ g of O} \times \frac{1 \text{ mole of O}}{16.0 \text{ g O}} = 0.925 \text{ mol of O}$$

b) The law of Dulong and Petit asserts that the product of a solid element's atomic weight and specific heat capacity is about 25 J $mol^{-1}K^{-1}$. The heat capacity of this sample of X is 9.29 J K^{-1}. Therefore its *specific* heat capacity is (9.29/35.2) J $g^{-1}K^{-1}$. The approximate molar weight of X is 25 divided by (9.29 / 35.2) = 94.7 g mol^{-1}.

c) The original sample contains 0.925 mol of O (part a). It also contains some X. The chemical amount of X is its weight divided by its molar weight. That is:

$$35.2 \text{ g X} \times \frac{1 \text{ mole X}}{94.7 \text{ g X}} \simeq 0.37 \text{ mol X}$$

The approximately-equal sign is used because the molar weight of X is only an estimate. Because there is 0.925 / 0.37 $\simeq$ 2.5 mol of O per mol of X the empirical formula is X_2O_5.

---

43. **a)** Start with the weight of the AgBr produced by the reaction which, together with the balanced equation, tells the chemical amount of XBr_2 present:

$$1.0198 \text{ g of AgBr} \times \frac{1 \text{ mol AgBr}}{187.779 \text{ g AgBr}} \times \frac{1 \text{ mol XBr}_2}{2 \text{ mol AgBr}}$$

$$= 0.0027154 \text{ mol of XBr}_2$$

But this chemical amount weighs 0.5000 g. So the molar weight is (0.5000 g / 0.0027154 mol) = 184.13 g mol^{-1}.

b) The molecular weight of XBr_2 (184.13) is the sum of the contribution of 2 Br's (at 79.909 each) and 1 X so the atomic weight of X is 24.31. Referring to the periodic table reveals that X must be magnesium.

---

45. ***a)*** It is straightforward to calculate the molar weight of the gas because the density is given at STP:

$$(3.418 \text{ g L}^{-1}) \times 22.4 \text{ L (at STP) mol}^{-1} = 76.6 \text{ g mol}^{-1}$$

More than three place precision is misleading. Real gases occupy slightly varying volumes per mole at STP, not exactly 22.4 L mol^{-1}.

b) Because water is a combustion product, one of the two constituents of the gas is obviously hydrogen: Compute the chemical amount of this H:

$$\frac{0.8031 \text{ g H}_2\text{O}}{500 \text{ cm}^3 \text{ gas}} \times \frac{1 \text{ cm}^3}{1 \text{ mL}} \times \frac{1000 \text{ mL}}{\text{L}} \times \frac{22.4 \text{ L(STP)}}{\text{mol of gas}} \times$$

$$\frac{1 \text{ mol H}_2\text{O}}{18.02 \text{ g H}_2\text{O}} \times \frac{2 \text{ mol H}}{1 \text{ mol H}_2\text{O}} = 4.00 \text{ mol H}$$

Since one mole of the unknown binary gas contains four moles of H the unknown's maximum molar weight is 76.6 − 4(1.00797) = 72.57 g mol^{-1}. This value is very close to that of Ge (germanium). Other possible values of the atomic weight are 72.57 / 2, 72.57 / 3, etc. But the only value of this series which is sufficiently close to the atomic weight of a real element is the first. Compare to **problems 3d and 39c.** The molecular formula is therefore GeH_4.

---

47. ***a)*** The path to an answer follows one guiding light: whatever the units in which they are expressed the *actual* masses of atoms of ^{32}S and of P are the *same* in the distant galaxy as here. The problem then becomes one of proper conversion of units. By *our* definitions there are N_0 (6.0220×10^{23}) atoms of ^{32}S per mole of ^{32}S. Therefore:

$$\frac{N_0 \text{ atom } ^{32}\text{S}}{31.972 \text{ g } ^{32}\text{S}} \times \frac{4.213 \text{ g } ^{32}\text{S}}{1 \text{ marg } ^{32}\text{S}} \times \frac{32.000 \text{ marg } ^{32}\text{S}}{1 \text{ elom of } ^{32}\text{S}}$$

$$= \frac{2.539 \times 10^{24} \text{ atoms } ^{32}\text{S}}{1 \text{ elom of } ^{32}\text{S}}$$

"Ordagova's number" is 2.539×10^{24}.

b) A series of unit-conversions works well in attacking this problem:

$$\frac{30.9738 \text{ g P}}{1 \text{ mole P}} \times \frac{1 \text{ marg P}}{4.213 \text{ g P}} \times \frac{1 \text{ mole P}}{N_0 \text{ atom of P}} \times \frac{N_{or} \text{ atom P}}{1 \text{ elom P}}$$

$$= \frac{(30.9738)\, N_{or}}{(4.213)\, N_0} \text{ marg P elom}^{-1}$$

But note from part a) that:

$$N_{or} = [\, N_0(4.213)(32.000) / 31.972 \,] \text{ atom elom}^{-1}$$

Upon substituting *this* expression for N_{or} into the equation just preceding, N_0 and "4.213" cancel away to give:

$$\frac{30.9738\ (32.000)\ \text{marg P}}{31.973\ \text{elom P}} = 31.000 \text{ marg P elom}^{-1}$$

The cancellation is important. It allows five significant figures in the answer. It shows that the answer is *independent* of the definition of "marg" in terms of a gram and of a "elom" in terms of a mole.

---

49. *a)* Write an unbalanced equation to represent what the problem tells about the process.

$$C_{12}H_{22}O_{11} + O_2 \rightarrow C_6H_8O_7 + H_2O$$

Balance this equation first as to carbon by inserting the coefficient 2 in front of the citric acid. Then balance the hydrogens by putting a 3 in front of the water (of the 22 H's on the left, 16 appear in the citric acid and the rest appear in the water). Consider now the oxygen. The right side has $(2 \times 7) + (3 \times 1) = 17$ O's. On the left side the sucrose furnishes 11 O's so the remaining 6 must come from 3 molecules of oxygen.

$$C_{12}H_{22}O_{11} + 3\ O_2 \rightarrow 2\ C_6H_8O_7 + 3\ H_2O$$

b) The balanced equation provides essential information:

$$15.0 \text{ kg sucrose} \times \frac{1 \text{ kilomol sucrose}}{342.3 \text{ kg sucrose}} \times \frac{2 \text{ kmol citric acid}}{1 \text{ kmol sucrose}} \times \frac{192.13 \text{ kg citric acid}}{\text{kmol citric acid}}$$

$$= 16.8 \text{ kg citric acid}$$

It is perfectly acceptable to create and use factors like "1 kilomole sucrose / 342.3 kg sucrose". Also, it was *not* necessary to have the entire balanced equation to work part b). Knowing that the reaction produces two moles of citric acid for every one mole of sucrose that it consumes is sufficient. The O_2 and H_2O could have been left unbalanced.

---

51. The equations show that in the Solvay process each mole of NH_3 takes up a mole of CO_2 and in effect holds it for NaCl to attack. Each mole of $NaHCO_3$ thus formed gives ½ mole of Na_2CO_3. Thus, for each mole of NH_3 ½ mole of Na_2CO_3 can form. In the following a capital M stands for mega or 10^6:

$$\frac{1 \text{ megamol } NH_3}{22.4 \text{ megaL } NH_3} \times \frac{1 \text{ Mmol } Na_2CO_3}{2 \text{ Mmol } NH_3} \times \frac{105.98 \text{ Mg } Na_2CO_3}{\text{Mmol } Na_2CO_3}$$

$$= 2.37 \text{ Mg } Na_2CO_3 \text{ per megaliter of ammonia}$$

A metric ton equals a megagram.

---

53. 7.32 g of M takes up (36.18 − 7.32) = 28.86 g of Cl as the MCl_3 is formed. The chemical amount of the Cl is 28.86 g × (1 mol Cl / 35.453 g) = 0.8140 mol. The formula (MCl_3) shows that the chemical amount of the M in the compound is $^1/_3$ that of the Cl. This chemical amount is 0.8140 mol / 3 or 0.2713 mol. Now, both the chemical amount (in moles) of M and the mass (in grams) of M are known. The molar weight of M is the ratio of the second to the first or 7.32 g / 0.2713 mol = 26.98 g mol^{-1}. Consulting the periodic table shows that the element is aluminum.

---

55. ***a)*** For every 1 mol of O_2 consumed 2 mol of H_2O evolves because H_2O is the only place oxygen can go in this reaction. If x mol of H_2 forms, then the total chemical amount of H atoms (not molecules) on the right is [4 (from the water) + $2x$ (from the H_2)] mol. This hydrogen all comes from the HBr and HCl on the left. Because BrCl, which contains Br and Cl in a 1:1 ratio, is formed and there are no other Br or Cl containing products, the chemical amounts of HBr and HCl on the left must equal each other. Their sum is $(4 + 2x)$, the total chemical amount of the H. Therefore the chemical amount of each is one-half of $(4 + 2x)$:

$$(2 + x)\,HCl + (2 + x)\,HBr + 1\,O_2 \rightarrow (2 + x)\,BrCl + 2\,H_2O + x\,H_2$$

b) BrCl is a product of the reaction. It can only come from the combination of HCl and HBr on the left side of the equation. Therefore one of the simpler reactions is the combination of HCl and HBr to give BrCl and H_2:

$$2\,HCl + 2\,HBr \rightarrow 2\,BrCl + 2\,H_2$$

The other simpler reaction must produce water and consume oxygen, the left-over members of the system. Only one reaction is possible which also involves a substance in common with the first simple reaction:

$$O_2 + 2\,H_2 \rightarrow 2\,H_2O$$

c) If hydrogen is consumed by the second simpler reaction at exactly the rate that it is produced by the first then the following equation results:

$$2\,HCl + 2\,HBr + O_2 \rightarrow 2\,BrCl + 2\,H_2O$$

Chapter 2

CHEMICAL PERIODICITY AND INORGANIC REACTIONS

2.1 *THE PERIODIC TABLE*

The periodic table summarizes a large number of important chemical facts and relationships. It can be used to predict the physical and chemical properties of unfamiliar elements and compounds. **See Problem 1.** It tells the number of *valence electrons* that each element can furnish and in that way provides the basis for writing Lewis structures. **See problems 3 and 5.** Columns in the table are *groups* and rows are *periods.* Elements in the table are categorized as follows:

- *The Representative Elements.* These elements comprise 8 groups, each headed by a Roman numeral:

I	II	III	IV	V	VI	VII	VIII
H							He
Li	Be	B	C	N	O	F	Ne
Na	Mg	Al	Si	P	S	Cl	Ar
K	Ca	Ga	Ge	As	Se	Br	Kr
Rb	Sr	In	Sn	Sb	Te	I	Xe
Cs	Ba	Tl	Pb	Bi	Po	At	Rn
Fr	Ra						

About half of the representative elements are metals and half non-metals. The step line in the above figure marks the boundary between the two. Metals lie to the left. Elements adjoining the line have intermediate properties. It is important to identify an element as a metal or non-metal because the distinction is used in naming compounds. **See problem 9.** Among the representative elements the group number equals the number of valence electrons (see below).

- *Transition Metal Elements.* These include 10 groups of three metals each. The groups appear in the table between Group II and III in the 4th, 5th, and 6th rows.

- *The Lanthanide Elements.* These are lanthanum, element 57 in the table, and the 14 elements immediately subsequent. All are metals.

- *The Actinide Elements.* These are actinium, element 89 in the table, and the 14 elements that follow.

2.2 *LEWIS STRUCTURES*

Diagrams in which dots represent the bonding electrons are Lewis structures. Bonding electrons, or *valence* electrons, are the electrons in the outermost regions of an atom. An atom's position in the periodic table tells its number of valence electrons. The number of valence electrons of atoms in groups I through VII equals the group number. The noble gases (group VIII) usually have zero valence electrons (do not bond). In the rare cases when group VIII elements bond they are counted as having 8 valence electrons.

Whenever more than two kinds of atoms are bonded in a molecule, the question of *connectivity* (order in which atoms are linked) arises. A Lewis structure *always* states the connectivity of atoms. The Lewis structure of water shows that the order of the three atoms is H O H and not H H O. Writing the three atoms in a straight line is equivalent in a Lewis structure to writing them with an angle at the oxygen atom. One does *not* read bond angles from a Lewis structure.

Lewis structures are formulated in the following steps:

1. Sum up the total number of valence electrons in the molecule or molecule-ion under consideration.

Example: N_3^-. Each N has 5 valence electrons (N is in Group V). The -1 charge on the group as a whole adds one electron. $3(5) + 1 = 16$ valence electrons.

2. Count the number of connnections among the atoms. Joining p atoms together requires at least $(p-1)$ connections. The actual structure may use more than this many links but it can never use fewer.

3. Determine the "skeleton" of the molecule or molecule-ion. This is the pattern by which the atoms are joined together.

A molecular or empirical formula does not reveal which atoms are located next to which others. Indeed some sets of atoms may arrange themselves in more than one skeleton. However, there are ways to get a skeleton in most cases:

a) The chemical formula of the molecule as written may specify the order of the connections among the atoms.

Example: The linkage of the atoms in the compound dichlorine heptoxide, Cl_2O_7, is obscure. Writing the formula as $O_3ClOClO_3$ suggests:

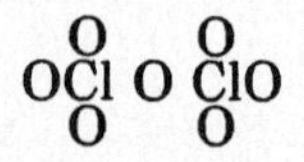

b) Hydrogen atoms are always on the outside of Lewis structures. In Lewis structures H atoms can form only one covalent bond. Therefore they are incapable of lying between two other atoms in a structure.

Example: H_2O_2. The skeleton is [H O O H], not [H O H O] or anything else with interior hydrogen atoms.

c) Molecules and molecular ions tend to be clusters with any unique atom at the center of the cluster.

Example: SF_4. The sulfur atom is the unique atom. Therefore, place it at the center of a cluster with all four fluorine atoms bonded to it. Naturally, various chain arrangements are possible.

4. Place the valence electrons around the symbols for the elements in the skeleton in such a way that all the elements attain a noble gas electronic configuration.

For most elements this means that the elements should "see" an octet of electrons -- four pairs. For hydrogen, each atom should "see" a single pair of electrons. If there were no sharing of electrons, satisfaction of the octet rule would require eight times the number of non-hydrogen atoms plus two times the number of hydrogen atoms. However:

• • Every single bond reduces the number of electrons required because shared electrons are counted twice.

The minimum number of connections among the atoms is given by step 2. It is one less than the number of atoms. Therefore, the number of valence electrons required for single-bonding with octets (duets for the hydrogens) becomes:

$$e^{-}\text{'s needed} = [8 \times N_{non\text{-}H}] + [2 \times N_H] - [2 \times (p - 1)]$$

where p is the number of atoms of all kinds. For example, for HNO_3, substitution in the formulas gives:

$$e^{-}\text{'s needed} = [8 \times 4] + [2 \times 1] - [2 \times (4 - 1)] = 26$$

Three outcomes are possible:

a) the number of valence electrons by required by the theory may *equal* the actual number available.

b) the number of valence electrons required by the theory may *exceed* the actual number of valence electrons available. The species is "electron-deficient."

c) the number of valence electrons required by the theory may be *less than* the number of valence electrons that must be used. The species has an "electron excess."

Case a) presents no problems.

In case b) the answer is to invoke multiple bonding, sharing four or six electrons between some atoms. The more that atoms share electrons, then the fewer electrons are required to give every atom an octet. The formation of *rings* of atoms also allows the use of fewer electrons.

In case c) there is no easy answer. There is an excess of electrons according to the octet rule. All of the electrons must be accommodated. Therefore the octet rule *fails.* Examples are SF_6, which has a central sulfur atom with 12 electrons around it, and ICl_7, which has a central iodine with no fewer than 14 electrons around it.

Shortcomings of Lewis Structures

- Some electron deficient molecules (BF_3, BeF_2) do *not* have octets on the central atom despite the fact that by invoking multiple bonding the octet rule could be satisfied.

- In electron-excess Lewis structures, the octet rule must be broken for at least one atom.

- Satisfactory Lewis structures cannot be drawn unless the number of valence electrons is an even number. Despite this, many "odd molecules" exist.

Formal Charges

To determine the formal charge on an atom in a molecule first draw a Lewis structure. Then:

Formal charge = Group No. − No. of nonbonding electrons − No. of electron pair bonds

The formal charge on an atom in a molecule is thus the number of electrons the atom originally contributed to the molecule minus the number of valence electrons the atom would get if all of its bonds were *homolytically cleaved.* Homolytical cleavage simply means a 50:50 split of all bonding electrons. In the Lewis structure:

:Ö::S̈:Ö̤:

the oxygen on the left has a formal charge of zero. It has 4 electrons in lone pairs, and if its bonds to S are homolytically cleaved it gets 2 more electrons. It originally had 6 electrons: $6 - (2 + 4) = 0$ formal charge. The oxygen on the right has 6 electrons in three lone pairs and, if its bond to S is cleaved, it gets 1 more electron. It originally had 6 electrons: $6 - (6 + 1) = -1$ formal charge. The central S came to the compound with 6 valence electrons. If its three covalent bonds are homolytically cleaved it gets back only 3 electrons to go with the 2 in its lone pair. Because it is 1 electron short it has a $+1$ formal charge: $6 - (2 + 3) = +1$. So the Lewis structure of SO_2 represented along with the various formal charges is:

0 +1 −1
:Ö::S̈:Ö̤:

The algebraic sum of the formal charges on all of the atoms in a molecule or ion must equal the actual charge on that species.

Resonance

Often it is possible to draw distinct Lewis structures based on the same skeleton and having a different distribution of electrons. Lewis structures drawn on the same skeleton but with different arrangements of electrons are resonance structures. The rules for constructing Lewis structures oblige a decision as to the exact location of every electron: localized in a lone pair on the first atom, localized as part of a double bond between two other atoms, etc. Resonance is invoked when there is no reason for preferring one such location over another. The structures:

:Ö::Ö:Ö: ↔ :Ö:Ö::Ö:

are resonance structures for the O_3 molecule. Both localizations of the double bond (to the right and to the left) are inadequate representations of the bonding. Neither is preferable. The actual bonding is a *hybrid*, a mixture, of the resonance contributors. There is a 1½ bond on each side.

Not all resonance structures need contribute equally to the final picture of the bonding in a molecule or molecular ion. Structures that violate the octet rule or require the build-up of large formal charges on the atoms are then not so much wrong as low-percentage contributors. Every Lewis structure that displays the correct number of valence electrons is, in principle, defensible as at worst a low-percentage resonance contributor.

2.3 *INORGANIC NOMENCLATURE AND OXIDATION NUMBERS*

Ions and Ionic Compounds

Many inorganic compounds are *ionic*, and *binary*, composed of one type of *cation* and one type of *anion*. Cations are positively charged and derive from elements losing one or more electrons. Anions are negatively charged and come from elements gaining one or more electrons. In comparing two elements in a compound, the more *electronegative* tends to gain electrons and form anions. The more electropositive tends to lose electrons and form cations. By definition, the more electronegative elements have a greater tendency in compounds to draw electrons to themselves. Electronegativity increases going across the periodic table to the right and up the periodic table.

Monatomic cations are named after the element from which they are derived. In the case of some metals which have more than one ion, Roman numerals in parentheses after the name of the metal indicate the ion charge. Cations are mostly derived from metals. Simple anions are mostly derived from non-metals. A monatomic anion is named by adding the suffix *-ide* to the stem of the name of the element from which it derives. In names of binary ionic compounds the name of the cation comes first followed by the name of the ion, separated by a space. Neither is capitalized unless used at the beginning of a sentence.

There are many *polyatomic* anions, particularly oxyanions. It is best to *memorize* the names of the common inorganic polyatomic anions. See Table 2-2 (text p. 47).

Binary Covalent Compounds

The method of naming binary covalent compounds resembles that used for binary ionic compounds. The bonding is covalent because the electronegativities of the bonded atoms do not differ enough to establish clear-cut formation of ions.

Often, a given pair of elements forms more than one binary covalent compound. In such cases either:

- Specify the number of atoms of each element in the molecular formula with Greek prefixes. *Di-* stands for two. *Tri*, *tetra*, *penta*, *hexa*, *hepta* stand for three through seven, in order. **See problem 9.**

- Treat the compound as if it were ionic and place the charge accorded to the first-named element in parentheses after the name of that element.

Oxidation Numbers

In a compound, an atom's *oxidation number* is generally equal to the charge that atom would have if all of the electrons involved in bonds to it were assigned arbitrarily to the more electronegative element. The term *oxidation state* is synonymous with oxidation number.

2.4 *INORGANIC REACTION CHEMISTRY*

Dissolution and Precipitation Reactions

Many common inorganic substances dissolve in water to give solutions that are good conductors of electricity. The mechanism of these dissolution reactions is the formation of aquated ions. Solubilities vary widely. $LiClO_3$ is tremendously soluble (500 g / 100 mL) in cold water. AgCl is essentially insoluble. Potassium perchlorate is an intermediate case. It dissolves only to a limited extent (0.75 g / 100 mL of water).

There is always an upper limit to solvents' capacity for any solute. Whenever a solute's limit of solubility is exceeded the solute tends to *precipitate* from the solution. Precipitation reactions are the reverse of dissolution reactions. The direction taken by the reaction depends on the concentration of the ions and their identities. Placing one mole of NaCl in a liter of water gives complete dissolution. Placing one mole of AgCl in a liter of water gives mostly undissolved solid.

The formation of a precipitate is most striking when two solutions containing ions which form an insoluble combination are mixed. The insoluble product forms, often quite quickly, and settles out of the solution.

Acid-Base Reactions

Acids are substances that increase the concentration of the H_3O^+ ion when they are placed in water. H_3O^+ ion is the *hydronium* ion. Bases are substances that increase the concentration of the *hydroxide* ion when they are placed in water.

Even the purest water contains small concentrations of H_3O^+ and OH^- because:

$$2\ H_2O \rightarrow H_3O^+(aq) + OH^-(aq)$$

When an acidic solution and a basic solution are mixed, *neutralization*, the reverse of the above reaction, occurs.

Acid and base are more broadly defined in the Lewis definition. A Lewis acid is an electron-pair acceptor and a Lewis base is an electron-pair donor. The generalization avoids the dependence on aqueous solvent and allows convenient use of Lewis structures.

Oxidation-Reduction Reactions

Oxidation reduction equations represent chemical reactions in which at least one element changes its oxidation number.

2.5 *BALANCING OXIDATION-REDUCTION EQUATIONS*

The criteria for balancing these equations are the same as for other chemical equations: the number of each kind of atom shown may not change from left to right nor may the net electrical charge change from left to right.

Two situations are common:

- All of the reactants and products are specified and the entire task is to find a set of coefficients.

- The reaction takes place in aqueous solution where $H_2O(l)$, H^+ *(aq)*, and OH^- *(aq)* may enter the equation as either reactants or products. Such cases require *completion* in addition to balancing.

When adding reactants or products to equations to complete them, bear certain ideas in mind:

- The formula H^+ *(aq)* is equivalent to H_3O^+ *(aq)*. The two are both valid representations of the hydrogen ion in aqueous solution. The second is the first with an H_2O added in. Use whichever is more convenient.

- In acid solution the concentration of OH^- *(aq)* is low. It is therefore not realistic to use it as a reactant in a balanced equation. Instead use H_2O as a source of O in the -2 oxidation state. If OH^- *(aq)* is formed in an oxidation-reduction reaction in acidic solution, then it will react with the plentiful H^+ *(aq)* ion to give water. Recognize this fact in balanced equations.

- In basic solution, H_3O^+ *(aq)* is scarce and should not be used as a source of hydrogen in the $+1$ oxidation state. Instead use H_2O. If H_3O^+ *(aq)* forms as a product, then it will react with the plentiful OH^- *(aq)* to give water. Recognize this in balanced equations. **See problems 19, 21 and 37.**

The text gives a clear summary of the steps to take to balance the oxidation-reduction equation in either case. Note certain common errors:

- Inserting H_2, O_2, H_2O_2 and other species in which H and O have oxidation numbers other than $+1$ and -2 respectively in attempting to complete equations. This always causes failure.

- Failing to recognize that two or three or even more elements in a compound can change oxidation number in the same reaction. **See problem 37a.**

- Not understanding that a single substance can *disproportionate*, that is, be both oxidized and reduced in the same reaction. **See problem 21d.**

DETAILED SOLUTIONS TO ODD-NUMBERED PROBLEMS

---

1. Make the predictions for the properties of scandium by taking the means of the corresponding values for Ca and Ti.

	m.p.(°C)	b.p. (°C)	density (g cm^{-3})
Sc(predicted)	1250	2386	3.02
Sc(observed)	1539	2730	3.02

Sc lies between Ca and Ti in the 4th row of the periodic table. Its properties are intermediate between those of Ca and Ti.

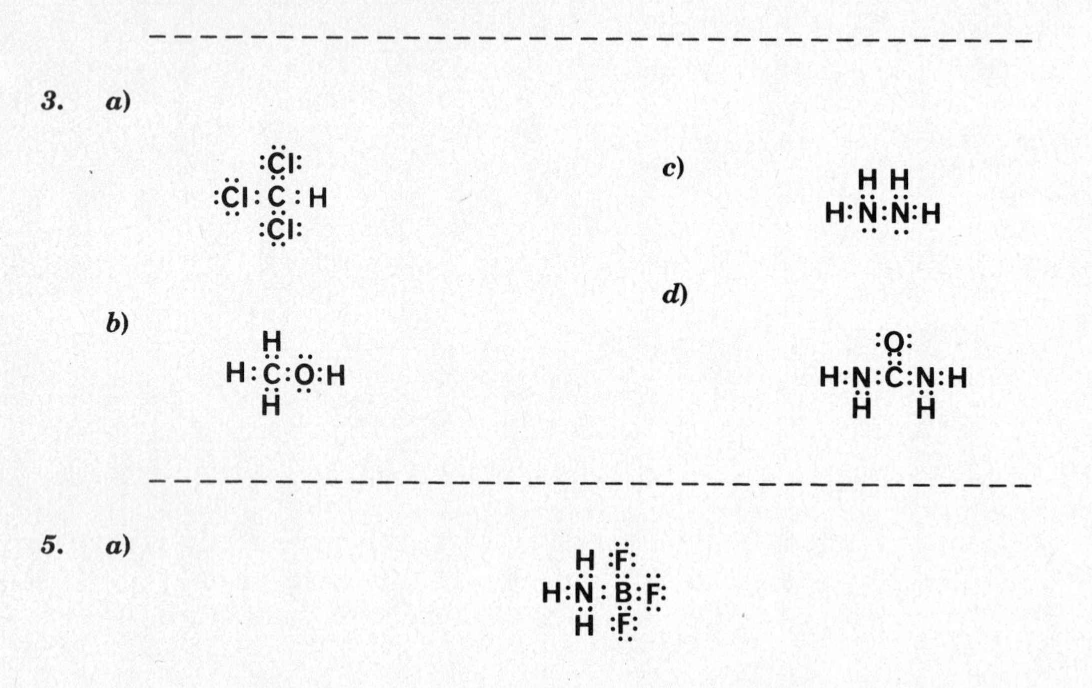

The formal charge on N is −1. It is +1 on B, and zero on all of the other atoms.

b)

[H:C:C:O:]$^{1-}$ (Lewis structure: H above and below the first C; double-bonded O above the second C)

The formal charge on the single-bonded O is −1. All other formal charges are zero.

c) The azide ion is linear. This fact is given to rule out *cyclic* Lewis dot structures in which each N is bonded to both others.

[:N̈::N::N̤̈]$^-$ ↔ [:N̤̈:N:::N:]$^-$ ↔ [:N:::N:N̤̈:]$^-$

In the left-most Lewis structure the formal charges, from left to right, are 0, +1, −1; in the center structure they are −2, +1 and 0; in the right-most structure 0, +1, −2.

d) The bicarbonate ion has 24 valence electrons. The C contributes 4, the H contributes 1, the 3 O atoms contribute 6 each, and a final electron is added to make the net charge -1. The resonance Lewis structures are:

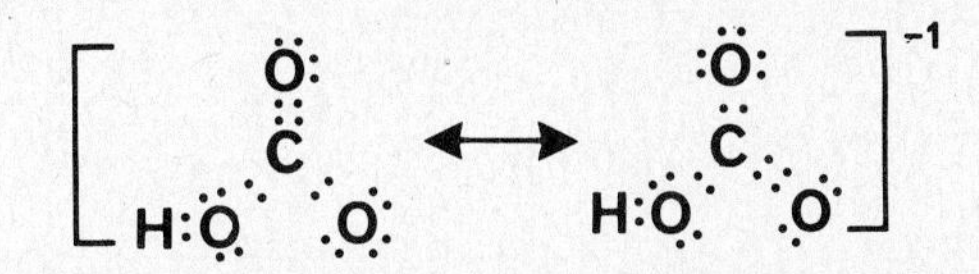

The double headed arrow links the two resonance structures. The -1 outside the brackets is the net charge on the ion. Formal charges on all atoms are zero with the exception of the O atoms which are singly bonded (one pair of dots) to C atoms and *not* bonded to H. These oxygen atoms have a formal charge of -1. The sum of the formal charges in either of the resonance structures is -1, equal to the net charge on the HCO_3^- ion.

Resonance structures differ *only* in the positions of the electrons. One common unacceptable answer to this problem presents a third resonance structure in which the oxygen atom on left shares two pairs of electrons with the central carbon atom. Such a structure forces a $+1$ formal charge onto the oxygen on the left, which is undesirable. If the H atom is moved to one of the other two oxygen atoms to avoid this outcome, then the formal charge arrangement is fine, but the structure is *not* a true resonance structure, because an atom as well as electrons has moved.

---

7. *a)* SiO_2

b) $(NH_4)_2CO_3$

c) PbO_2

d) P_2O_5

e) CaI_2

f) $Fe(NO_3)_3$

---

9. *a)* Copper(I) sulfide and copper(II) sulfide

b) sodium sulfate

c) tetraarsenic hexoxide or arsenic(III) oxide

d) zirconium(IV) chloride

e) dichlorine heptoxide or chlorine(VII) oxide

f) gallium(I) oxide

---

11. The oxidation numbers are determined by the standard rules: Sr $+2$, Br -1; Zn $+2$, O -2, H $+1$; Si $+4$, H -1; Ca $+2$, Si $+4$, O -2; Cr $+6$, O -2; Ca $+2$, P $+5$, O -2; K $+1$, O $-\frac{1}{2}$.

---

13. *a)* The Lewis acid *accepts* a pair of electrons. The Lewis base *donates* a pair of electrons. Ag^+ is the Lewis acid. It accepts one electron pair from each ammonia, the Lewis base.

b) $N(CH_3)_3$ is the Lewis base. BF_3 is the Lewis acid.

c) $AlCl_3$ is the Lewis acid. Cl^- is the Lewis base.

---

15. The balanced equation is:

$$Na_2O(s) + H_2O(l) \rightarrow 2NaOH(aq)$$

The oxidation number of Na on the left-hand side is +1. It is also +1 on the right-hand side. Similarly, the oxidation numbers of O and H remain unchanged by the reaction (at −2 and +1 respectively). The reaction is the displacement of a very weak Lewis acid (Na^+) from its association with O^{2-} by H^+, a stronger Lewis acid.

---

17. **a)** H is oxidized from 0 to +1 and O is reduced from +2 to −2. F retains the −1 oxidation number. To reduce one OF_2 then requires 4 electrons. This means 4 H atoms or 2 H_2 molecules. H_2 and OF_2 therefore must react in a 2 to 1 ratio:

$$2\ H_2(g) + OF_2(g) \rightarrow H_2O(g) + 2\ HF(g)$$

b) Titanium is reduced from +4 to 0. Mg is oxidized from 0 to +2, and Cl stays at −1. Therefore there must be 2 Mg atoms oxidized to furnish the electrons to reduce one Ti:

$$TiCl_4(l) + 2\ Mg(s) \rightarrow 2\ MgCl_2(s) + Ti(s)$$

c) Lead is reduced from +4 to +2; iodide is oxidized from −1 to 0. There must be two iodide's oxidized for every one Pb reduced. Some iodide is *not* oxidized:

$$PbO_2(s) + 4\ HI(g) \rightarrow PbI_2(g) + I_2(g) + 2\ H_2O(l)$$

d) The oxygen in hydrogen peroxide is reduced from −1 to −2. The carbon in CH_3OH is oxidized from −2 to +4. The H's and O in CH_3OH do not change oxidation number. Therefore:

$$3\ H_2O_2(l) + CH_3OH(l) \rightarrow CO_2(g) + 5\ H_2O(l)$$

---

19. **a)** The $VO_2^+(aq)$ ion is reduced and $SO_2(g)$ is oxidized:

$$2\ VO_2^+(aq) + SO_2(g) \rightarrow 2\ VO^{2+}(aq) + SO_4^{2-}(aq)$$

b) Copper metal is oxidized and nitrate ion is reduced. Because each copper atom loses two electrons and each nitrate ion gains one electron, the ratio of Cu to NO_3^- in the final balanced equation is 1 to 2:

$$Cu(s) + 4\ H^+(aq) + 2\ NO_3^-(aq) \rightarrow Cu^{2+}(aq) + 2\ NO_2(g) + 2\ H_2O(l)$$

c) Dichromate ion is reduced and neptunium(IV) is oxidized to Np(VI). Each of two Cr(VI)'s is reduced to Cr(III). Hence it takes six electrons from neptunium(IV) to achieve the reduction. Each Np(IV) loses two electrons so the ratio of Np(IV) to dichromate must be 3:1 in the final equation:

$$Cr_2O_7^{2-}(aq) + 3\ Np^{4+}(aq) + 2\ H^+(aq) \rightarrow 2\ Cr^{3+}(aq) + 3\ NpO_2^{2+}(aq) + H_2O(l)$$

d) Formic acid is oxidized and permanganate ion is reduced:

$$5\ HCOOH(aq) + 2\ MnO_4^-(aq) + 6\ H^+(aq) \rightarrow 5\ CO_2(aq) + 2\ Mn^{2+}(aq) + 8\ H_2O(l)$$

e) The reduction is the disproportionation of Pb_3O_4. Lead starts in the $+^8/_3$ oxidation state. Some is reduced to +2. Some is oxidized to +4:

$$4\ H^+(aq) + Pb_3O_4(s) \rightarrow 2\ Pb^{2+}(aq) + PbO_2(s) + 2\ H_2O(l)$$

The same answer comes from regarding Pb_3O_4 as $PbO_2 \cdot 2PbO$ partially dissolving in acid. The PbO_2 remains behind and the base PbO picks up H^+ to form Pb^{2+} and water. **See problem 33.**

f) Gold is oxidized and mercury is reduced:

$$3\ Hg_2HPO_4(s) + 2\ Au(s) + 8\ Cl^-(aq) + 3\ H^+(aq) \rightarrow 6\ Hg(l) + 3\ H_2PO_4^-(aq) + 2\ AuCl_4^-(aq)$$

---

21. When balancing equations for reactions taking place in basic aqueous solution $OH^-(aq)$ and H_2O may appear at will as products or be consumed as reactants, but $H^+(aq)$ may not. If it were generated then it would instantly react with $OH^-(aq)$ to give H_2O. Because the solution is basic the concentration of $H^+(aq)$ is low and the ion is too scarce to join in most reactions.

a) $AsO_3^{3-}(aq) + Br_2(aq) + 2\ OH^-(aq) \rightarrow AsO_4^{3-}(aq) + 2\ Br^-(aq) + H_2O(l)$

b) $ZrO(OH)_2(s) + 2\ SO_3^{2-}(aq) \rightarrow Zr(s) + 2\ SO_4^{2-}(aq) + H_2O(l)$

c) $7\ HPbO_2^-(aq) + 2\ Re(s) \rightarrow 7\ Pb + 2\ ReO_4^-(aq) + 5\ OH^-(aq) + H_2O(l)$

In this reaction rhenium is oxidized from 0 to +7 and lead is reduced from +2 to 0. This accounts immediately for the coefficients of 2 and 7. Once this ratio is fixed some provision must be made to carry off the extra oxygen from the seven $HPbO_2^-(aq)$ ions. Only 8 of the 14 O's are used to create the $2ReO_4^-(aq)$ ions. This is the reason for the OH^-'s and H_2O.

d) $4\ HXeO_4^-(aq) + 8\ OH^-(aq) \rightarrow 3\ XeO_6^{4-}(aq) + Xe(g) + 6\ H_2O(l)$

e) $3\ Ag_2S(s) + 2\ Cr(OH)_3(s) + 10\ OH^-(aq) \rightarrow 6\ Ag(s) + 3\ S^{2-}(aq) + 2\ CrO_4^{2-}(aq) + 8\ H_2O(l)$

f) $N_2H_4(aq) + 2\ CO_3^{2-}(aq) \rightarrow N_2(g) + 2\ CO(g) + 4\ OH^-(aq)$

---

23. Aluminum(III) oxide is Al_2O_3. To prepare elemental aluminum this oxide must be *reduced.* The carbon is therefore oxidized:

$$2\ Al_2O_3 + 3\ C \rightarrow 4\ Al + 3\ CO_2$$

---

25. ***a)*** There are only two elements in the compound. Indium and oxygen have the weight percentages 82.71 and 17.29 respectively. Take a random portion of the oxide, say, 92.53 g. In a 92.53 g sample of the indium oxide there are (0.1729 × 92.53) g of O or:

$$16.0\ \text{g O} \times \frac{1\ \text{mol O}}{16.0\ \text{g O}} = 1.00\ \text{mol O}$$

and (0.8271 × 92.53) g of In or:

$$76.53\ \text{g In} \times \frac{1\ \text{mol In}}{76} = 1.01\ \text{mol indium}$$

The empirical formula is therefore InO *if* indium has an atomic weight of 76.

The weight of substance to consider is a matter of choice. The "random" choice of 92.53 g in this case happens to give one mole of each element. If 100.0 g of the indium oxide had been considered, then the chemical amounts of oxygen and indium would have been 1.08 and 1.09 respectively...still in a 1 to 1 ratio.

b) The law of Dulong and Petit allows an estimate of the atomic weight of indium. The law states that the product of the molar weight of a metallic element and its specific heat capacity is about 25 J $mol^{-1}K^{-1}$. Therefore the molar weight of indium is about 107 g mol^{-1}. This number is approximately $^3/_2$ of 76. If the atomic weight of indium is $^3/_2$ times larger than 76, then $^2/_3$ as many moles of it are present in the sample of the oxide. The empirical formula is $In_{2/3}O_1$ or, with whole number subscripts, In_2O_3.

Take In_2O_3 as the correct formula of the indium oxide and suppose there is a 100.0 g sample of it. The sample contains 17.29 g of O or 17.29 g/15.9994 g mol^{-1} = 1.0807 mol of O. The chemical amount of indium in the sample is two-thirds that of the O:

$$^2/_3 \times (1.0807)\ \text{mol} = 0.72044\ \text{mol of indium.}$$

The mass of indium in the 100.0 g sample is 82.71 g. Therefore 82.71 g of indium is the same as 0.72044 mol of indium, and there is 82.71 g / 0.72044 mol = 114.8 g mol^{-1} of indium.

---

27. The binary compounds would have the formulas SbH_3, HBr, SnH_4, H_2Se. Antimony is in Group V, bromine in Group VII, tin in Group IV and selenium in Group VI. In each case the number of hydogens in the compound is 8 minus the group number. The formulas are given with the customary order of the elements. The compounds are named stibine, hydrogen bromide, stannane and hydrogen selenide, respectively.

---

29. The Lewis structure of ammonia, NH_3, has a central N surrounded by 4 pairs of valence electrons. Three of these pairs are shared severally with the three H atoms. The last pair is entirely owned by the N. The formal charges on all atoms are 0. There are no important resonance structures.

In methane, CH_4, the central C is surounded by 4 pairs of valence electrons. Each pair is shared with an H atom. The formal charges on all atoms are 0. There are no important resonance structures.

In carbon dioxide, CO_2, the central C is bonded with double bonds to each of 2 O atoms. Both O atoms have two pairs of nonbonding electrons of their own. There are thus 16 valence electrons in the Lewis structure. No resonance structures are necessary. All atoms have formal charges of 0.

Nitric acid has the skeleton $H-O-NO_2$. The central N atom shares one pair of electrons with the O atom which is attached to the H atom. It shares *two* pairs of electrons with one of the other O atoms and one pair with the second. The central N atom has a formal charge of $+1$. The O atom which shares only one pair of electrons has a -1 formal charge. Since it is immaterial which of the 2 O atoms this is, two resonance Lewis structures should be written. The only difference between them is the location of the $N=O$ double bond.

Sulfur trioxide, SO_3, has a central S surrounded by the O atoms. Two of the $S-O$ bonds are single bonds (one shared pair of electrons). The third is a double bond. The central S has a formal charge of $+2$. The single-bonded O atoms both have -1 formal charges, and the double-bonded O atom has a formal charge of 0. There are three resonance structures each showing a different O atom to have the double bond.

---

31. Three resonance Lewis structures, each of which satisfies the octet rule, can be constructed for BF_3. In each of the structures one of the fluorines has a double bond to the central B while the other two F's retain single bonds. The double-bonded F's have a formal charge of $+1$ in these structures, and the B has a formal charge of -1.

---

33. ***a)*** In PbO lead has a $+2$ oxidation number. In PbO_2 lead has $+4$ oxidation number. In Pb_2O_3 lead has $+3$ oxidation number. In Pb_3O_4 lead has $+{}^8/_3$ oxidation number.

PbO_2 conceivably could be "lead(II) peroxide" in which the oxygens would have -1 oxidation numbers and Pb a $+2$ oxidation number. The assignment of oxidation numbers is intrinsically arbitrary. Assigning the Pb a $+2$ oxidation number in PbO_2 could make no difference whatever in an practical proceeding such as the balancing of an oxidation-reduction equation. A $+4$ oxidation number for Pb in PbO_2 is the conventional choice.

b) Pb_2O_3 can be rewritten as $PbO \cdot PbO_2$. Pb_3O_4 reformulates as $PbO_2 \cdot (PbO)_2$.

---

35. ***a)*** $CO_2(g) + H_2O(l) \rightarrow H_2CO_3(aq)$ (carbonic acid)

b) $SO_2(g) + H_2O(l) \rightarrow H_2SO_3(aq)$ (sulfurous acid)

c) $SO_3(g) + H_2O(l) \rightarrow H_2SO_4(aq)$ (sulfuric acid)

d) $N_2O_5(g) + H_2O(l) \rightarrow 2\ HNO_3(aq)$ (nitric acid)

e) $P_4O_{10}(g) + 6\ H_2O(l) \rightarrow 4\ H_3PO_4(aq)$ (phosphoric acid)

f) $Cl_2O_7(l) + 2\ H_2O(l) \rightarrow 2\ HClO_4(aq)$ (perchloric acid)

---

37. ***a)*** Both Cr(III) and I(−I) are oxidized. That is, the compound CrI_3 loses electrons from both of its constituent elements. Also, all of the oxygen on the right-hand side must come from the solvent. Both facts are shown in the balanced oxidation *half-equation*:

$$25\ H_2O + 2\ CrI_3 \rightarrow Cr_2O_7^{2-} + 6\ IO_3^- + 50\ H^+ + 42\ e^-$$

Two Cr's go from +3 to +6 and six I's go from −1 to +5 for a loss of 2(3) + 6(6) = 42 electrons. A single Br_2 absorbs 2 electrons as it is reduced to 2 Br^-. At this rate it takes 21 Br_2 molecules to consume the 42 electrons from the oxidation. The balanced equation becomes:

$$25\ H_2O(l) + 2\ CrI_3(s) + 21\ Br_2(l) \rightarrow 42\ Br^-(aq) + Cr_2O_7^{2-}(aq) + 6\ IO_3^-(aq) + 50\ H^+(aq)$$

b) In this equation vanadium is oxidized. The H_2 gas comes from the reduction of water. Consider the oxidation separately:

$$33\ OH^-(aq) + 6\ V(s) \rightarrow HV_6O_{17}^{3-}(aq) + 16\ H_2O(l) + 30\ e^-$$

It is an error to insert hydrogen in the form of H_2 (oxidation number zero) among the oxidation products when it starts (in the OH^- ion) with an oxidation number of +1. Each separate species that gives or takes electrons should be dealt with separately. The H_2 comes from the reduction of water. The overall balanced equation:

$$3\ OH^-(aq) + 6\ V(s) + 14\ H_2O(l) \rightarrow HV_6O_{17}^{3-}(aq) + 15\ H_2(g)$$

c) On the left-hand side, Te has oxidation numbers of 6 and zero. On the right-hand side, Te has oxidation number +4 and +7. Clearly the elemental tellurium is oxidized. Suppose for the moment the Te(0) gives rise to the Te(VII) product. The oxidation is then:

$$8\ OH^-(aq) + Te(g) \rightarrow TeO_4^-(aq) + 4\ H_2O(l) + 7\ e^-$$

and the reduction half-equation is:

$$2\ e^- + H_6TeO_6(g) \rightarrow TeO_2(s) + 2\ OH^-(aq) + 2\ H_2O(l)$$

The overall balanced equation:

$$7\ H_6TeO_6(s) + 2\ Te(s) \rightarrow 7\ TeO_2(s) + 2\ TeO_4^-(aq) + 20\ H_2O(l) + 2\ H^+(aq)$$

If, instead of the above, the oxidation is imagined to go Te $\rightarrow$ TeO_2, then the other change in oxidation number, $H_6TeO_6 \rightarrow TeO_4^-$, is *also* an oxidation. This is impossible because there must be a reduction for every oxidation.

d) The reaction is the oxidation of thiocyanate ion to give cyanide and sulfate ions. It is not important to decide which elements in the CNS^- ion gain exactly how many electrons. The ion as a whole definitely loses 6 electrons and that is what counts:

$$4\ H_2O(l) + CNS^-(aq) \rightarrow CN^-(aq) + SO_4^{2-}(aq) + 8\ H^+(aq) + 6\ e^-$$

The iodate ion is reduced. The overall balanced equation:

$$2\ CNS^-(aq) + 2\ H^+(aq) + 6\ Cl^-(aq) + 3\ IO_3^-(aq) \rightarrow$$
$$2\ CN^-(aq) + 2\ SO_4^{2-}(aq) + 3\ ICl_2^-(aq) + H_2O(aq)$$

e) The reduction is of a compound. In the compound two elements go to lower oxidation numbers. The reduction is:

$$16\ e^- + 24\ H^+(aq) + 18\ I^-(aq) + Hg_5(IO_6)_2(s) \rightarrow 5\ HgI_4^{2-}(aq) + 12\ H_2O(l)$$

The 18 I^- in this half-equation only help to make the HgI_4^{2-} ion and do not change oxidation number. Meanwhile some additional iodide *is* oxidized:

$$2I^-(aq) \rightarrow I_2(aq) + 2e^-$$

The overall balanced equation:

$$24\ H^+(aq) + 34\ I^-(aq) + Hg_5(IO_6)_2(s) \rightarrow 5\ HgI_4^{2-}(aq) + 8\ I_2(aq) + 12\ H_2O(l)$$

Chapter 3

KINETIC-MOLECULAR THEORY OF GASES

3.1 *THE CHEMISTRY OF GASES*

Unlike solids and liquids, gases expand to fill completely any container they occupy. There are only weak interactions between the particles of real gases, and the particles on the average are far apart from each other.

Gases form chemically:

- By thermal decomposition of solids (such as mercury(II) oxide and ammonium hydrogen carbonate).
- By the direct reaction of many elements with oxygen to give oxides.
- By the action of acids on many ionic solids. When ionic solids are dissolved in water and then mixed the escape of a water-insoluble gas tends to drive the reaction toward the products.

3.2 *IDEAL GASES*

Three variables find great use in describing the physical state of gases: pressure, temperature and volume. At sufficiently low densities, all gases follow the *ideal gas law,* an equation relating these variables:

$$PV = nRT$$

In this equation V represents the volume of the gas, P its pressure, T its absolute temperature and n its amount. R is a constant.

The absolute temperature must be used whenever any form of the ideal gas law is used. **See problems 5 and 7b.**

Volume is the three-dimensional region available to the sample of gas under consideration. The natural SI unit of volume is the cubic meter. Because this unit is inconveniently large, additional units of volume are in common use:

$$^{1}/_{1000}\ m^3 = 1\ \text{liter (L)}$$

$$^{1}/_{1000}\ L = 1\ mL = 1\ cm^3$$

Pressure is force divided by area. Intuitively, a force is a push. Even a weak push driving a sharp pin against the skin causes pain because the area of the pin's point is small and enormous pressure exists directly under it. The SI unit of force is the *newton* (kg m s^{-2}). The natural SI unit of pressure is the newton $meter^{-2}$, also called the *pascal.* The pascal is unsatisfactorily small for common applications. This is not because the newton is such a small unit of force but rather because a square meter is quite a large area. Many other pressure units are defined and in use. Some important ones are related to each other and the pascal as follows:

$$101{,}325 \text{ pascal} = 1 \text{ atmosphere} = 760 \text{ torr} = 1.01325 \text{ bar} = 14.7 \text{ psi}$$

A torr is the pressure exerted by a column of mercury one millimeter high when the temperature of the mercury is 0° C. The diameter of such a column does not matter. A large diameter column of mercury exerts a large force but spreads it over a large area. A small column of mercury exerts less force, but because the area at its base is proportionately less, the pressure there is the same. This fact is the basis for the operation of a barometer.

The everyday pressure of the earth's atmosphere fluctuates in the neighborhood of one atmosphere (atm). A psi is a pound per square inch.

The pressure *of* a gas is equal to the pressure *on* that gas. When the internal pressure of a sample of gas differs from the external pressure, then the gas either expands or is compressed.

The constant, R, is the *universal gas constant.* It is as fundamental a constant of nature as the speed of light. The units of the gas constant are:

$$\text{energy}\cdot\text{temperature}^{-1}\,(\text{chemical amount})^{-1}.$$

In the SI system, $R = 8.314 \text{ J K}^{-1}\text{mol}^{-1}$. Another useful value of R is 0.082057 $\text{L atm K}^{-1}\text{mol}^{-1}$. The first value is mostly used in calculations involving the energy of molecular motion. The second is more useful in calculations involving different states of gases.

- It is legitimate and often advantageous to rewrite R as, for example, 8.314×10^{-3} $\text{kJ K}^{-1}\text{mol}^{-1}$ or $82.057 \text{ cm}^3 \text{atm K}^{-1}\text{mol}^{-1}$. **See problems 15 and 31.**

A comparison of the different values of R shows that the L atm is a unit of energy. It is (force/area) × volume which is force × distance. Think of a piston being forced along a cylinder by the expansion of a gas.

The ideal gas equation includes several important physical relationships which have names of their own:

- Boyle's law states that the pressure of a quantity of gas is inversely proportional to its volume when the gas is held at a constant temperature.

$$P = nRT \times (1 / V)$$

where n is the quantity of gas, and T is the temperature.

- Charles' law states that the volume of a quantity of gas is linearly dependent on its temperature when the gas is held at a constant pressure. All gases (at low enough pressures) expand by 1 / 273.15 of their volume at 0° C for every degree Celsius they are heated. This finding by Charles suggested the definition of the absolute scale of temperature upon which the volume of the gas is directly proportional to the temperature:

$$V = (nR / P) \times T$$

The absolute scale has degrees (named kelvins) of the same size as Celsius degrees but has its zero at -273.15° C:

$$T = t + 273.15$$

• Avogadro's hypothesis (Chapter 1) is also implicit in the ideal gas law. At a specified temperature and pressure, the volume of the gas is directly proportional to the chemical amount of the gas:

$$V = (RT / P) \times n$$

There is a limitless number of possible problems that use the ideal gas equation. Many of these have their roots in practical laboratory work and others are more fanciful. Whatever the problem, it will fall into one of two broad categories:

• *Problems Involving a Change of Conditions.* The temperature, pressure and volume of a gas are stated. One or two of the variables are then changed and the resulting value of the third variable is required. Such problems are usually easy. A reading of the problem allows construction of a table of P_1, V_1, T_1 and P_2, V_2, T_2. Then, because:

$$R = PV / nT$$

it follows that:

$$P_1 V_1 / n_1 T_1 = P_2 V_2 / n_2 T_2$$

and the rest is substitution. **See problems 3 and 5.** Problems in which the chemical amount of the gas changes along with or instead of P, V and T also occur. **See problems 43 and 45.**

• *Problems Requiring Computation of an Unknown Variable.* The problem will in some guise give the values of three out of four of the variables in the ideal gas equation and ask the computation of the fourth. All that is required is access to a suitable value for R, the universal gas constant, substitution, and some care with the units. Important variations on this theme do not give explicit values of variables but instead give a quantity relating two of them. For example, knowing the *ratio* (n/V), the molar density of a gas, is just as good as knowing both n and V if P or T must be computed.

3.3 *DALTON'S LAW OF PARTIAL PRESSURES*

According to Dalton's law, each gas in a mixture of gases exerts a *partial pressure* equal to the pressure it would exert if it were present by itself under the same conditions. The sum of the partial pressures of the components of a mixture of gases equals the total pressure:

$$P_{tot} = P_1 + P_2 + P_3 + \ldots$$

where P_1 is the partial pressure of the first component, P_2 is the partial pressure of the second, etc. Dalton's law implies that a molecule in a mixture of gases interacts with *unlike* molecules to just the same extent it does with *like* molecules.

A common application of Dalton's law occurs when gases are collected over volatile liquids, *e.g.*, the collection of oxygen gas over water. The O_2 gas is admixed with water vapors. The total pressure of the sample is then:

$$P_{tot} = P_{oxygen} + P_{water\ vapor}$$

To get the pressure that the oxygen would exert if it were pure, the contribution of the water vapor must be subtracted. Fortunately this depends only on the temperature (see Chapter 4) and is widely tabulated.

3.4 *KINETIC MOLECULAR THEORY*

The postulates of the kinetic molecular theory are brief:

1. A pure gas has many identical molecules of negligible size. That is, the molecules are approximately point masses.

2. The molecules of a gas are in chaotic motion with a distribution of speeds.

3. There are no interactions among the molecules except during *elastic* collisions. An elastic collision is a collision in which no translational energy is lost.

From these postulates it is possible to derive Boyle's law (See text p. 81-83). The derivation uses classical mechanics.

An important aspect of the derivation is the difference between *speed* and *velocity*. The velocity, v, of an object is a vector. It has both magnitude and direction. The object's speed, u, is how fast it moves without regard to direction, that is, speed is a magnitude alone. A car proceeding too fast gets a speeding ticket. A car proceeding (even slowly) the wrong way on a one-way street gets a velocity ticket.

Like any vector, velocity can be expressed in terms of a triple of components, v_x, v_y, and v_z, relative to a set of Cartesian coordinate axes. The square of the speed is:

$$u^2 = v_x^2 + v_y^2 + v_z^2$$

Consider two cars. The first is going south at 50 m s^{-1} and has Cartesian velocity components (0, −50, 0). The two zeros in the triple mean the car is not headed east or west or up or down. The second is going north at 50 m s^{-1}. Its velocity components are (0, 50, 0). The average speed of the cars is 50 m s^{-1}, but their average velocity is 0.0 m s^{-1}.

¤ The addition of velocities must take their directions into account. **See problem 19.**

Distribution of Molecular Speeds

The speeds of the molecules in a sample of gas are ordinarily distributed across a wide but statistically predictable range. The prediction is made by the *Maxwell-Boltzmann equation*:

$$f(u) = 4\pi\left[\frac{m}{2\pi k_B T}\right]^{3/2} u^2\, e^{-mu^2/2k_B T}$$

To study and understand this complex expression it is best to take it apart. Two terms involve u, the molecular speed: $e^{-mu^2/2k_BT}$ and u^2. The quantity $(-mu^2/2k_BT)$ gets more and more negative as u increases and m and T remain unchanged. (The remaining factor, k_B, is a constant.) Therefore $e^{-(mu^2/2k_BT)}$ gets smaller and smaller as u goes up. If this exponential term alone controlled the distribution of molecular speeds, the molecules in a gas would scarcely be moving. But this term is multiplied by u^2 which of course increases as u increases. As a result, f(u) at first grows as u rises from zero, then reaches a maximum, and finally diminishes as the effect of the exponential term becomes dominant. The following table shows how this works out for N_2 at 273.15 K:

u (m s^{-1})	u^2 (m^2 s^{-2})	$e^{-mu^2/2k_BT}$	f(u) (s m^{-1})
0	0	1.000	0.0
50	2.5×10^{3}	0.985	8.5×10^{-5}
100	1.0×10^{4}	0.940	3.3×10^{-4}
200	4.0×10^{4}	0.781	1.1×10^{-3}
300	9.0×10^{4}	0.574	1.8×10^{-3}
400	1.5×10^{5}	0.373	2.1×10^{-3}
500	2.5×10^{5}	0.214	1.9×10^{-3}
1000	1.0×10^{6}	2.10×10^{-3}	7.3×10^{-5}
3000	9.0×10^{6}	7.81×10^{-25}	2.4×10^{-25}

At low speeds u^2 dominates. At high speeds the exponential term crushes the total function f(u) to zero. The two leading terms in the Maxwell-Boltzmann distribution are constants for a given gas at a given temperature. In this example their combined value is 3.457×10^{-8} s^3 m^{-3}.

The units of f(u) are s m^{-1}, the reciprocal of the units of speed. When f(u) is multiplied by Δu, a range of speed, all of the units cancel out. This reveals f(u)Δu as a *probability distribution*, a pure number telling the chance that a molecule will have a speed between u and u + (Δu).

Example. What is the chance that an N_2 molecule at 273.15 K has a speed between 390 and 410 m s^{-1}?

Solution. Δu is 20 m s^{-1}, f(u) at 400 m s^{-1} is 2.1×10^{-3} (from the table). Therefore, f(u)Δu = 0.042. This is the chance that an individual N_2 molecule is in the desired range of speeds. It is as well the *fraction* of all the molecules with speeds in the 390 to 410 m s^{-1} range. The answer assumes that f(u) is constant across the 390 to 410 m s^{-1} range. It can be improved by computing f(u) at various speeds between 390 and 400 m s^{-1}, using shorter Δu's which bracket these speeds, and adding up the subtotals.

The Maxwell-Boltzmann distribution function is formed in such a way that the sum of all possible probabilities is 1.

The root-mean-square speed is the square root of the mean of the squares of the speeds of a collection of molecules. It is defined by the equation:

$$u_{rms} = (3k_BT / m)^{1/2}$$

The *Boltzmann constant*, k_B, has the units of (energy / temperature). It equals 1.38×10^{-23} J K^{-1}. T and m stand for the absolute temperature of the gas and the mass of a molecule. If m is in kg then u_{rms} comes out in m s^{-1}. Common gases at room temperature have root-mean-square molecular speeds on the order of 10^3 m s^{-1}.

The Boltzmann constant is equal to the universal gas constant divided by Avogadro's number:

$$k_B = R / N_0$$

This means that k_B / m can be replaced by R / M where M is the mass of a mole (Avogadro's number) of molecules. Therefore the root-mean-square speed is also:

$$u_{rms} = (3RT / M)^{1/2}$$

If R in this equation is taken as 8.314 J $K^{-1}mol^{-1}$ then M must have the units kg mol^{-1} (and not the more usual g mol^{-1}) in order for the units of the speed to come out as m s^{-1}. To prove this, recall that a joule is a kg $m^2 s^{-2}$.

The Maxwell-Boltzmann distribution curve is not symmetrical about its maximum. Consequently two other representative speeds of molecules, differing from u_{rms}, arise:

$$\bar{u} = (8k_BT / \pi m)^{1/2} = (8RT / \pi M)^{1/2}$$

$$u_{mp} = (2k_BT / m)^{1/2} = (2RT / M)^{1/2}$$

$\bar{u}$ is the arithmetic mean of the speeds of the molecules in a sample of gas. It is always less than u_{rms} because squaring the speeds, the procedure used in computing a root-mean-square speed, gives extra emphasis to larger speeds. The third type of u is u_{mp}, the most probable speed. It is always less than both u_{rms} and $\bar{u}$. It is the speed at which the Maxwell-Boltzmann distribution curve reaches its maximum.

The average translational kinetic energy is related simply to u_{rms}:

$$E_{avg} = (\tfrac{1}{2} mu^2)_{avg} = \tfrac{1}{2} m (u_{rms})^2$$

How a Velocity Selector Works

Imagine two disks set parallel to each other, arranged to spin on the same shaft, and separated by a distance of one meter. Each disk has a narrow slit so that a beam of atoms can get in and out of the region between the disks. The slits are offset by a 90° angle so one lags the other at the shaft spins. If the disks are set spinning at a rate of 250 revolutions per second then it takes 1/250 of a second for a complete revolution to take place and 1/1000 of a second elapses between the time an atom entering through the leading slit can possibly exit through the lagging slit. Atoms in the beam going 1000 m s^{-1} will be able to pass both slits. These fast atoms will just make it through the interior of the velocity selector in the time it takes the second slit catch up to the angular position the first slit had when the beam of atoms entered. Most other atoms will not pass the selector, although atoms with a speed of $^1/_5$ of 1000 m s^{-1} will, in their more sluggish way, get to the second disk just as it has completed $^5/_4$ revolutions and will also be able to exit.

3.5 *APPLICATIONS OF THE KINETIC THEORY*

Wall Collisions, Effusion and Diffusion

The rate of collisions of the molecules of a gas with a section of a wall of its container is:

$$Z_w = \tfrac{1}{4}\,(N/V)\,\bar{u}\,A = \tfrac{1}{4}\,(N/V)\,[8k_BT/\pi m]^{\frac{1}{2}}\,A$$

where A is the wall area under consideration and (N/V) is the *number density* of the gas. The number density of any gas is equal to its molar density multiplied by Avogadro's number (N_0). The units of Z_w are time^{-1}, that is, Z_1 tells collisions per unit time.

Graham's law of effusion follows from this expression. Effusion is the escape of a gas from a container through a small hole into a vacuum.

- Graham's law states that the rate of effusion of a gas is inversely proportional to the square root of its molecular weight at constant temperature and pressure.

Graham's law could also be stated in terms of gas *densities* because the density of a gas is directly proportional to its molecular weight. Diffusion, which is physically a different process, is also governed by Graham's law.

The typical problem involving Graham's law require the computation of the molecular weight of an unknown gas by comparison of its rate of effusion to the rate of effusion of a known gas under the same conditions. **See problem 51.**

Frequency of Molecular Collisions

If the molecules of a gas were mathematical points, then they could collide only with the walls of their container. They would always miss each other. In real gases the molecules have a small effective diameter, *d.* Then Z_1, the rate at which a typical molecule experiences collisions, is:

$$Z_1 = \sqrt{2}\,(N/V)\,\pi d^2\,\bar{u} = \sqrt{2}\,(N/V)\,\pi d^2\,[8k_BT/\pi m]^{\frac{1}{2}}$$

Z_1 has the SI units s^{-1}. The derivation of this expression assigns a diameter, *d,* to the molecules of the gas and imagines an individual molecule sweeping out a cylinder in some fixed time. The frequency of collisions of this molecule with others varies with the square of the diameter of its cylinder. A bigger molecule is harder to avoid. Because a faster moving molecule sweeps out a longer cylinder in the fixed time, the frequency of collisions is also directly proportional to the root-mean-square speed of the molecules. Finally, the greater the density of the gas, then the more encounters the test molecule experiences. The frequency of molecular collisions is therefore directly proportional to the number density of the gas.

Mean Free Path and Diffusion

The mean free path of a gas (symbolized λ) is the average distance its molecules travel between collisions. It is the product of the average speed of the molecules ($\bar{u}$) and the average time between their collisions. The average time between collisions is the reciprocal of the rate of collision. Therefore:

$$\lambda = \bar{u} / Z_1$$

It is a good idea to study expressions like this in the above simple form, despite the fact that some further manipulation is possible. After substituting for Z_1, $\bar{u}$ cancels out:

$$\lambda = \frac{1}{\sqrt{2}\,(N/V)\pi d^2}$$

This result shows that the mean free path of a gas is independent of its temperature and its molecular mass. As the mean free path of a gas becomes comparable to its molecular diameter, deviations from ideality (next section) become important.

3.6 *EQUATIONS OF STATE AND NONIDEAL GASES*

The ideal gas equation is an equation of state. It relates some of the state variables that describe the physical behavior of a sample of gas. It is not the only possible equation of state, although it is certainly the simplest.

More elaborate equations of state take into consideration the definite, if small, size of the molecules comprising the system and the existence of intermolecular attractions. Such equations of state fit observed P-V-T data better. One comparatively simple equation of state is the van der Waals equation:

$$\left(P + \frac{an^2}{V^2}\right)(V - nb) = nRT$$

The van der Waals equation is the ideal gas equation elaborated with two correction factors. An additive correction ($a\, n^2/V^2$) operates on the observed pressure of the gas. Intermolecular attractions tend to reduce pressures from what they would be in the ideal case (no attractions) and the a-factor correction makes up for the reduction.

A subtractive correction (nb) is applied to the actual volume available to the motion of the molecules of the gas to account for the volume excluded to free motion by the existence of other molecules.

The strength of the intermolecular attractions in a gas and the size of the excluded volume both depend on the identity of the gas. Therefore the values of the van der Waals constants a and b differ from gas to gas. They are experimentally determined.

DETAILED SOLUTIONS TO ODD-NUMBERED PROBLEMS

1. Because water is considerably less *dense* than mercury it takes a longer column of it to balance the pressure of the atmosphere. 1.00 atm of air requires a barometric mercury column 76.0 cm high. Mercury's density is 13.6 g cm^{-3}, and water's is only 1.00 g cm^{-3}. Therefore the height of the column of water would be 76.0 cm × (13.6 g cm^{-3}/1.00 g cm^{-3}) = 13.6 × 76 cm = 1.03 × 10^3 cm, nearly 34 feet. Such water barometers have been built.

---

3. Assume that Kr is an ideal gas and then write the ideal gas equation for the system before and after the change, using subscripts to distinguish the two states:

$$P_1 V_1 = n_1 R T_1 \quad \text{(before compression)}$$

$$P_2 V_2 = n_2 R T_2 \quad \text{(after compression)}$$

The quantity of krypton gas is not changed in the process, so $n_1 = n_2$. The process goes on at a constant temperature so $T_1 = T_2$. Therefore:

$$P_1 V_1 = P_2 V_2$$

In this case P_1 is 1.00 atm because the compression starts at the standard pressure. V_1 is 5.00 L and V_2 is 2.30 L. Substitution gives:

$$P_2 = 1.00 \text{ atm} \times (5.00 \text{ L} / 2.30 \text{ L}) = 2.17 \text{ atm}$$

---

5. The actual pressure inside the tire is 30 psi + 14.7 psi = 44.7 psi. Assuming that the air in the tire behaves ideally:

$$\frac{P_1 V_1}{P_2 V_2} = \frac{n_1 T_1}{n_2 T_2}$$

The subscripts refer to the variables' values before and after the change from 0° C to 32° C. $V_1 = V_2$ because the tire is non-expandable. Heating does not change the quantity of air inside the tire, so $n_1 = n_2$. Cancelling these quantities:

$$P_1 / P_2 = T_1 / T_2$$

The temperatures are T_1 = 273 K and T_2 = 305 K. P_1 = 44.7 psi. Substituting and rearranging:

$$P_2 = P_1 (305 \text{ K} / 273 \text{ K}) = 44.7 \text{ psi } (305 / 273) = 49.9 \text{ psi}$$

It was unnecessary to convert the pressure to metric units. The units of P_2 are automatically the same as the units of P_1. The conversion to absolute temperature *was* necessary. If T_1 had not been converted in this case it would have meant dividing by zero.

Gauge pressure is 14.7 psi *less* than the actual pressure. The gauge pressure of the tire at 32° C is thus 49.9 − 14.7 = 35.2 psi. The answer is larger than the original gauge pressure, which fulfills common-sense expectation.

---

7. **a)** Assume that argon follows the ideal gas equation. Then PV = nRT, and all of the variables are known except the chemical amount (n) of argon. The weight of argon is the chemical amount multiplied by 39.948 g mol^{-1}, argon's molar weight. First, the chemical amount:

$$n = \frac{PV}{RT} = \frac{0.395 \text{ atm} \times 0.750 \text{ L}}{0.08206 \text{ L atm K}^{-1}\text{mol}^{-1} \times 298.15 \text{ K}}$$

The volume of the argon has been converted to L and its temperature to K. Doing the arithmetic gives n = 0.01211 mol, and the weight of Ar (molar weight 39.948 g mol^{-1}) is 0.484 g. An additional significant digit was carried in the calculation of n but the final answer was rounded off to three significant figures.

b) Qualitatively, the decrease in volume and increase in temperature both tend to *increase* the pressure of the gas. From the ideal gas equation:

$$\frac{P_1 V_1}{P_2 V_2} = \frac{n_1 T_1}{n_2 T_2}$$

where the subscripts distinguish the before and after values. The amount of argon gas does not change, so n_1 equals n_2. Furthermore:

T_1 = 298.15 K T_2 = 373.15 K
V_1 = 750 cm^3 V_2 = 500 cm^3
P_1 = 0.395 atm P_2 is the only unknown

Solving the above equation for P_2:

$$P_2 = \frac{T_2}{T_1} \times \frac{V_1}{V_2} \times P_1$$

This means that the new pressure equals the old pressure multiplied first by a factor reflecting the temperature change and then by a similar factor reflecting the volume change:

$$P_2 = \frac{373.15 \text{ K}}{298.15 \text{ K}} \times \frac{750 \text{ cm}^3}{500 \text{ cm}^3} \times 0.395 \text{ atm}$$

Both of the multiplying factors exceed one, as expected. It was not necessary to change the units of the volumes. The ratio of 750 cm^3 to 500 cm^3 is after all the same as the ratio of 0.75 L to 0.50 L. The conversion to Kelvin temperatures *was* necessary. The ratio 100° C / 25° C is *not* the same as 373.15 / 273.15. The new pressure is 0.742 atm.

c) Density is mass divided by volume. The initial gas density is therefore 0.484 g / 750 cm^3 or 0.645×10^{-3} g cm^{-3}, or 0.645 g L^{-1}. After the compression the density is 0.484 g / 500 cm^3 or 0.968 g L^{-1}.

When the gas is compressed to $^2/_3$ of its original volume (from 750 to 500 cm^3), the gas density must increase by a factor of $^3/_2$ to 0.968 g L^{-1}. The temperature change has no effect on the density of the argon.

---

9. ***a)*** The balanced equation:

$$2\ Na(s) + 2\ HCl(g) \rightarrow H_2(g) + 2\ NaCl(s)$$

b) Use a series of conversion factors. The key factor is the second. It is supplied by the balanced equation.

$$6.24\ g\ Na \times \frac{1\ mol\ Na}{23.0\ g\ Na} \times \frac{1\ mol\ H_2}{2\ mol\ Na} = 0.136\ mol\ H_2$$

Assuming that the hydrogen behaves ideally:

$$V = nRT / P$$

where n = 0.13565 mol, T = (273.15 + 50) K, P = 0.850 atm and R = 0.08206 L atm $K^{-1}mol^{-1}$. Note the use of extra significant figures in n. Inserting these numbers in the equation and completing the arithmetic gives V = 4.23 L.

It is a mistake to try to compute the volume of H_2 by multiplying the 0.136 mol H_2 by 22.4 L mol^{-1}. Even though the units come out right the 22.4 L mol^{-1} applies only at STP.

---

11. The hydrocarbon contains *only* C and H. We can compute the mass of C and the mass of H in the original sample from the amounts of CO_2 and H_2O that the sample furnishes:

$$1.930\ g\ CO_2 \times \frac{12.01\ g\ C}{44.01\ g\ CO_2} = 0.5267\ g\ C$$

$$1.054\ g\ H_2O \times \frac{2.016\ g\ H}{18.015\ g\ H_2O} = 0.1179\ g\ H$$

Next calculate the chemical amounts of C and H:

$$0.5267\ g\ C \times \frac{1\ mol\ C}{12.01\ g\ C} = 0.04386\ mol\ C$$

$$0.1179\ g\ H \times \frac{1\ mol\ H}{1.00797\ g\ H} = 0.1170\ mol\ H$$

The C to H molar ratio is 1 to 2.667, and the empirical formula is C_3H_8.

---

13. Assume that for the mixture PV = nRT. This expression is readily solved for n, the total chemical amount of mixed gas in the vessel. Substituting T = (273.15 + 50) K, V = 50 L, P = 2.5 atm, and taking R as 0.08206 L atm $K^{-1}mol^{-1}$ gives n = 4.714 mol.

If 0.40 of this 4.714 mol of gas is Ar, then n_{Ar}= 1.8856 mol and the mass of argon is 1.8856 mol × 39.948 g mol^{-1} = 75.326 or 75 g of argon. Rounding off to two significant figures is reserved for the last step.

Because hellium comprises 0.60 of the molecules in the mixture it contributes 0.60 of the total pressure or 0.60 × 2.5 = 1.5 atm.

---

15. ***a)*** Assume the water vapor mixed with air can be treated as a mixture of ideal gases. Then for the water vapor:

$$P_{water} = n_{water}RT / V$$

holds, a consequence of Dalton's law of partial pressures. The volume is 1.0 cm^3, the temperature is (20 + 273.15) K, and P_{water} is 0.0230 atm. Substituting these values of the variables together with R = 82.057 cm^3 atm $K^{-1}mol^{-1}$ gives n_{water} = 9.56 × 10^{-7} mol.

Each mole of H_2O contains Avogadro's number, N_0, of molecules so the 1.0 cm^3 of saturated air contains (9.56 × 10^{-7})mol × (6.022 × 10^{23})mol^{-1} = 5.76 × 10^{17} molecules of water. Rounding off to two significant figures, this becomes 5.8 × 10^{17} molecules.

b) Because 1.0 cm^3 holds only about 10^{-6} mole of water, it will take much more than 1 cm^3 to hold 0.50 mol of water:

$$0.50 \text{ mol } H_2O \times \frac{1.0 \text{ cm}^3 \text{ sat. air}}{9.56 \times 10^{-7} \text{ mol } H_2O} = 5.23 \times 10^5 \text{ cm}^3$$

After converting to liters and rounding off to two significant figures, the answer is 5.2 × 10^3 L.

---

17. ***a)*** The product of the pressure exerted by the Ar atoms and the volume of the system is $^2/_3$ of the total kinetic energy of translation of the argon (according to the derivation of Boyle's law from the kinetic-molecular theory):

$$P_{Ar} V = {}^2/_3 \, E_k$$

Because 35 percent of the molecules in the methane argon mixture are argon:

$$P_{Ar} = 0.35 \times 5.00 = 1.75 \text{ atm}$$

The total kinetic energy of the Ar molecules is therefore:

$${}^3/_2 \times P_{Ar} \times V = {}^3/_2 \times 1.75 \text{ atm} \times 100 \text{ L} = 262.5 \text{ L atm.}$$

The problem now consists in converting from the less familiar energy unit L atm to joules:

$$262.5 \text{ L atm} \times \frac{10^{-3} \text{ m}^3}{\text{L}} \times \frac{101{,}325 \text{ pascal}}{1 \text{ atm}}$$

$$\times \frac{1 \text{ newton m}^{-2}}{1 \text{ pascal}} = 2.66 \times 10^4 \text{ N m}$$

A newton meter (N m) is a joule.

b) The temperature is 100° C which equals 373.15 K. Assuming that the argon behaves ideally:

$$n_{Ar} = (1.75 \text{ atm} \times 100 \text{ L}) / (0.08206 \text{ L atm K}^{-1}\text{mol}^{-1} \times 373.15 \text{ K})$$

$$n_{Ar} = 5.72 \text{ mol}$$

The molar weight of Ar is 39.948 g mol^{-1} which makes the 5.72 mol equal to 228 g of argon.

c) The force per unit area exerted by the CH_4 molecules is just the pressure exerted by the CH_4 molecules. This is the mole fraction of CO_2 times the total pressure:

$$0.65 \times 5.00 \text{ atm} = 3.25 \text{ atm.}$$

Because 1 atm = 101,325 pascal, 1.75 atm = 3.29×10^5 pascal or 3.29×10^5 newton m^{-2}.

---

19. The root-mean-square speed of the molecules of a gas depends on the absolute temperature of the gas:

$$u_{rms} = [3k_BT / m]^{\frac{1}{2}}$$

For O_2, $m = 5.314 \times 10^{-26}$ kg. The mass of an oxygen molecule is the mass of a mole of O_2 molecules (0.032 kg) divided by the number of O_2 molecules in a mole, N_0. The temperature as given in the problem is 400 K.

The Boltzmann constant $k_B = 1.381 \times 10^{-23}$ J K^{-1}. Hence, the root-mean-square speed is:

$$u_{rms} = [3(1.381 \times 10^{-23} \text{ J K}^{-1})(400 \text{ K}) / 5.314 \times 10^{-26} \text{ kg}]^{\frac{1}{2}}$$

$$= [3.118 \times 10^5 \text{ J / kg}]^{\frac{1}{2}}$$

A joule is a kg $m^2 s^{-2}$ so:

$$u_{rms} = [3.118 \times 10^5 \text{ m}^2\text{s}^{-2}]^{\frac{1}{2}} = 5.58 \times 10^2 \text{ m s}^{-1}$$

The average speed is given by the expression:

$$\bar{u} = [8k_BT / \pi m]^{\frac{1}{2}}$$

Comparing this to the expression for u_{rms} shows that:

$$\bar{u} = u_{rms} \, (^{8}/_{3}\, \pi)^{1/2} = 0.9213 \, u_{rms}$$

Accordingly, the average speed is 5.14×10^{2} m s^{-1}. The average speed is 0.9213 times the root-mean-square speed regardless of the identity of the gas or the temperature, as long as the Boltzmann distribution is followed.

To calculate the average momentum of an oxygen molecule, start with the definition of the momentum of a particle:

$$p = m\,v$$

where m is its mass and v is its velocity. Before leaping to an answer by multiplying the mass of an oxygen molecule by its average speed, remember the difference between velocity and speed. Although the average speed of the oxygen molecules is 5.14×10^{2} m s^{-1}, their average velocity is zero. Velocity has both direction and magnitude. The oxygen molecules are moving in random directions so their velocities have a vector sum of zero. If their average momentum were non-zero the sample of gas would move as a whole. The average product of the speed and mass of the oxygen molecules can of course be computed. The mass of an oxygen molecule is:

$$m_{O_2} = 0.0320 \text{ kg mol}^{-1} \times 6.022 \times 10^{23} \text{ mol}^{-1} = 5.31 \times 10^{-26} \text{ kg}$$

Multiplying this and the average speed gives mu_{avg}, 2.73×10^{-23} kg m s^{-1}.

---

21. Until the Maxwell-Boltzmann distribution of speeds is attained, the collection of He atoms does not *have* a temperature. After this distribution of speeds is reached, the total translational kinetic energy of the He atoms must be the same as when the He ions exited the cyclotron, ignoring the very slight changes caused by neutralizing the electrical charge on the helium ions. Translational energy is conserved during the re-distribution of speeds because the collisions among the He atoms are elastic.

The original translational kinetic energy is E_k:

$$E_k = \tfrac{1}{2}\, N m u^2$$

where u is 2.00×10^{4} m s^{-1}, N is the number of He atoms, and m is a single He atom's mass. This mass is $(4.003 \times 10^{-3} / N_0)$ kg. For simplicity work with one mole of He, that is, let $N = N_0$. Then:

$$E_k = \tfrac{1}{2}\, N_0 (4.003 \times 10^{-3} \text{ kg} / N_0)(2.00 \times 10^{4})^2 \text{ m s}^{-1}$$

$$= 8.01 \times 10^{5} \text{ kg m}^2 \text{s}^{-2} = 8.01 \times 10^{5} \text{ J}$$

From the kinetic-molecular theory $E_k = {}^{3}/_{2}$ nRT. Therefore:

$$T = {}^{2}/_{3}\, E_k / nR$$

Substituting n = 1 mol and the proper values for E_k and R, and completing the arithmetic:

$$T = 6.42 \times 10^4 \text{ K}$$

The helium atoms all exit the cyclotron with the same speed. Their speed distribution curve consists of a single sharp peak at 2.00×10^4 m s^{-1}, and the average speed, the root-mean-square speed and the most probable speed obviously all are 2.00×10^4 m s^{-1}. But recall:

$$\bar{u} = (8k_BT / \pi m)^{1/2}$$

$$u_{rms} = (3k_BT / m)^{1/2}$$

$$u_{mp} = (2k_BT / m)^{1/2}$$

If these three different speeds are all equal to each other then the gas has three different temperatures each of which is as good as the next. This absurdity vanishes when it is realized that the gas really has *no* temperature until its molecules redistribute their speeds to fit the Maxwell-Boltzmann curve.

---

23. Consider the equation giving the rate of collision of the molecules of a gas with an area, A, of its walls:

$$Z_w = \frac{1}{4} \times \frac{N}{V} \times \sqrt{\frac{8k_B T}{\pi m}} \times A$$

The chemical amount of air which has leaked into the 500 cm^3 bulb one hour after it was sealed can be computed using the ideal gas equation and the pressure, temperature and volume observed at that time:

$$n = (1.00 \times 10^{-7} \text{ atm})(0.500 \text{ L}) / (0.082057 \text{ L atm K}^{-1}\text{mol}^{-1})(300 \text{ K})$$

The bulb contains 2.0311×10^{-9} mol of air. Because a mole contains 6.022×10^{23} particles this is 1.22×10^{15} molecules.

The collisions of the molecules of the air *outside* the vessel with the area of the hole let in 1.22×10^{15} molecules in 3600 s or $1.22 \times 10^{15} / 3600 = 3.4 \times 10^{11}$ molecules s^{-1}.

The density of the air outside the vessel is computed using the ideal gas equation and the known temperature and pressure of the outside air:

$$n/V = (1.00 \text{ atm}) / (0.082057 \text{ L atm mol}^{-1}\text{K}^{-1})(300 \text{ K}) = 4.062 \times 10^{-2} \text{ mol L}^{-1}$$

This is 2.46×10^{22} molecules L^{-1} or 2.46×10^{25} molecules m^{-3}.

From $\bar{u} = (8k_BT / \pi m)^{1/2}$ the average velocity of the molecules in the outside air is 469.5 m s^{-1}. Solving the original equation for *A* and substituting:

$$A = \frac{4(3.4 \times 10^{11} \text{ molecule s}^{-1})}{(2.46 \times 10^{25} \text{ molecules m}^{-3}) \times (469.5 \text{ m s}^{-1})}$$

which is 1.184×10^{-16} m^2. Because the hole is circular, its radius can be computed from the formula $r = (A/\pi)^{1/2}$. The radius is 6.14×10^{-9} m or 6.14 nanometers.

---

25. The rate of diffusion of a gas is inversely proportional to the square root of its molecular weight. One stage of diffusion of a mixture of $^{235}UF_6$ and $^{238}UF_6$ enriches the product gas by a factor of $\sqrt{352/349}$ or 1.0043 in the lighter gas, $^{235}UF_6$. 352 is the molecular weight of $^{238}UF_6$ (the atomic weight of ^{238}U + 6 times the atomic weight of fluorine). 349 is the molecular weight of $^{235}UF_6$. Fortunately, F has only one naturally occurring isotope so these are the only molecular weights to consider.

The problem calls for enrichment from 0.72 percent to 95 percent, a factor of 131.9. Each stage achieves only very slight enrichment, by a factor of 1.0043. But in a train of such stages each improves on the previous. If there are n stages then:

$$131.9 = 1.0043^n$$
$$\log 131.9 = n \log 1.0043$$
$$n = 1138$$

---

27. The mean free path of a molecule in a gas is inversely proportional to the number density (N/V) of the molecules and to the square of the diameter of the molecule (d^2).

$$\lambda = [\sqrt{2}\,\pi\, d^2\, (N/V)]^{-1}$$

Meanwhile, the collision frequency Z_w of the molecules of the gas with a specified area A of the walls of its container is

$$Z_w = {}^1/_4\, (N/V)\, \bar{u}\, A$$

where $\bar{u}$ is the average molecular speed. Solving for N/V from the first equation and substituting in the second gives:

$$Z_w = (4\sqrt{2}\,\pi d^2 \lambda)^{-1}\, \bar{u}\, A$$

or:

$$Z_w = (4\sqrt{2}\,\pi d^2 \lambda)^{-1}\, (3k_B T / 8\pi m)^{1/2}\, A$$

All the quantities on the right-hand side of this equation are known. Summarizing them (and converting them to SI units as necessary): $A = 1.00 \times 10^{-4}$ m^2, $m = 3.348 \times 10^{-27}$ kg (the mass of a single H_2 molecule), $k_B = 1.381 \times 10^{-23}$ J K^{-1}, T = 400 K, $\lambda = 3.00 \times 10^{-8}$ m, $d = 2.34 \times 10^{-10}$ m. Substitution of these values gives $Z_w = 7.02 \times 10^{24}$ s^{-1}.

---

29. ***a)*** The average distance an atom in a gas travels before colliding with another atom is the mean free path, λ, of the atom:

$$\lambda = [\sqrt{2}\,\pi d^2 (N/V)]^{-1}$$

In this problem the gas is atomic hydrogen in interstellar space for which:

$$d = 1.2 \times 10^{-10}\ \text{m};\quad N/V = 10\ \text{cm}^{-3} = 10 \times 10^{6}\ \text{m}^{-3}$$

λ is therefore 1.56×10^{12} m, or, to two significant figures, 1.6×10^{12} m.

b) The longer the mean free path the more time it takes a molecule to traverse it. The faster a molecule moves the less time will elapse between collisions. Therefore the average time between collisions is directly proportional to the mean free path of the molecules and inversely proportional to the average speed. The mean free path was computed in part a). The average speed is:

$$\bar{u} = [8k_BT / \pi m]^{\frac{1}{2}} = 1449.3\ \text{m s}^{-1}$$

Then:

$$\lambda / \bar{u} = 1.08 \times 10^{9}\ \text{s} = 34\ \text{yr}$$

This answer is the reciprocal of the collision frequency.

c) Because collisions among H atoms in this region of interstellar space are so infrequent the diffusion coefficient should be large. H atoms are diverted only rarely from their straight-line paths. Text equation **3.21** (text p. 92) gives the diffusion constant in terms of the mean free path and the average speed:

$$D = (3\pi / 16)\,\lambda\,\bar{u}$$

where D is the diffusion constant, λ is the mean free path (see part a) and $\bar{u}$ is the average speed (part b). Simple substitution, followed by rounding off to two significant figures, gives $D = 1.3 \times 10^{15}\ \text{m}^2\ \text{s}^{-1}$.

---

31. The key is to recognize that the pressure of the krypton is directly proportional to its number density. This fact follows from the ideal gas law:

$$P = (n/V)\,RT = (1 / N_0)\,(N/V)\,RT$$

where the quantity on the right reflects the fact that the number of molecules of gas (N) divided by Avodagro's number (N_0) equals the number of moles of gas. Because:

$$\lambda = [\sqrt{2}\,\pi d^2 (N/V)]^{-1}$$

it follows that:

$$N/V = [\sqrt{2}\,\pi d^2 \lambda]^{-1}$$

where d is the molecular diameter and λ is the mean free path. Substituting for N/V in the first equation:

$$P = [N_0 \sqrt{2}\,\pi d^2 \lambda]^{-1}\ RT$$

The volume of the spherical vessel is 1.00 L (1.00×10^{-3} m^3) and $V = {}^4/_3\,\pi r^3$ where r is the vessel's radius. The radius of the vessel is therefore 0.0620 m and its diameter is 0.124 m.

Now set λ equal to 0.124 m, fulfilling the condition that the mean free path be comparable to the diameter of the vessel. From the statement of the problem, T = 300 K and $d = 3.16 \times 10^{-10}$ m. R is 8.206×10^{-5} m^3 atm mol^{-1} K^{-1} (note the carefully chosen units of R). Substitution gives $P = 7.42 \times 10^{-7}$ atm.

---

33. ***a)*** Rearranging the ideal gas law gives P = nRT / V. In this case n = 50.0 / 44.0 = 1.136 mol, T = 298.15 K, V = 1.00 L and R = 0.08206 L atm mol^{-1} K^{-1}. Substitution gives P = 27.8 atm.

b) The van der Waals equation includes terms, a and b, which depend on the identity of the gas. For CO_2 *(g)* these are a = 3.592 atm L^2 mol^{-2} and b = 0.04267 L mol^{-1}.

The equation is:

$$\left(P + \frac{an^2}{V^2}\right)(V - nb) = nRT$$

so:

$$P = \frac{nRT}{V - nb} - \frac{an^2}{V^2}$$

Substituting, P = 24.6 atm.

---

35. The earth is covered by an ocean of air that exerts a pressure of 730 mm Hg all over its surface. Imagine the air replaced by an ocean of mercury. To exert the same pressure the mercury ocean would need to be only 730 mm deep. The volume of such a mercury ocean would be, to a close approximation, the surface area of the earth times its depth d:

$$V = 4\pi r^2 d$$

When $r = 6.370 \times 10^6$ m (converting from km) and $d = 730 \times 10^{-3}$ m, the mercury has a volume of 3.72×10^{14} m^3. The mass of a substance is its volume times its density. Mercury's density is about 13.6 g cm^{-3} which, remembering that 10^6 cm^3 equals 1 m^3 and that 1000 g equals 1 kg, converts to 13.6×10^3 kg m^{-3}. The mass of the mercury ocean is therefore 5.06×10^{18} kg.

This is the mass of the atmosphere, too, because the atmosphere exerts the same pressure on the earth's surface as this much mercury would.

---

37. The question amounts to asking what value of t_F in the equation:

$$V = 209.4 + 0.456\, t_F$$

causes V to become equal to 0. Substituting V = 0 and solving for t_F gives t_F = −459.2. The other data are not needed.

---

39. There is no doubt that the same chemical amount of C_2H_2 gas will evolve in the two experiments because the same mass of starting material is used under similar conditions; the only difference between the first and second runs of the reaction is the temperature at which the C_2H_2 gas is captured. Charles' law states that the volume occupied by a given amount of gas at constant pressure is directly proportional to the absolute temperature. The volume of gas collected at 400° C is therefore:

$$64.5 \text{ L} \times \frac{(273.15 + 400) \text{ K}}{(273.15 + 50) \text{ K}} = 134 \text{ L}$$

The balanced equation is not needed as long as it is known that the *same* equation applies to both runs of the reaction.

---

41. Collect P-V-T data on a precisely measured quantity of each gas. Then use the ideal gas equation to calculate n, the chemical amount of gas. The molar weight of the gas is its mass divided by n, its chemical amount. For greater precision, fit the P-V-T data to the van der Waals equation. That is, determine *a* and *b* in addition to n. Three sets of P-V-T values would allow determination of the three parameters.

---

43. Use the ideal gas equation. Enough gas is withdrawn to exert a pressure of 1 atm in a volume of 3.50 cm³:

$$n_{withdrawn} = 1 \text{ atm } (3.50 \text{ cm}^3) / RT$$

The original amount of gas equals the amount withdrawn plus the amount left:

$$n_{total} = n_{withdrawn} + n_{left}$$

Solving the ideal gas equation for each of these n's and then substituting:

$$(0.980 \text{ atm}) \, V / RT = (1 \text{ atm})(3.50 \text{ cm}^3) / RT + (0.855 \text{ atm}) \, V / RT$$

where V is the unknown volume. RT is the same in all 3 terms so:

$$V = \frac{1 \text{ atm } (3.50 \text{ cm}^3)}{(0.980 - 0.855) \text{ atm}} = 28.0 \text{ cm}^3$$

---

45. Assume that the gases behave ideally before the catalyst is introduced and that they behave ideally *after* the reaction goes to completion, too. To *react* on the other hand is quite non-ideal behavior.

A simplifying step is to imagine that the temperature and volume of the system are such that the total chemical amount of gases starts at 0.100 moles. Under this assumption, the reaction:

$$C_2H_2\,(g) + 2\,H_2\,(g) \rightarrow C_2H_6\,(g)$$

decreases the total chemical amount of gas to 0.042 mol. This is true because chemical amount is directly proportional to the pressure if T and V do not change.

Let x represent the original chemical amount of $C_2H_2\,(g)$ and y the original chemical amount of $H_2\,(g)$. Before reaction there is no $C_2H_6\,(g)$ so:

$$x + y = 0.100 \text{ mol}$$

The reaction produces exactly x mol of $C_2H_6\,(g)$ as it consumes $2x$ mol of $H_2\,(g)$ and x mol of $C_2H_2\,(g)$. Since $C_2H_2\,(g)$ is the limiting reagent, the reaction stops after the x mol of C_2H_2 gas is used up. At this point, the vessel contains x mol of $C_2H_6\,(g)$, the product, and $(y - 2x)$ mol of left-over $H_2\,(g)$ Hence:

$$x + (y - 2x) = 0.042 \text{ mol}$$

Solving the two simultaneous equations for x, we find it to equal 0.029. There is 0.029 mol of $C_2H_2\,(g)$. The original mole fraction of $C_2H_2\,(g)$ is therefore 0.029 / 0.100 = 0.29. Had we assumed the system to be big enough to hold for example 100 moles of gases all of the numbers in this computation, except the answer, would have been 1000 times bigger.

---

47. ***a)*** The mean speed of He atoms is given by the formula:

$$\bar{u} = [8k_BT / \pi m]^{1/2} \qquad \textit{hence} \qquad T = \pi m(\bar{u})^2 / 8\,k_B$$

From the problem, $\bar{u} = 11.2 \times 10^3$ m s^{-1}. The mass of He molecules is $(4.00 \times 10^{-3} / 6.022 \times 10^{23})$ kg and k_B is 1.381×10^{-23} J K^{-1}. Therefore T = 2.37×10^4 K.

b) For Ar and Xe the only change from the above calculation will be to employ a value of m which is for Ar (39.95 / 4.00) and for Xe (131.3 / 4.00) times larger. 39.95 and 131.3 are the atomic weights of Ar and Xe, respectively. The temperatures are 2.37×10^5 K and 7.78×10^5 K for Ar and Xe respectively.

c) The mean speed of a gas molecule is:

$$\bar{u} = [8k_BT / \pi m]^{1/2}$$

For He, Ar and Xe at T = 2000 K, the $\bar{u}$ values are 3254 m s^{-1}, 1030 m s^{-1} and 568 m s^{-1}, respectively.

The ratios of terrestrial escape velocity (11.2×10^3 m s^{-1}) to these speeds are 3.44, 10.9 and 19.7 respectively. These values rise in proportion to the square root of the atomic weights of He, Ar and Xe.

---

49. *a)* The average translational kinetic energy of the molecules in an ideal gas depends only on the absolute temperature of the gas. At 200° C *all* of the species (from $^{16}O^{16}O$ to $^{18}O^{18}O$) have the *same* average translational kinetic energy. The heavier ones move more slowly. At 400° C the average translational kinetic energies for each species are much higher, higher by the factor (400 + 273.15) / (200 + 273.15), but still all equal to each other.

b) The heavier molecules move, on the average, less rapidly at any temperature. This occurs in proportion to the square root of the molecular masses. Hence an average $^{18}O^{18}O$ molecule (molecular weight 36) moves $(^{32}/_{36})^{1/2}$ as fast as an average $^{16}O^{16}O$ (molecular weight 32). The same ratio holds at both 200° C and 400° C.

---

51. Convert the rates of effusion of the two gases from g min^{-1} to mol min^{-1} so that Graham's law can be applied. Oxygen's molar weight is 32.0 g mol^{-1}, so its rate of effusion becomes 3.25 g min^{-1} / 32.0 g mol^{-1}. Similarly, the rate of effusion of the unknown becomes 1.96 g min^{-1} / x g mol^{-1}. where x g mol^{-1} is the molar weight of the unknown.

Writing Graham's law for both gases gives:

$$3.25 \text{ g min}^{-1} / 32.0 \text{ g mol}^{-1} = k\,(1/32)^{1/2}$$

$$1.96 \text{ g min}^{-1} / x \text{ g mol}^{-1} = k\,(1/x)^{1/2}$$

In these equations k is a constant of proportionality. Both experiments are carried out under identical conditions of temperature and pressure so k is the same in the two. Dividing the first equation by the second allows cancellation of k and the rate units and ready computation of x. x = 11.6. It was not necessary to know T and P as long as they were the same in the two experiments.

---

53. Assume that the rate of escape of the natural gas is governed by the rate of collision of its molecules with the small hole that developed at the weld. The rate of collision, Z_w, is:

$$Z_w = \tfrac{1}{4}\,(N/V)\,(8k_BT/\pi m)^{1/2}\,A$$

The number density N/V of the natural gas depends on its temperature and pressure. Assuming that the natural gas behaves ideally:

$$N/V = P\,N_0/(RT)$$

The pressure in the tank is 136.05 atm (converting from 2000 psi). R is 0.08206 L mol^{-1}K^{-1}, T is 293.15 K, and N_0 is 6.022 × 10^{23} mol^{-1}. Hence:

$$N/V = 3.406 \times 10^{24} \text{ L}^{-1} = 3.406 \times 10^{27} \text{ molecules m}^{-3}$$

The quantity $(8k_BT/\pi m)^{1/2}$ is the average speed of the natural gas molecules. The mass is (0.01604 / 6.022 × 10^{23}) kg, k_B is 1.381 × 10^{-23} J K^{-1} and T is again 293.15 K. Substitution gives:

$$\bar{u} = 622.1 \text{ m s}^{-1}$$

It is useful (although not necessary) to compute intermediate values in this kind of problem as a check. If $\bar{u}$ for example had come out too far from 1000 m s^{-1}, it would have triggered some re-calculation; $\bar{u}$ values around room temperature are known to be on the order of 10^3 m s^{-1}.

Combining the intermediate values with $A = 1.00 \times 10^{-6}$ m^2 gives $Z_w = 5.29 \times 10^{23}$ s^{-1} as the rate of collision. After 86400 s (1 day):

$$5.29 \times 10^{23} \text{ s}^{-1} \times 86400 \text{ s} = 4.58 \times 10^{28} \text{ molecules}$$

will have leaked out.

The tank has a volume of 1.78×10^3 m^3 (based on $V = \pi r^2 h$ and computed after converting to SI units of length). This is 1.78×10^6 L. It contains 6.06×10^{30} molecules of gas at the start. (This number comes from the ideal gas equation applied to the contents of the tank.) Therefore a 0.0076 part or 0.76 percent of the gas escapes in a day.

The escape of gas during the one day lowers the number of molecules inside the storage tank to about 99.2 percent of its original value. This reduction of the interior number density during the course of the day means less frequent collisions with the area of the leaky spot by day's end. Therefore the answer is in principle slightly high although still correct within the precision of the other numbers in the problem.

---

55. ***a)*** The diffusion constant for self-diffusion of a gas is given by:

$$D = (3\pi / 16)\,\lambda\,\bar{u}$$

where $\bar{u}$ is the average speed and λ is the mean free path. By assuming the ammonia to have the same molecular radius (r) as the major components of air (O_2 and N_2) the problem allows the use of this equation. The mean free path of NH_3:

$$\lambda = \frac{1}{\sqrt{2}\,(N/V)\pi d^2}$$

where $d = 2r = 6 \times 10^{-10}$ m, and N/V is the number density of the NH_3, computed from the ideal gas equation:

$$N/V = \frac{N_0 P}{RT} = \frac{6.022 \times 10^{23} \times 1.00 \text{ atm}}{(8.206 \times 10^{-5} \text{ m}^3 \text{ atm K}^{-1} \text{mol}^{-1})(293.15 \text{ K})}$$

$$N/V = 2.50 \times 10^{25} \text{ m}^{-3}$$

Combining the intermediate numbers:

$$\lambda = 2.50 \times 10^{-8} \text{ m}$$

The average speed of the NH_3 molecules comes from:

$$\bar{u} = (8\,k_B T / \pi m)^{1/2}$$

With $k_B = 1.381 \times 10^{-23}$ J K^{-1}, T = 293.15 K, $m = (0.0170 / 6.022 \times 10^{23})$ kg:

$$\bar{u} = 604.1 \text{ m s}^{-1} \qquad \textit{and} \qquad D = 8.9 \times 10^{-6} \text{ m}^2 \text{s}^{-1}$$

This is only an estimate under the assumptions of self-diffusion and ideality.

b) Some time, *t*, after the spill the rms displacement of the NH_3 molecules is:

$$r_{\text{rms}} = (6\,Dt)^{\frac{1}{2}}$$

In this case r_{rms} is 100 m:

$$(100 \text{ m})^2 = 6\,Dt$$

Therefore:

$$t = 10^4 \text{ m}^2 / [6 \times (8.9 \times 10^{-6} \text{ m}^2 \text{s}^{-1})] = 1.9 \times 10^8 \text{ s}$$

This is about 5.9 years. Diffusion is slow. Rapid mixing of gases depends on convection.

Chapter 4

CONDENSED PHASES AND SOLUTIONS

4.1 *PHASE DIAGRAM OF A SUBSTANCE*

Phase transitions are conversions from solid to liquid, solid to gas and liquid to gas and their reverses. The different types of phase transitions have names. Melting and freezing refer to solid-liquid transitions; sublimation and condensation refer to solid-gas transitions; vaporization and condensation refer to gas-liquid transitions. Many substances have more than one solid phase, but none has more than one liquid or gaseous phase.

A *phase diagram* of a pure substance is a graph with T on the horizontal axis and P on the vertical axis. Lines (generally curved) on this graph define combinations of temperature and pressure at which two phases coexist at equilibrium. If two phases are at equilibrium, then no further macroscopic changes are detectible, although microscopic exchange of molecules (or atoms) between the phases continues.

A typical diagram (right) displays the relationships among three phases of a pure substance. It has three different coexistence lines radiating from a central point (T, the *triple point*). The liquid-gas coexistence line terminates at the *critical point* (labelled C).

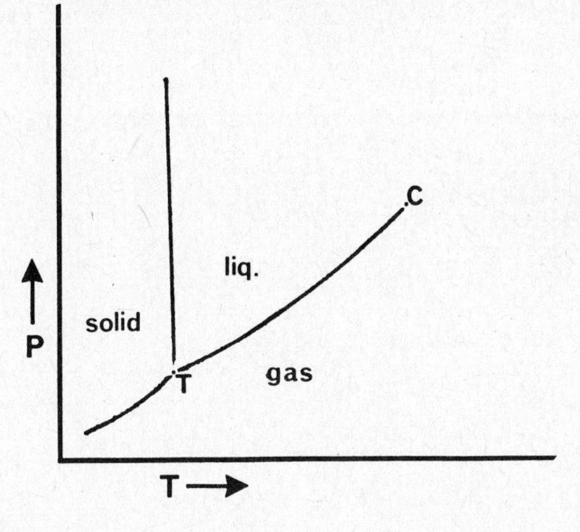

Freezing, Boiling and Sublimation Points

The normal freezing point of a pure substance is the temperature at which solid and liquid are in equilibrium when the pressure is one atmosphere. Similarly, the normal boiling point of a pure substance is the temperature at which liquid and gas are in equilibrium when the pressure is one atmosphere. Finally, the normal sublimation point of a pure substance is the temperature at which solid and gas are in equilibrium at one atmosphere pressure. All three, in diagrammatic terms, are intersections of the P = 1 atm line and the appropriate phase coexistence line. No pure substance can have all three of these points. After all, a solid heated slowly at a constant pressure of 1 atm either turns to a liquid or a gas, but not both. Some substances have *none* of these points. They simply decompose chemically before the temperature is high enough for them to melt, boil or sublime.

Phase Equilibria and Triple Points

Consider just the three phases, solid, liquid and gas. There are exactly three possible equilibrium lines (solid-liquid, solid-gas and liquid-gas) along which two of these three phases coexist. Regions between these lines are sets of P-T values at which only *one* phase exists at equilibrium. If two phase coexistence lines intersect, then, at the point of intersection, *three* phases coexist. This means that the third phase coexistence line must pass through the intersection of the first two. In general the coexistence lines on the P versus T diagram *do* intersect. The resulting specific combination of pressure and temperature at which three phases coexist is a *triple point* of a pure substance.

Critical Point

The essence of a phase transition is an abrupt discontinuous change in physical properties (*e.g.* density or viscosity) as the transition occurs. The distinction between liquid and gas *disappears* at temperatures and pressures that exceed the T and P of the *critical point*, a certain P-T combination on every substance's phase diagram. Distinguishable, abrupt liquid to gas transitions are no longer observed.

At temperatures above its *critical temperature* a gas cannot be liquefied no matter how great the pressure. The *critical pressure* is the pressure just necessary to liquefy a gas at its critical temperature. The critical pressure exceeds normal atmospheric pressure for all common substances. For this reason the "normal" pattern of phase change is the melting of a solid to a liquid followed by boiling to a gas. The observed behavior of substances at and beyond their critical points seems paradoxical.

Many interesting and thought-provoking problems are constructed around the facts presented on the phase diagrams of pure substances. For example:

- Description of the events within a container of a pure substance when the temperature is changed at constant pressure or when the pressure is changed at constant temperature.

  A pressure change at constant temperature corresponds to a vertical line on a phase diagram. A temperature change at constant pressure corresponds to a horizontal line. When such lines cut through phase coexistence lines then phase transitions occur. Keep in mind that heating, for example, a sample of a pure liquid substance at constant pressure would mean coping with enormous changes in volume when vaporization occurs.

- Construction of a phase diagram given the temperature and pressure at the triple point and critical point and perhaps the densities of the solid and liquid phases. **See problem 29.**

- Determination of the equation defining one or more of the phase coexistence lines. Although these lines are in general *not* straight lines, segments of them (over limited T and P ranges) are often approximated as straight. **See problem 31.**

A generic P versus T phase diagram has three curved lines radiating from the central triple point. The solid-liquid coexistence line usually goes almost straight up, because freezing points are not usually strongly dependent on pressure. Its slight slope is generally to the left (a negative slope) because, for most substances, the density of the solid phase exceeds the density of the liquid phase. Water is the outstanding exception. Ice floats in liquid water proving that it is less dense than the liquid. The solid-liquid coexistence line for H_2O has a slight positive slope.

The liquid-gas coexistence line of all substances slopes to the right on the P-T diagram because the gas phase is always less dense than the liquid phase. Increases in pressure favor the more dense liquid phase. This line terminates at the critical point. The solid-vapor coexistence line also always has a positive slope and rises from somewhere in the lowest reaches of temperature and pressure to terminate in the triple point.

Qualitative discussions of phase diagrams often deal with tidy-looking diagrams that attain their apparent symmetry by virtue of severe distortion of the scales on the T and P axes. When phase diagrams are drawn on axes with regularly spaced increments of T and P, areas of interest (triple points, critical point, etc.) are commonly forced onto an edge of the diagram.

4.2 *COMPOSITION OF SOLUTIONS*

Homogeneous systems that contain two or more substances are solutions. Solutions have ranges of composition. Therefore a new variable (in addition to pressure, temperature and volume) is needed fully to describe them.

The following measures are used to describe the composition of solutions:

• **Mole fraction.** The mole fraction of a substance in a mixture is its number of moles divided by the total number of moles of all the different substances present. Mole fraction is symbolized with a capital *X*:

$$X_1 = n_1 / n_{tot}$$

The sum of the mole fractions of all of the components of a solution must equal 1. Two component systems are common in problems and the relationship $X_1 + X_2 = 1$ is often crucial in solving these problems. **See problems 21, 23, 25.**

• **Molality.** The molality of a solution is the number of moles of solute divided by the number of kilograms of solvent. When more than one solute is present then the solution has a molality in the first, another molality in the second and so forth. The *solvent* is the component of the solution that is present in the greatest chemical amount. Solutes are the components present in lesser chemical amounts. The symbol for molality is *m* and the SI unit is mol kg^{-1}.

All that is needed to convert freely between mole fractions and molalities are the molar weights of the components of the solution. **See problem 3.** Conversely, if the composition of a solution is expressed both in terms of molalities and mole fractions then molar weights can be calculated.

The case of knowing the molality of a two component system and its composition by weight is particularly common. These facts allow computation of the solute's molar weight. **See problems 13, 37, 38.**

- **Molarity.** The molarity of a solution is the number of moles of the solvent per liter of solution. The symbol for molarity is M and its unit is mol L^{-1}. Because this unit of concentration has a volume in the denominator, conversion between it and either molality or mole fraction requires a knowledge of the density of the *solution* (not of the pure components). **See problems 3 and 5.** Densities fluctuate with temperature and it follows that molarities do too. In *dilute* aqueous solutions the molarity is approximately the same as the molality because the density of the solution is approximately 1 g cm^{-3}.

Many problems involve the conversion among different units of concentration. It is *not* wise to memorize lists of specific formulas for these conversions. Instead learn the definitions of the different units of concentration.

The units of concentration fall into two categories, those having a *mass* in the denominator (mole fraction, molality, weight percent) and those having a *volume* in the denominator (molarity, volume percent). To convert between these two categories requires knowledge of the density of the solution.

There are other ways of expressing concentration. Weight percent (mass percent) and volume percent are two examples.

Preparation of Solutions

The preparation and manipulation of solutions have enormous practical importance. To prepare a solution that is for example 1.20 M in a certain solute one takes 1.20 mol of the solute and adds enough solvent to make the total volume exactly 1 L. If only 517 mL of such a solution is required, then one takes (0.517 × 1.20) mol of the solute and adds enough solvent to bring the total volume exacty to 517 mL. This is *not* the same as adding 517 mL of solvent. The difference stems from the volume that the solute occupies. **See problem 7.** *Volumetric flasks*, calibrated to hold exactly known volumes at specified temperatures, are used in preparing solutions of accurately known concentration.

Doubling the total volume of a solution by adding more solvent obviously cuts the concentrations of all solutes in half. It does not change the chemical amount of solutes present. In general, for the dilution of a solution:

$$C_f = C_i \, V_i \, / \, V_f$$

where the subscripts refer to the final and initial values. C and V stand for concentration and volume respectively.

The preparation and dilution of solutions are the subjects of many exercises. **See problems 7 and 35.** A potentially puzzling variation of such dilution problems involves the *concentrating* of a solution from a lower to a higher concentration. The same formula would apply except that V_f would be smaller than V_i instead of larger.

4.3 *COLLIGATIVE PROPERTIES OF SOLUTIONS*

There are four colligative properties of solutions. Colligative properties do not depend on the identity of the solute but instead only on the effect of the number of solute particles.

To understand these properties one must first understand *Raoult's law*:

$$P_1 = X_1 P_1^\circ$$

This law states that the vapor pressure of a component in a solution depends on its mole fraction times the vapor pressure the pure component exerts. When P_1° is 0 then the component has no vapor pressure and is non-*volatile.*

Solutions that follow Raoult's law are *ideal* solutions. Contrast them with ideal gases. In an ideal gas, there are negligible forces of attraction among the molecules of the gas. In an ideal solution the intermolecular attractions are not negligible. Instead the solvent to sol*ute* attractions are the *same* as the solvent-sol*vent* interactions.

Vapor Pressure Lowering

The case of a volatile solvent and a non-volatile solute is common. The addition of the solute to the solvent leads to a reduction of the total vapor pressure of the solution. The reduction is easily measured and is proportional to the mole fraction of the solute. Thus vapor pressure measurements give solution composition:

$$-\Delta P = \text{vapor pressure lowering} = P^\circ\,(\text{solvent})\; X_2$$

See problem 9. An observed vapor pressure lowering, taken in combination with the molar weight of the solvent and the masses of the solute and solvent, gives the solute's molar weight. **See problem 11.** An obvious variation is to give the molar weight of the *solute*, the masses of solute and solvent, and ask the molar weight of the solvent. The approach is essentially unchanged in such a case.

Boiling Point Elevation

Although the vapor pressure of a volatile solvent is lowered by the presence of a non-volatile solute, its boiling point is raised. The approximate equation representing this is:

$$\Delta T_b = K_b\, m$$

where K_b is a constant that depends only on the solvent, m is the molality of the solute, and ΔT is the final boiling temperature minus the original boiling temperature. The quantity ΔT_b is always positive, by the nature of the phenomenon. Tables of K_b values are available. The units of K_b are K m^{-1}, that is, "kelvins per molal", which is the same as K kg mol^{-1} where it is understood that the kg refers to the mass of the solvent and the mol to the chemical amount of the solute.

The simple form of the boiling point elevation formula lends itself to many practical problems involving once again the determination of the molar weight of unknown non-volatile solutes. **See problems 13 and 15.**

Freezing Point Depression

Freezing point depression is analogous to boiling point elevation. The formula for the amount of change in the freezing point of a volatile solvent occasioned by the presence of a non-volatile solute is:

$$-\Delta T = K_f m$$

where K_f is the solvent's freezing point depression constant. K_f and K_b values differ for any solvent. The negative sign on ΔT appears because T_2 is always less than T_1 in a temperature lowering.

Typical applications again involve the determination of the molar weights of unknown solutes. **See problem 37.**

Osmotic Pressure

The colligative property that is most used in biochemical applications is the osmotic pressure. All solutions have an osmotic pressure. In the case of dilute solutions:

$$\Pi = c\,R\,T$$

where Π is the osmotic pressure, c is the concentration of the solution, T is its absolute temperature and R is the gas constant. Osmotic pressures are measured in set-ups using semi-permeable membranes, materials that allow passage of molecules of solvent but not of solute.

Solutes That Dissociate

A complication arises in the case of solutions (nearly always aqueous) of dissociating solutes. Such solutes give rise to two or more particles in solution for every particle added, corespondingly affecting the magnitude of the several colligative properties. If a colligative property is enhanced in this way the apparent molality, not actual molality, is increased. The theoretical number of particles is often evident from the formula and name of the solute: Na^+Cl^-, 2; $Ca^{2+}(Cl^-)_2$, 3; $(K^+)_2(SO_4{}^{2-})$, 3; etc. **See problems 15 and 17.**

4.4 *PHASE DIAGRAMS OF MIXTURES*

Solutions of two (or more) components have phase diagrams, just as do pure substances. A full phase diagram of a two-component solution would require three axes, one for pressure, one for temperature, and the third for composition. A three-component solution would require *four* axes because two different composition variables require specification. Evidently, full phase diagrams for many-component solutions are at best difficult and often impossible to draw. Instead *cross-sections* of phase diagrams are sketched. Text Figure 4-9 (text p. 122) is a constant temperature cross-section of the phase diagram of a two-component ideal solution. Pressure and composition vary on the vertical and horizontal axes respectively. Figure 4-12, text p. 125, shows composition on the horizontal axis but temperature on the vertical axis. The pressure is constant.

Real solutions' phase diagrams are complex and differ sharply from the simple predictions of Raoult's law. Nevertheless, at a sufficiently low concentration of solute (even in nonideal solutions) there is still some simplicity:

$$P(\text{solute}) = k\,X(\text{solute})$$

This is Henry's law. The constant k is different for every solute-solvent combination. Henry's law often occurs in problems involving solutions of gases in liquids. **See problem 21.**

Distillation

When a solution of volatile components is heated, the total vapor pressure above the solution increases. The vapors in equilibrium with the liquid are in general richer in the more volatile components than the liquid. If the solution is ideal, the degree of enrichment can be computed. **See problems 23, 25 and 43.** Raoult's law gives the partial vapor pressures of the components, and Dalton's law gives the mole fractions of the components in the vapor.

4.5 *COLLOIDAL SUSPENSIONS*

Colloids are mixtures of two or more substances in which one phase is suspended as small particles in a second. The particles of the dispersed phase are larger than single molecules. They are from 10^{-9} to 10^{-6} m in diameter and too small to be distinguished by eye. Colloids are metastable. In principle, the dispersed phase in a colloid will settle out...eventually. In practice, this sedimentation is *slow.* When colloids are *flocculated* by adding soluble salts to the dispersing phase, sedimentation speeds up.

DETAILED SOLUTIONS TO ODD-NUMBERED PROBLEMS

1. Treat the mercury vapor above the liquid mercury in a barometer as an ideal gas. The problem asks for the *number density* (N/V) of this mercury vapor. The number of molecules (N) in a sample of gas is Avogadro's number (N_0) times the chemical amount of the gas (n) in moles:

$$N = n\,N_0$$

Combining this with the ideal gas equation:

$$(N/V) = N_0 P / RT$$

The temperature is 300 K, P = 2.87×10^{-6} atm, R = 82.06 cm^3 atm $mol^{-1} K^{-1}$ and N_0 = 6.022×10^{23} mol^{-1}. Substitution gives (N/V) = 7.02×10^{13} cm^{-3}, that is, 7.02×10^{13} atoms of Hg in every cubic centimeter above the liquid mercury. This, the *Torricellian* vacuum, is far from perfect.

---

3. Consider a convenient arbitrary mass of the solution, say, 100.0 g. This weight of the nitric acid solution contains 40.00 g of pure HNO_3, but has a volume of only 80.238 cm^3:

$$V = m / d = 100.0 \text{ g} \;/\; 1.173 \text{ g cm}^{-3} = 80.238 \text{ cm}^3$$

Since the molar weight of HNO_3 is 63.013 g mol^{-1}, 40.00 g HNO_3 is 0.6348 mol HNO_3. If the solution has 0.6348 mol HNO_3 in its 80.238 cm^3 then it has 7.911 mol HNO_3 per 1000 cm^3. But moles of solute per 1000 cm^3 (1000 cm^3 is 1 L) of solution *defines* molarity. Hence the molarity of the solution is 7.911 M.

The *molality* of a solution is the number of moles of solute per kilogram of *solvent*. This solution has 0.6348 mol HNO_3 for every 60.00 g of H_2O. Therefore it has a proportionately larger quantity of acid in 1000 g (1 kg) of water: 10.58 mol HNO_3. It is 10.58 *m*.

The *mole fraction* of nitric acid is the number of moles of HNO_3 divided by the total number of moles of all substances in the mixture. The chemical amount of HNO_3 in 100 g of solution is 0.6348 mol (see above). The chemical amount of H_2O, the only other component, is (60.00 g / 18.015 g mol^{-1}) = 3.330 mol. Let X_2 represent the mole fraction of the solute nitric acid:

$$X_2 = 0.6348 \text{ mol} / (0.6348 + 3.330) \text{ mol} = 0.1601$$

---

5. Suppose there is 1000 cm^3 of the solution. A molarity of 13.137 means there is 13.137 mol of HCl in such a sample. The molar weight of HCl is 36.461 g mol^{-1} so:

$$13.137 \text{ mol HCl} \times 36.461 \text{ g mol}^{-1} = 478.99 \text{ g HCl}$$

This weight of HCl is 40.00 percent of the total weight of the solution: which therefore must be (478.99 g / 0.4000) = 1197.5 g. Density is mass divided by volume so the density of the solution is:

$$1197.5 \text{ g} / 1000 \text{ cm}^3 = 1.1975 \text{ g cm}^{-3}$$

The molality of the solution is the number of moles of solute in a kilogram of *solvent* (not solution). The 1000 cm^3 of solution weighs 1.1975 kg of which 0.47899 kg is HCl. The rest, 0.7185 kg, must be water. The molality then is:

$$13.137 \text{ mol HCl} / 0.7185 \text{ kg solvent} = 18.28\ m$$

---

7. ***a)*** To mix 600 mL of the solution requires only $^6/_{10}$ as much NaCl as mixing 1000 mL of the solution would require. To make 1000 mL (1 L) of solution would require 0.1500 mol NaCl or 0.1500 mol × 58.44 g mol^{-1} = 8.766 g NaCl. Therefore, take 5.26 g NaCl, which is $^6/_{10}$ of 8.766, and add enough water to bring the total volume to 600 mL. This is *not* the same as adding 600 mL of water. Adding 600 mL of water to 5.26 g NaCl of NaCl would give a final volume more than 600 mL.

b) The 600 mL of solution is to be diluted from 0.150 M to 0.0400 M. This dilution is by a factor of 0.150 M / 0.0400 M, or 3.75. The final volume therefore is 3.75 × 600 mL = 2250 mL.

---

9. The total vapor pressure above the carbon tetrachloride is lowered by the addition of a solute which, being non-volatile itself, contributes nothing to that vapor pressure. Let the subscript 1 refer to the CCl_4, the solvent, and subscript 2 to the solute. Raoult's law states:

$$P_1 = X_1 P_1^\circ$$

In this problem P_1 = 0.411 atm and P_1°, the vapor pressure of pure CCl_4, is 0.437 atm. Therefore X_1 = 0.941 and X_2 = 1 − 0.941 = 0.0590. The mole fraction of the solute, X_2, is:

$$X_2 = \frac{n_2}{n_1 + n_2}$$

where n_2 is the chemical amount of solute and n_1 is the chemical amount of solvent. In this case:

$$n_1 = 100 \text{ g} / 153.82 \text{ g mol}^{-1} = 0.650 \text{ mol}$$

because 153.82 g mol^{-1} is the molar weight of CCl_4. It is straightforward to substitute and solve for n_2:

$$n_2 = X_2\, n_1 / X_1 = 0.0408 \text{ mol}$$

Because 7.42 g of the solute amounts to 0.04081 mol, the molar weight of the solute is 7.42 g / 0.0408 mol = 182 g mol^{-1}

The above calculation exhibits rounded-off intermediate values for the sake of clarity. If no rounding-off is done until the last step, then n_2 is 0.041126 and the molar weight of the solute comes out to 180 g mol^{-1}. This is a more nearly correct answer.

---

11. The vapor pressure of the acetone is lowered by addition of the solute. The vapor pressure *lowering* is directly proportional to the mole fraction of the non-volatile solute (component 2):

$$\Delta P = X_2 P_2^\circ$$

$$X_2 = 0.0249 \text{ atm} / 0.3720 \text{ atm} = 0.066936$$

The solution contains 50.0 g / 58.0807 g mol^{-1} = 0.86087 mol of acetone. The mole fraction of acetone is 1.0000 − 0.066936 or 0.9330064. The solvent acetone contributes 0.9330064 of the total chemical amount of all substances in the mixture. The *total* chemical amount of solute + solvent is therefore equal to 0.86087 mol / 0.9330064 or 0.92263 mol. The difference between the total moles (0.92263 mol) and the moles of solute (0.86087 mol) is the moles of solute. It is 0.06176 mol. The molar weight of the solute is 15.0 g / 0.06176 mol = 242.88 g mol^{-1}. Additional significant figures have been carried in the intermediate values in this computation. The final answer must be rounded off to 3 significant digits. It is 243 g mol^{-1}.

---

13. ***a)*** The presence of a solute elevates the boiling temperature of the toluene. Assuming that the solute is non-volatile, the boiling point elevation of this dilute solution follows the equation:

$$\Delta T_b = K_b\, m$$

The data on the anthracene in toluene solution allow the computation of K_b. ΔT_b for this solution is 2.27 K. The chemical amount of anthracene is 7.80 g / 178.23 g mol^{-1} = 0.0438 mol. The molality of the anthracene solution is 0.0438 (the number of moles of anthracene) divided by 0.100 (the number of kg of toluene) and is 0.438 *m*. By substitution, K_b equals 5.187 K kg mol^{-1}.

For the unknown solution ΔT_b is 111.94° C − 110.60° C or 1.34° C. This *difference* in temperature is also 1.34 K because the two units of temperature are the same size. Recall that the Celsius and absolute temperature scales differ only in the location of the zero point. Putting 1.34 K into the boiling point elevation formula for toluene gives a concentration of 0.258 *m* for the unknown.

Since there is 0.258 mol of unknown in a kilogram of toluene there must be 0.120 × 0.258 mol = 0.0310 mol of unknown in the 120 g of toluene used in the experiment. For the unknown then, 4.80 g *is* 0.0310 mol which means its molar weight is 4.80 / 0.0310 or 155 g mol^{-1}.

b) In the separate experiment, 0.2580 g of unknown is burned to give 450 cm^3 of CO_2 *(g)* at STP. This is, by the ideal gas equation, 0.0201 mol of CO_2. The 0.0201 mol of CO_2 contains 0.0201 mol of C which weighs 0.2411 g. The rest of the unknown is H which, by subtraction from the total weight, weighs 0.0169 g . This weight of H is 0.0167 mol of H. The number of moles of the two elements, 0.0201 and 0.0167, lie in a 6:5 ratio. That is, 0.0201 / 0.017 is

1.20. Since C and H are present in a 6:5 molar ratio, the *empirical* formula is C_6H_5. But this empirical formula corresponds to a molecular weight of only 77.11. Doubling it, to $C_{12}H_{10}$, gives a molecular weight closely in accord with the results of part a).

---

15. The $CaCl_2$ solution has 0.030 mol / 0.400 kg = 0.075 mol $CaCl_2$ per kg of solvent. Because $CaCl_2$ dissociates to three particles when dissolved in water:

$$CaCl_2\,(s) \rightarrow Ca^{2+}(aq) + 2\,Cl^-(aq)$$

the solute particles in the solution are 3 × 0.075 molal. The freezing point depression is:

$$\Delta T_f = -K_f m$$

where K_f is a constant characteristic of the solute and m is the molality of the solution. For water, K_f is 1.86 K kg mol^{-1}. ΔT_f in this problem then is $-1.86 \times 3 \times 0.075 = -0.42$ K. This is the same as $-0.42°$ C because kelvins and degrees Celsius are the same magnitude. The change in temperature is the final temperature minus the original temperature:

$$\Delta T_f = T_2 - T_1 = T_2 - 0.00° \text{ C}$$

$$T_2 = -0.42° \text{ C}$$

Ice crystallizes from the solution at this temperature. As the first ice leaves the solution, what remains becomes more concentrated in $CaCl_2$ and has an even lower freezing point. Pure substances have sharp freezing points. Mixtures like this one freeze over a temperature range.

---

17. The 10.4 g of $NaHCO_3$ is 0.124 mol of $NaHCO_3$ so the aqueous solution has 0.124 mol solute per 0.200 kg solvent and is 0.619 *m*. A 0.619 solution of a non-dissociating solute would change the freezing point of water:

$$-1.86 \text{ K kg mol}^{-1} \times 0.619 \text{ mol kg}^{-1} = -1.15 \text{ K.}$$

$NaHCO_3$ changes the freezing point by almost exactly twice this ($-2.30°$ C). The $NaHCO_3$ must dissociate into 2 ions: Na^+ and HCO_3^-.

---

19. The osmotic pressure, Π, of a dilute solution is:

$$\Pi = cRT$$

where c is the concentration of the solution and the other symbols have their usual meanings. If the units of R are L atm K^{-1}mol^{-1} then Π must be in atm, and c in mol L^{-1}.

The concentration of the solution in this problem is:

$$c = 0.0105 \text{ atm} / [(0.08206 \text{ L atm K}^{-1}\text{mol}^{-1}) \times 300 \text{ K}]$$

$$c = 4.27 \times 10^{-4} \text{ mol L}^{-1}$$

The 200 mg of the polypeptide dissolved in 25.0 cm^3 of water has a concentration of 8.00 g L^{-1}. The concentration of the solution is thus known both in moles per liter and in grams per liter. The second divided by the first has units of g mol^{-1} and is the molar weight of the polypeptide: 1.88×10^4 g mol^{-1}.

---

21. ***a)*** Henry's law is similar in form to Raoult's law:

$$P_1 = X_1 P_1^\circ \qquad \text{Raoult's law}$$

$$P_1 = X_1 k \qquad \text{Henry's law}$$

The Henry's law constant, k, replaces the P° of Raoult's law. Solutions obeying Henry's law are not ideal solutions but are still quite simply described.

In this problem $k = 1.65 \times 10^3$ atm and $P_1 = 5.0$ atm so the mole fraction of CO_2 *(aq)* is 3.03×10^{-3}.

Consider a sample of this carbonated water containing a total of 100.000 mol of CO_2 and water. The chemical amount of CO_2 is 0.303 mol and H_2O 99.697 mol. The total mass of the sample is the sum of the mass of the CO_2:

$$0.303 \text{ mol } CO_2 \times 44.010 \text{ g mol}^{-1} = 13.3 \text{ g } CO_2$$

and the mass of the H_2O:

$$99.687 \text{ mol } H_2O \times 18.0153 \text{ g mol}^{-1} = 1796.1 \text{ g } H_2O$$

$$1796.1 \text{ g} + 13.3 \text{ g} = 1809.4 \text{ g}$$

Since the mass of the solution is 1809.4 g the volume of the solution is 1.8094 L (because its density is 1.00 g cm^{-3}, equivalent to 1000 g L^{-1}). The molarity of the CO_2 is then:

$$0.303 \text{ mol } CO_2 \,/\, (1.8094 \text{ L}) = 0.167 \text{ M}$$

b) When the bottle of carbonated water is uncapped the pressure of CO_2 above the liquid is suddenly reduced. The product $k\, X_{CO_2}$ exceeds P_{CO_2}, and bubbles of CO_2 *(g)* leave the solution spontaneously.

---

23. ***a)*** The two components are both volatile and both contribute to the vapor pressure above the solution. According to Raoult's law:

$$P_A = X_A P_A^\circ \quad \textit{and} \quad P_B = X_B P_B^\circ$$

P_A° and P_B° are 0.200 atm and 0.330 atm respectively. X_A and X_B are 0.400 and 0.600 respectively. Substitution gives $P_A = 0.080$ atm and $P_B = 0.198$ atm. The vapor pressure of the solution is the sum of these two partial vapor pressures: 0.278 atm.

b) Assume that the mixture of vapors above the solution obeys the ideal gas law and Dalton's Law. Component A contributes 0.080 atm of a total pressure of 0.278 atm above the solution. Therefore 0.080 / 0.278 of the molecules in the vapor are molecules of A, and the mole fraction of A in the vapor is 0.288. The mole fraction of A in the liquid was substantially more than this. The vapors have been enriched in the component with the *greater* vapor pressure (the more volatile component), which is B. The phenomenon is general and underlies all distillative separations.

---

25. *a)* The total pressure of the solution is the sum of the partial vapor pressures of the two volatile compounds that make it up. The vapor pressures of each of the two are given by Raoult's law. Therefore:

$$P_{tot} = X_I P_I^\circ + X_{II} P_{II}^\circ$$

In this problem $P_I^\circ = 0.500$ atm, $P_{II}^\circ = 0.660$ atm, $P_{tot} = 0.590$ atm and X_I and X_{II} are the mole fractions of the two components in the liquid phase. Because:

$$X_I + X_{II} = 1$$

it is straightforward to solve for X_{II}:

$$0.590 \text{ atm} = (1 - X_{II})(0.500 \text{ atm}) + X_{II}(0.660 \text{ atm})$$

$$X_{II} = 0.563$$

b) The total vapor pressure above the solution is 0.590 atm. Of this 0.437×0.500 atm is exerted by molecules of compound I. The mole fraction of compound I in the vapors:

$$X_I(g) = (0.437 \times 0.500) \text{ atm} / 0.590 \text{ atm} = 0.370$$

The mole fraction of compound I is less in the vapors than in the liquid because it is less volatile than compound II. The molar weights of the compounds are not needed.

---

27. Assume that the water vapor and the air in the room are ideal gases and also obey Dalton's law. The humidifier goes to work and saturates the air in the room with water vapor. Then for the water vapor:

$$n_{water} = P_{water} V / RT$$

In this equation $V = 110 \text{ m}^3 = 110 \times 10^3$ L, $T = 298.15$ K, $R = 0.082057$ L atm $K^{-1} mol^{-1}$ and $P_{water} = 0.03126$ atm. Substitution quickly gives the chemical amount of water. The mass of water is the molar weight of water (18.015 g mol^{-1}) times n_{water}; the mass is 2.53×10^3 g.

29.

The figure (*right*) presents the data for O_2 as given in the problem. The scales on both axes are non-linear. Despite the distortions, the slopes of the coexistence lines are correctly portrayed. The fact that the density of the solid exceeds the density of the liquid requires a negative slope for the solid/liquid coexistence line. Similarly, the slope of the liquid/gas line confirms that the density of the gas is less than the density of the liquid. Finally, an ideal gas at the critical point of oxygen would have a density of 0.126 g cm^{-3}. Oxygen gas is 3 ½ times denser than this at its critical point.

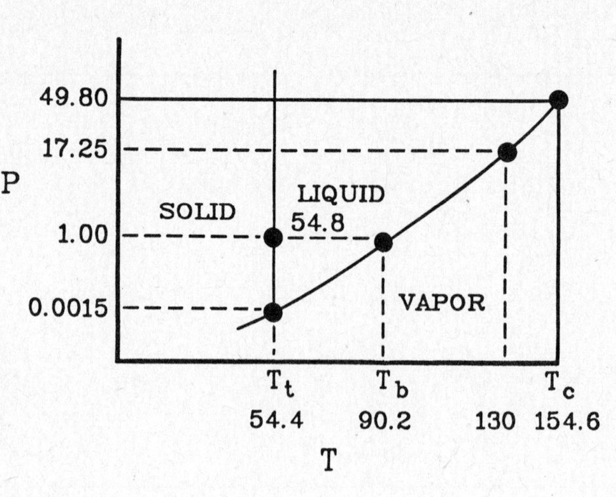

---

31. We know two points, call them 1 and 2, on the liquid-solid coexistence curve of water:

$$t_1 = 0.000^\circ\text{ C},\ P_1 = 1.000 \text{ atm}$$

$$t_2 = 0.010^\circ\text{ C},\ P_2 = 0.0060 \text{ atm}$$

These two points define a straight line on a P-T graph for water. The "two-point" form of the equation of a straight line is common in analytical geometry. It is:

$$\frac{P - P_1}{t - t_1} = \frac{P_1 - P_2}{t_1 - t_2}$$

After substitution of the numbers for the two known points and rearrangement, the equation becomes:

$$P = -99.4\,t + 1.000$$

Assume that this equation holds at P=50 atm and compute the temperature corresponding to this pressure:

$$t = (50 - 1.000) / -99.4 = -0.49^\circ\text{ C}$$

Converting all temperature values to the absolute scale would change the equation (corresponding to a shift in the origin) but not the answer.

b) A skater of mass 75 kg exerts a force on the ice under his or her skates given by:

$$F = mg$$

where g, the acceleration of gravity, equals 9.80 m s^{-2}. The force falls on a blade area of 0.80 cm^2 which is 0.80×10^{-4} m^2. The pressure under the blades of the skates is the force divided by the area:

$$P = F / A = mg / A$$

$$P = 75 \text{ kg} \times 9.80 \text{ m s}^{-2} / 0.80 \times 10^{-4} \text{ m}^2 = 9.2 \times 10^6 \text{ pascal}$$

Because 1 atm is equal to 101,325 pascal the pressure under the skates is 91 atm. The freezing point of water under this pressure can be estimated using the straight-line equation developed in part a. It is −0.9° C.

c) The freezing point of water is lowered by the pressure of the skater's blades but not enough to explain the fact that skates work even at very low temperatures. The probable explanation of the smooth action of ice skates is that the friction of the runners over the ice melts a lubricating layer of water.

---

33. Liquid N_2 has a density at 77.3 K of 0.808 g cm^{-3}. Because the molar weight of N_2 is 28.01 g mol^{-1}, the *molar* density of N_2 *(l)* is 0.0288 mol cm^{-3}. The reciprocal of this is 34.7 $cm^3 mol^{-1}$. It is the *molar volume* (volume occupied by one mole) of liquid nitrogen at 77.3 K. One mole of N_2 contains N_0 molecules, so:

$$(34.7 \text{ cm}^3 \text{mol}^{-1}) / (N_0 \text{ mol}^{-1}) = 5.76 \times 10^{-23} \text{ cm}^3$$

In liquid N_2, 1 molecule has the volume 5.76×10^{-23} cm^3 available to it.

In the gas phase the ideal gas equation allows estimation of the molar volume:

$$V / n = RT / P$$

where P = 1 atm (normal boiling point means P = 1 atm), T = 77.3 K and R = 0.08206 L atm $K^{-1} mol^{-1}$:

$$V / n = 6.34 \text{ L mol}^{-1} = 6.34 \times 10^3 \text{ cm}^3 \text{mol}^{-1}$$

which amounts to 1.05×10^{-20} cm^3 available per single molecule.

The average distances between molecules are on the order of the cube roots of these volumes: 3.9×10^{-8} cm in the liquid; 22×10^{-8} cm in the gas.

The approximate molecular diameter is 3.15×10^{-8} cm. Apparently in N_2(*l*) at 77.3 K molecules are packed together nearly as tightly as possible whereas in N_2 *(g)* at 77.3 K vapor they have 5 or 6 times more room.

---

35. The concentration of the liquid mixture is the total chemical amount of HNO_3 divided by the total volume of solution. The first solution contributes:

$$0.0800 \text{ mol L}^{-1} \times 0.600 \text{ L} = 0.048 \text{ mol HNO}_3$$

and the second:

$$0.0600 \text{ mol L}^{-1} \times 0.900 \text{ L} = 0.054 \text{ mol HNO}_3$$

Because there is 0.102 mol of HNO_3 in 1.500 L, the resulting solution is 0.0680 M HNO_3.

---

37. The Rast method uses the measurement of a freezing point depression to determine molecular weights. For camphor as a solvent in a Rast determination:

$$\Delta T_f = -K_f m$$

In this problem, the camphor has its freezing point sharply lowered, from 178.4° C to 170.8° C. This means that ΔT_f is −7.6° C or −7.6 K.

Simple substitution gives the molality of the solution:

$$-7.6 \text{ K} / -37.7 \text{ K kg mol}^{-1} = 0.202 \text{ mol kg}^{-1} = 0.202\ m$$

0.840 g of the unknown *solute* in 25.0 g of camphor corresponds to 33.6 g of it in 1 kg of camphor. But there is 0.202 mol of unknown per kg of camphor. Therefore 33.6 g of unknown comprises 0.202 mol of unknown, and the molar weight of the unknown is 166 g mol^{-1}. The difference in temperature is known to only two significant figures. Therefore, the molar weight of the unknown must be rounded off to two significant figures also. It becomes 1.7×10^2 g mol^{-1}. The rounded-off answer implies a precision of 1 part in 17 which is more nearly consistent with the precision of the ΔT_f (1 part in 7.6) than an implied precision of 1 part in 166.

---

39. Sap rises in trees because it contains solutes and is enclosed by semi-permeable membranes (in the trees' roots) that allow entry of water from the ground but no exit of solutes to the ground. The osmotic pressure of the sap depends not only on the height h it rises, but also on the concentration of solutes in the sap:

$$\Pi = \rho g h \quad \textit{and} \quad \Pi = n/V\ RT$$

where ρ is the density of the sap, g is the acceleration of gravity and n/V is the concentration of the sap. Eliminating Π between the two equations and rearranging gives:

$$n / V\rho = gh / RT$$

But the volume of a solution multiplied by its density is its mass, that is, $V\rho = M$. With this in mind the equation becomes:

$$n / M = gh / RT$$

If SI units are exclusively used on the right-hand side of this equation the units of the quantity on the left will be mol kg^{-1}. Substituting g = 9.80 m s^{-2}, h = 24.4 m, R = 8.314 J mol^{-1}K^{-1} and T = 298.15 K gives a sap concentration of 0.0965 mol kg^{-1}. If the sap is approximated as a solution of sucrose (molar weight 342.3 g mol^{-1}) then it contains 33.0 g of sucrose per kilogram.

---

41. The vapor pressure of water above the first beaker is greater than it is above the second beaker because the second beaker has more non-volatile solute per liter of solution ($0.25\ mol\ L^{-1}$ versus $0.100\ mol\ L^{-1}$).

Water will tend to transfer from the first to the second beaker until vapor pressures above the two become equal. The vapor pressures above the two beakers become equal when the concentrations of NaCl in the two become equal. The first beaker contains:

$$0.100\ \text{mol L}^{-1} \times 0.400\ \text{L} = 0.0400\ \text{mol of NaCl}$$

and the second contains:

$$0.25\ \text{mol L}^{-1} \times 0.200\ \text{L} = 0.0500\ \text{mol of NaCl}$$

If the concentrations of NaCl in the two beakers are to be equal then the second must end up with 1.25 times the volume of solution of the first inasmuch as it holds 1.25 times more solute. The total volume of water is 600 mL so:

$$600\ \text{mL} = x + 1.25x$$

where x is the final volume in the first beaker. Clearly:

$$x = 266.7\ \text{mL}$$

The final volume in the second beaker is 333.3 mL, by difference. The final concentration in both beakers is 0.150 M.

0.150 M is a *weighted* mean. (See p. 10 of this Guide.) It is the average of the concentrations in the two beakers weighted by the relative volume of their contents. The fact that the container is small saves us from having to consider the quantity of water that ends up in the vapor state.

---

43. ***a)*** The problem is similar to problem 25. Call methyl *n*-butyrate component 1 and ethyl acetate component 2. By Raoult's law:

$$P_{tot} = X_1 P_1^\circ + X_2 P_2^\circ$$

and obviously:

$$X_1 + X_2 = 1$$

The vapor pressure of pure methyl *n*-butyrate, P_1° is 0.1443 atm at 50.0° C and the vapor pressure of pure ethyl acetate, P_2°, at the same temperature is 0.3713 atm. P_{tot} is 0.2400 atm. Hence, numerical solution of the two simultaneous equations is possible:

$$0.2400 = (1 - X_2)(0.1443) + X_2(0.3713)$$

$$0.2400 = 0.1443 + 0.2270\ X_2$$

$$X_2 = 0.4216$$

b) The ethyl acetate contributes:

$$X_2 P_2^o = 0.4216 \times 0.3713 \text{ atm} = 0.1565 \text{ atm}$$

to the total vapor pressure above the solution. The total pressure above the solution is 0.2400 atm. Since 0.1565 atm of this comes from the ethyl acetate, the mole fraction of ethyl acetate in the vapor is (0.1565 atm / 0.2400 atm) = 0.6522.

Chapter 5

CHEMICAL EQUILIBRIUM

5.1 *CHEMICAL REACTIONS AND EQUILIBRIUM*

A chemical system left to itself ultimately comes to a state of *equilibrium* in which macroscopic properties like pressures and concentrations no longer change. Mechanical analogies are easy to find: the unwinding of a spring, the slump of a pile of gravel to a final angle of repose. Such analogies fail in a crucial respect. Chemical equilibria are *dynamic*. Although there is never a visible sign of change in a chemical system at equilibrium, there is continuous exchange between the reactants and products on the molecular level. There is a dynamic balance between forward and reverse reactions.

The Law of Mass Action

Experiment and theory both show that the concentrations of the reactants and products present in a chemical system at equilibrium are related to each other. Consider a system in which all of the reactants and products are gases:

$$a\,\mathrm{A}(g) + b\,\mathrm{B}(g) \rightleftarrows c\,\mathrm{C}(g) + d\,\mathrm{D}(g)$$

The *reaction quotient* (Q) for this reaction is:

$$Q = \frac{P_C^c\, P_D^d}{P_A^a\, P_B^b}$$

The reaction tends to occur either from left to right or from right to left to adjust the value of Q to equal a constant K that differs for every reaction. *K is the reaction's equilibrium constant.* It depends upon the temperature at which the reaction occurs.

Large values of K correspond to nearly complete equilibrium conversion of reactants to products. Small values of K correspond to only slight forward reaction at equilibrium and slight conversion to products.

Many problems give an initial set of partial pressures of reactants and products. Substitution of these values into the mass-action expresssion gives the initial values of Q. If Q exceeds K, then the reaction proceeds to the left, the reverse of the direction in which it is written. If K exceeds Q, then the reaction proceeds toward the right. If by some chance $Q = K$, then the system is at equilibrium as originally mixed and no perceptible reaction will occur. Thus it is easy to predict the direction of reaction given a set of initial pressures and a value of K. **See problem 11a.**

Identifying the *tendency* to react in one direction or the other does not tell how *fast* the reaction will be. Many chemical systems remain without apparent change for years in states which are far from equilibrium.

Activities

Both K and Q are *dimensionless* quantities, that is, K and Q are pure numbers that have no units. The law of mass action is strictly true only when the partial pressures of gases (given in atm) are replaced by the unitless *activities* of the gases.

- The activity of a dilute gas is the *ratio* of its partial pressure to some reference pressure, P_{ref}

The impact of the term "activity" is that the greater the partial pressure of a gas taking part in a reaction, the more active it is in determining the exact nature of that reaction's equilibrium. It makes sense to choose the reference pressure to be exactly 1 atm. That way, as long as partial pressures are in atm their conversion to activities requires nothing more than the dropping of the unit. Numerical values of K obviously depend on the P_{ref} One atmosphere is the usual reference choice, but not the only one.

The partial pressure of a gas is directly proportional to its concentration, assuming it is ideal. Indicate a substance's concentration in mol L^{-1} by a set of brackets around its formula. Thus, for the gas A:

$$[A] = n_A / V = P_A / RT$$

In view of this equation, a mass action expression in terms of partial pressures is easy to rewrite employing the concentrations of the various gases taking part in the reaction. Doing this, of course, exerts no effect on events in a reaction system which is coming to equilibrium. However, the numerical value of the equilibrium constant in general differs when concentrations are used. For the general reaction:

$$a\,A + b\,B \rightleftarrows c\,C + d\,D$$

the equilibrium constant employing concentrations is:

$$K_c = \frac{[C]^c\,[D]^d}{[A]^a\,[B]^b}$$

Let K_p stand for the equilibrium constant employing partial pressures. Then:

$$K_c = K_p\,(RT / P_{ref})^{a + b - c - d}$$

The quantity $(c + d - a - b)$ is Δn_g, the change in the chemical amount of gas between the two sides of the reaction. Thus, the exponent in the above equation is $-\Delta n_g$:

$$K_c = K_p\,(RT / P_{ref})^{-\Delta n_g}$$

Relationships Among Equilibrium Expressions

If the coefficients of a balanced chemical equation are all multiplied by a constant, the equation is still balanced. The corresponding mass action expression is raised to a power equal to the multiplying constant, even if the multiplying constant is a fraction. Consider for example the chemical equation and equilibrium constant expression:

$$A(g) \rightleftarrows B(g) + 2\,C(g) \qquad K = [B][C]^2 / [A]$$

Compare it to the equation and equilibrium constant expression that follow:

$$\tfrac{1}{2}\,A(g) \rightleftarrows \tfrac{1}{2}\,B(g) + C(g) \qquad K = [B]^{1/2}[C] / [A]^{1/2}$$

The second chemical equation is the first multiplied through by one-half. The *K* of the second equation is the square root of the *K* of the first. Recall that raising to the ½ power means taking the square root.

- If two balanced chemical equations are added then the corresponding mass action expressions are multiplied. **See problem 23a.**

- If one chemical equation is subtracted from a second then the mass action expression of the resulting equation is the mass action expression of the second *divided* by the mass action expression of the first. **See problem 55.**

5.2 *CALCULATION OF GAS-PHASE EQUILIBRIA*

The law of mass action provides a reliable mathematical equation relating the equilibrium partial pressures of all the gases taking part in a reaction. It is however but a single expression. There may be three or four or even more different gases present at equilibrium. To calculate the partial pressures of all the gases in an equilibrium mixture requires finding additional relationships among the partial pressures. There are some important points to bear in mind in doing this.

- The partial pressures substituted into the mass action expression must be *equilibrium* partial pressures. Problems often quote initial partial pressures; final (equilibrium) partial pressures differ sharply. Use initial values to compute an initial reaction quotient. **See problem 11a.**

- The units of all partial pressures must be the same. Recall that numerous different units measure pressure. If a problem is presented in mixed units convert, preferably to atm, before going further.

- The initial partial pressures in a reaction mixture are determined by the person setting up the experiment. They may be given specifically in a problem. **See problems 23b and 39.** When they are not specified they are often equal to zero **(see problem 9)** or their values are given indirectly by the wording of the problem. **See problem 13.**

- The stoichiometry of the equation determines the *changes* in the partial pressures of reactants and products as a chemical reaction tends toward equilibrium.

- The sum of the partial pressures of all of the gases in a system equals the total pressure. In practical reactions the total pressure is usually quite easy to measure. Knowing it provides an additional relationship among the several partial pressures. **See problem 5 and 41.**

To solve gas-phase equilibrium problems:

1. Write down the balanced reaction and the corresponding mass action expression.

2. Write down the initial partial pressures of all reactants and products. Include zero partial pressures for gases not initially present.

3. Determine the changes required in the partial pressures to reach equilibrium. If a change is not known, symbolize it with x or y. Remember that changes made by the reaction as it goes to equilibrium are not independent, but are related by the coefficients in the balanced equation.

4. Substitute the expressions for the equilibrium partial pressures into the mass action expression and solve for x or y, the changes. If tangled and complex algebraic expressions arise, avoid the tedium and possible error of analytical solution by using the method of successive approximations. Easy access to electronic calculators makes this method the best in many cases. **See problems 11, 13 and 43.**

5. Calculate the equilibrium partial pressures by adding the appropriate change to the initial partial pressures.

LeChatelier's Principle

LeChatelier's principle states that if a stress is applied to a system at equilibrium, then the position of the equilibrium will shift in the direction that counteracts the stress. For example:

- A product is taken out of a system at equilibrium. The equilibrium shifts to the right and partially remedies the loss.

- A reactant is removed from an equilibrium system. The equilibrium shifts to the left to minimize the loss, producing more reactants.

- A product is added to an equilibrium system. The equilibrium shifts to the left, consuming some of that product.

- A reactant is added to an equilibrium system. The equilibrium re-establishes itself after shifting to the right.

- The volume of an equilibrium chemical system is changed. A reduction in volume causes an increase in the pressure and the system responds by shifting toward the side of the reaction having the fewer moles of gas. Fewer moles of gas require smaller volume. Increasing the volume has the reverse effect.

LeChatelier's principle is a handy qualitative guide in grappling with equilibrium problems. Its use avoids a troublesome difficulty. In problems, it is often possible inadvertently to set up the algebraic sense of the changes required to come to equilibrium so that they work out to be negative. If one expects that an equilibrium will shift to the right, getting a negative sign for changes in partial pressure simply means that it in fact shifts to the left. In practice, however, negative changes are not expected, especially if they come after some mathematical effort, and may have a depressing effect on the problem-solver's morale. LeChatelier's principle allows one to form a correct qualitative idea of the direction of prospective shifts and prevents this effect. **See problems 11c and 39c.**

5.3 *HETEREOGENEOUS EQUILIBRIA*

In general, chemical equilibria may involve solids, liquids and solutions in all combinations in addition to gases. The mass action expression in a general case is constructed in the same way as in a case involving only gases. As before, gases enter the expression as partial pressures, in atmospheres. Dissolved species enter as concentrations, in moles per liter.

- *Pure solids and pure liquids are omitted from mass action expressions.*

The activity of a pure solid or liquid is equal to 1. Even if it were raised to some large power in a mass action expression, an activity of 1 would make no numerical difference. **See problem 15.** Although the activities of pure solids and liquids does not appear explicitly in mass action expressions, the presence of the solids or liquids is still crucial to the existence of the equilibrium. **See problem 21b.**

In dilute solutions the activity of the solvent is quite close to 1. For this reason the solvent can be omitted from mass action expressions. **See problems 15, 17 and 23.**

5.4 *SOLUBILITY OF SALTS*

When a typical salt is dissolved in water the dissolved material is present as separate hydrated anions and cations. A *saturated* solution of a salt is a solution in which equilibrium exists between the solid salt and its dissolved ions. The equilibrium governing the dissolution of a salt in water is a simple heterogeneous equilibrium. For example:

$$PbSO_4\,(s) \rightleftarrows Pb^{2+}\,(aq) + SO_4^{2-}\,(aq)$$

This heterogeneous equilibria is described by the mass action expression:

$$K_{sp} = [Pb^{2+}]\,[SO_4^{2-}]$$

K_{sp} is a *solubility product* constant. It does not differ fundamentally from other equilibrium constants. The special name merely emphasizes that the mass action expression has the form of a product and not the more usual form of a quotient. **See problem 49** for an example of writing a K_{sp} expression in a difficult case.

The *solubility* of a substance is the number of grams or moles of it present at equilibrium in a given quantity of solvent or solution. Solubilities depend, often quite markedly, on the temperature. This means that K_{sp} values always have to be associated with a temperature. In problems that do not cite the temperature it is reasonable to assume that it is the same as the temperature at which the K_{sp} was measured.

Typical Problems Involving K_{sp}

- The calculation of K_{sp} from the solubility of a solid and the reverse. **See problems 25 and 27a.**

- The prediction of the concentrations of ions required in the solution for *precipitation* of a given solid upon mixing its constituent ions. Precipitation occurs whenever the product of the proper powers of the concentrations of the ions composing a salt exceeds the value of K_{sp}. **See problem 33a.**

- Evaluation of proposed selective precipitations in which, for example, a mixture of two cations is separated on the basis of the differing solubilities of their salts with an added anion. **See problems 31 and 51.**

The Common Ion Effect

If a salt is placed in water that is not pure but contains one of the ions produced by the dissolution of the salt (a common ion) then the solubility of the salt is *reduced.* This is a straight-forward conclusion from LeChatelier's principle. Although the common ion effect drastically changes salts' solubilities it does not change their K_{sp}'s, which are more fundamental quantities. **See problem 27.**

5.5 *EXTRACTION AND SEPARATION PROCESSES*

The separation of mixtures is important in research and industry. A type of equilibrium much used in separations is the *partition* of a substance between two immiscible liquids. Carbon tetrachloride and water quickly separate into two layers after being shaken together in a flask. A third substance, like I_2, partitions itself between the layers according to the ratio of its solubilities in the two solvents. An equilibrium is established:

$$K = [I_2](CCl_4)/[I_2](aq)$$

An impurity in the iodine will partition itself between the two liquids, too, but with some different *K*. Suppose the impurity has a smaller *K* than the I_2. Then it tends to concentrate in the water relative to the I_2. The I_2 tends to concentrate in the CCl_4 relative to the impurity. Even small differences in *K* can lead to effective separations if the equilibrium is repeatedly established in a cycled operation. **See problems 54 and 55.**

DETAILED SOLUTIONS TO ODD-NUMBERED PROBLEMS

1. *a)* The equilibrium expression is:

$$K = \frac{P_{CO_2} P_{H_2}}{P_{CO} P_{H_2O}}$$

b) The problem gives K and three of the four equilibrium partial pressures. Substitution in the equilibrium expression and solution for the partial pressure of H_2 is straighforward. The answer is 0.056 atm.

---

3. Benzyl alcohol has a molar weight of 108.14 g mol^{-1}. Place 1.20 g of it in the container. This is 0.011097 mol. Assume for a moment that the benzyl alcohol behaves as an ideal gas in the 2.00 L container at 523 K. Then its pressure in the container (based on PV = nRT) would be 0.23811 atm. Benzyl alcohol does *not* behave ideally; instead a portion of it reacts to give benzyaldehyde and hydrogen. When this reaction reaches equilibrium at 523 K:

$$0.558 = \frac{P_{H_2} P_{C_6H_5CHO}}{P_{C_6H_5CH_2OH}}$$

The partial pressure of benzyl alcohol is reduced from its ideal value. The stoichiometry of the equation and the fact that there are no other sources of either product require that:

$$P_{H_2} = P_{C_6H_5CHO}$$

Let $x = P_{H_2}$. Then the partial pressure of benzyl alcohol is $(0.23811 - x)$ atm. Substitution in the mass action expression gives:

$$0.558 = x^2 / (0.23811 - x)$$

which rearranges to give:

$$x^2 + 0.558x - 0.132865 = 0$$

Solving this quadratic equation and discarding the physically meaningless root:

$$x = 0.1800 \text{ atm} = P_{H_2}$$

Since the partial pressure of H_2 is 0.1800 atm, the partial pressure of benzyl alcohol is 0.23811 − 0.1800 atm = 0.05811 atm. The pressure of the benzyl alcohol is 0.2439 of what it would have been, absent the reaction. Therefore the fraction of the benzyl alcohol that is dehydrogenated at equilibrium is 1 − 0.2439 = 0.7561. The benzl alcohol is 75.6 percent dehydrogenated.

---

5. The reaction is the dimerization of acetic acid at 110° C:

$$2CH_3COOH(g) \rightleftarrows (CH_3COOH)_2(g)$$

Imagine that the process starts with pure acetic acid vapors (the reactant) present at a pressure P_0. Some of the acetic acid dimerizes to give the product which at equilibrium has a partial pressure x. For every 1 mole of dimer that appears 2 moles of monomer disappear. The equilibrium partial pressure of the reactant is therefore $P_0 - 2x$. Let the subscripts d and m stand for the monomer and dimer. Then:

$$K = \frac{P_d}{P_m^2} = \frac{x}{(P_0 - 2x)^2}$$

where $K = 3.72$. Meanwhile, the total pressure, the sum of the partial pressures, is 1.50 atm:

$$1.50 = (P_0 - 2x) + x$$

Eliminating P_0 between the two equations gives:

$$2.72\,x^2 - 11.16\,x + 8.37 = 0$$

The roots of this quadratic equation are $x = 3.12$ and 0.985. (A good way to get these roots is to use the quadratic formula.) The first root corresponds to a *negative* original pressure of monomer. It is rejected and the second root chosen. Thus, $x = 0.985$ which means that the pressure of the dimer is 0.985 atm at equilibrium.

---

7. The problem requires a grasp of what is meant by "fraction dissociated" and why the pressure in the vessel increases as the reaction goes to the right. The reaction is:

$$2\,A_2B(g) \rightleftarrows 2\,A_2(g) + B_2(g)$$

Imagine that the original pressure of A_2B in the vessel is 10 atm. This figure is entirely *arbitrary* and is *not* the answer. The original partial pressures of both products are 0 because the only way they form is *via* the reaction. Now imagine that various fractions of the A_2B molecules undergo dissociation. Tabulate the partial pressures of all components:

Fraction Dissociated	P_{A_2B}	P_{A_2}	P_{B_2}	P_{total}
0.0	10.0 atm	0.0 atm	0.0 atm	10.0 atm
0.10	9.0	1.0	0.5	10.5
0.30	7.0	3.0	1.5	11.5
0.50	5.0	5.0	2.5	12.5
0.60	4.0	6.0	3.0	13.0
1.00	0.0	10.0	5.0	15.0

The table is constructed on the principle that the pressure of a component of a gas mixture is proportional to its chemical amount. (This follows from the ideal gas law and Dalton's law.) As the table shows, the reaction of A_2B causes its own partial pressure to drop but forms enough products to make the total pressure rise. If *all* of the A_2B reacts then the total pressure is $1\frac{1}{2}$ times its original value. In mathematical form, with α standing for the fraction dissociated:

$$P_{tot} = P_{A_2B} + P_{A_2} + P_{B_2}$$

$$P_{tot} = P_0 (1 - \alpha) + P_0 \alpha + P_0 (\alpha / 2)$$

$$P_{tot} = P_0 (1 + \alpha / 2)$$

Having derived this formula, apply it to the situation described in the problem. In the problem, the final pressure is 30.0 percent more than the original pressure:

$$P_{tot} / P_0 = 1.30 \qquad \textit{hence} \qquad \alpha = 0.600$$

P_0 is what the pressure would have been if there had been no reaction. The original chemical amount of A_2B is 75.0 g / 150 g mol^{-1} = 0.500 mol. Substitution in the ideal gas equation with V = 22.4 L and T = 323.15 *K* gives P_0 = 0.5919 atm. The volume of the vessel is 22.4 L, the famous STP molar volume. This is just an accident and should not make one think that there is 1.00 mol of gas present or that P_0 is 1.00 atm.

The *actual* total pressure, P_{tot} in the vessel is not 0.5919 atm (P_0) but, instead, is 0.5919 × 1.30 or 0.7695 atm which, rounded to 3 significant digits, is 0.769 atm.

To calculate the equilibrium constant, substitute in the expression:

$$K = \frac{P_{A_2}^2 P_{B_2}}{(P_{A_2B})^2} = \frac{P_0^2 \alpha^2 P_0(\alpha/2)}{P_0^2 (1 - \alpha)^2}$$

Both P_0 and α are known so *K* is computed readily. It is 0.400.

---

9. ***a)*** Originally, PCl_5 *(g)* is the only substance in the bulb. To attain equilibrium some of the PCl_5 *(g)* dissociates to give PCl_3 *(g)* and Cl_2 *(g)*:

$$PCl_5 (g) \rightleftarrows PCl_3 (g) + Cl_2 (g)$$

The mass action expression for this reaction is:

$$K = 2.15 = P_{PCl_3} P_{Cl_2} / P_{PCl_5}$$

The reaction is the only source of the products so:

$$P_{Cl_2} = P_{PCl_3}$$

There is no way for Cl_2 to appear without PCl_3 appearing in equal molar measure. The total pressure, the sum of the components' partial pressures, is:

$$P_{PCl_5} + P_{PCl_3} + P_{Cl_2} = 0.79 \text{ atm}$$

There are now three equations relating three unknowns. Let $x = P_{PCl_3} = P_{Cl_2}$ and combine the three algebraically to give:

$$x^2 + 4.30x - 1.6985 = 0$$

Use of the quadratic formula gives $x = -4.664$ and 0.364. Therefore $P_{PCl_3} = P_{Cl_2} = 0.364$ atm and $P_{PCl_5} = 0.062$ atm.

b) Assuming ideal gas behavior for each component:

$$P = (n / V)\,RT = C\,RT$$

where C is the concentration of the component. Taking R = 0.08206 L atm $K^{-1} mol^{-1}$ and T = 523.15 K together with the three partial pressures gives:

$$[PCl_3] = [Cl_2] = 8.48 \times 10^{-3} \text{ mol L}^{-1}$$

$$[PCl_5] = 1.4 \times 10^{-3} \text{ mol L}^{-1}$$

In these expressions the square brackets specify the concentrations of the three species.

c) At equilibrium the glass bulb contains 8.48×10^{-3} mol L^{-1} of PCl_3 and 1.44×10^{-3} mol L^{-1} of PCl_5. Its volume is 0.100 L so it contains a total of 9.92×10^{-4} mol of the two phosphorus compounds. There is one mole of phosphorus per mole of each compound. Also, PCl_5 is the only source of phosphorus in the system. Therefore the original amount of PCl_5 was 9.92×10^{-4} mol. Converting this chemical amount to weight by multiplying by the molar weight of PCl_5 (208.23 g mol^{-1}) gives 0.207 g PCl_5.

---

11. ***a)*** Let P_{di} stand for the partial pressure of *di*phosphorus, P_2 and P_{tet} for the partial pressure of *tetra*phosphorus, P_4. The reaction quotient of the process is:

$$Q = P_{di}^2 / P_{tet}$$

Initially $P_{di} = 2.00$ atm and $P_{tet} = 5.00$ atm making $Q = 0.800$. Because $Q > K$ the equilibrium will tend to shift *to the left* (reducing the numerator and increasing the denominator in the above expression) until $Q = K$.

b) Let x equal the increase in the pressure of tetraphosphorus during the change. Then:

$$K = (2.00 - 2x)^2 / (5.00 + x) = 0.612$$

The equation can be solved with the quadratic formula. It is instructive however to get x by successive approximation. (See Appendix C of text.) Construct a table:

x	$(2.00 - 2x)^2$	$5.00 + x$	Q
0.00	4.00	5.00	0.800
0.100	3.24	5.10	0.635
0.120	3.10	5.12	0.605
0.115	3.13	5.115	0.612

As x increases Q decreases from 0.800. At $x = 0.12$ $Q = 0.605$, which is slightly over the true value of K. With an electronic calculator it takes only moments to get an accurate value of x. If $x = 0.115$ then $P_{di} = 1.77$ atm and $P_{tet} = 5.12$ atm.

c) If the volume of the system is increased, then, by LeChatelier's principle, there will be net *dissociation* of P_4. The system responds to its forcible rarefaction by producing more molecules to fill the larger volume.

---

13. The equilibrium constant expression for the synthesis of ammonia is

$$K = 6.78 \times 10^5 = \frac{P_{NH_3}{}^2}{P_{H_2}{}^3\ P_{N_2}}$$

This equation can be written effortlessly. The difficulty is in translating the other statements in the problem into mathematical terms. If the ratio of H to N atoms in the system is 3:1 then:

$$P_{H_2} = 3\ P_{N_2}$$

because the third component, NH_3, maintains, within itself, the required 3:1 ratio. The fact that the total pressure is 1 atm means:

$$P_{N_2} + P_{H_2} + P_{NH_3} = 1 \text{ atm}$$

Let $x = P_{N_2}$ and substitute in the mass action expression:

$$6.78 \times 10^5 = (1 - 4x)^2\ /\ (3x)^3\, x$$

$$6.78 \times 10^5 = 1 - 8x + 16x^2\ /\ 27x^4$$

This equation is hard to solve analytically. But x is expected to be small because at equilibrium the mixture is mostly NH_3 (K is big). Suppose $4x << 1$. Then:

$$6.78 \times 10^5 \simeq 1\ /\ 27x^4$$

$$x \simeq 0.0153$$

The best way to get a more exact answer is by successive approximations. The idea is to guess values of x and compute Q for each. Revise the guesses until Q becomes sufficiently close to K. The following table is a map of such a process. It starts with the x from the rough solution:

x	$(1 - 4x)^2$	$27x^4$	Q
0.0153	0.8813	1.480×10^{-6}	5.96×10^5
0.0145	0.8874	1.194×10^{-6}	7.43×10^5
0.0149	0.8843	1.331×10^{-6}	6.64×10^5
0.0148	0.8851	1.295×10^{-6}	6.83×10^5

K is given to three significant digits, so accept $x = 0.0148$ as close enough. If $x = 0.0148$ then:

$P_{N_2} = 0.0148$ atm $\quad P_{H_2} = 0.0444$ atm $\quad P_{NH_3} = 0.941$ atm

---

15. The activities of pure solids and liquids are 1 and are omitted from equilibrium constant expressions.

a)

$$K = \frac{P_{CO_2} P_{HF}^{\ 4}}{P_{CF_4}}$$

b)

$$K = \frac{P_{SO_2}^{\ 3}}{P_{O_2}^{\ 5}}$$

c)

$$K = \frac{[VO_3(OH)^{2-}][OH^-]}{[VO_4^{\ 3-}]}$$

d)

$$K = \frac{[HCO_3^-]^6}{P_{CO_2}^{\ 6}\,[As(OH)_6^{\ 3-}]^2}$$

---

17. ***a)*** In this heterogeneous equilibrium, the equilibrium constant expression is simplified because the activity of the pure $NH_4HSe(s)$ on the left is 1.

$$K = P_{NH_3} P_{H_2Se}$$

The total pressure above the solid NH_4HSe is 0.0184 atm. If the decomposition of the solid is the only source of $NH_3(g)$ and $H_2Se(g)$, then the partial pressures of both gases are $\frac{1}{2}(0.0184)$ atm and K is $[\frac{1}{2}(0.0184)]^2 = 8.46 \times 10^{-5}$.

b) Now the partial pressure of NH_3 has been adjusted by some outside means to 0.0152 atm. At equilibrium:

$$K = 8.46 \times 10^{-5} = P_{NH_3} P_{H_2Se} = 0.0152\ P_{H_2Se}$$

and the partial pressure of H_2Se is 5.57×10^{-3} atm.

---

19. The equilibrium constant expression is:

$$K = P_{CO} / P_{CO_2} P_{H_2}$$

and K is given as 3.22×10^{-4}. Assume that the $CO(g)$ is to be produced at 25° C. Let $P_{CO_2} = P_{H_2} = x$. Then, because $P_{CO} = 1$:

$$K = 1 / x^2 = 3.22 \times 10^{-4}$$

from which x = 55.7. At this high pressure deviations from ideality are important. The fact that the problem specifies the ideal gas assumption allows the answer: $P_{CO_2} = P_{H_2} = 55.7$ atm.

---

21. ***a)*** The mass action expression for the reaction is:

$$K = 4.0 = (P_{NH_3}\,P_{HCl})^{-1} \quad \textit{or} \quad 0.25 = P_{NH_3}\,P_{HCl}$$

at 340° C. If P_{NH_3} = 0.80 atm, then, by substitution, the equilibrium partial pressure of HCl*(g)* is 0.31 atm.

b) The equilibrium cannot occur before addition of NH_4Cl*(s)* to the container filled with ammonia. It is the only source of HCl*(g)*.

For every mole of HCl*(g)* produced one mole of NH_3*(g)* is added to the quantity responsible for the original 1.5 atm. Assuming ideality, the partial pressures of the two gases are directly proportional to their chemical amount (moles). Letting x equal the equilibrium partial pressure of HCl*(g)*:

$$0.25 = P_{NH_3}\,P_{HCl} = (1.5 + x)x$$

Solving the quadratic equation in x and rejecting the negative root gives x = 0.151. Therefore P_{HCl} = 0.15 atm, and P_{NH_3} = 1.5 + 0.15 or 1.65 atm.

---

23. ***a)*** The equation for the water gas reaction is the sum of the two other equations given in the problem. Therefore the equilibrium constant for the water gas equation is the *product* of K_1 and K_2:

$$K_1 = \frac{P_{CO}^{\,2}}{P_{CO_2}}$$

$$K_2 = \frac{P_{H_2}\,P_{CO_2}}{P_{CO}\,P_{H_2O}}$$

$$K_1 K_2 = \frac{P_{CO}^{\,2}\,P_{H_2}P_{CO_2}}{P_{CO_2}P_{CO}P_{H_2O}} = \frac{P_{CO}P_{H_2}}{P_{H_2O}} = K_3$$

K_3 then equals 2.611. Note that graphite, as a pure solid, has unit activity. It does not appear in the expressions.

b) In this part the value of K_3 and the equilibrium constant expression are used in an application:

$$K_3 = P_{CO}\,P_{H_2} / P_{H_2O}$$

where P_{H_2O} = 2.00 atm. If the only source of H_2*(g)* is the water gas reaction, then, letting $P_{CO} = P_{H_2} = x$:

$$x^2 = 2.611 \times 2.00 \quad \textit{and} \quad x = P_{CO} = 2.29 \text{ atm}$$

---

25. $AgBrO_3$ dissolves by dissociation:

$$AgBrO_3\,(s) \rightleftarrows Ag^+\,(aq) + BrO^{-}{}_3\,(aq)$$

In a saturated solution of silver bromate:

$$K_{sp} = [Ag^+][BrO_3^-]$$

and, in this case:

$$[Ag^+] = [BrO_3^-]$$

since the only source of the ions is the dissolution of the 0.358 g of $AgBrO_3$. The molar weight of $AgBrO_3$ is 235.78 g mol^{-1} so the 200 mL of water contains 1.518×10^{-3} mol of dissolved $AgBrO_3$ or 7.59×10^{-3} mol L^{-1}:

$$[Ag^+] = [BrO_3^-] = 7.59 \times 10^{-3} \text{ mol L}^{-1}$$

Substitution gives:

$$K_{sp} = (7.59 \times 10^{-3})^2 = 5.76 \times 10^{-5}$$

The K_{sp} has no units because the concentrations of the ions are replaced with their numerically equal but dimensionless *activities*.

---

27. ***a)*** Let S equal the molar solubility of $Ni(OH)_2\,(s)$. This ionic substance dissolves by the equilibrium:

$$Ni(OH)_2\,(s) \rightleftarrows Ni^{2+}\,(aq) + 2\,OH^-\,(aq)$$

If S mol L^{-1} dissolves then:

$$[Ni^{2+}] = S \qquad \textit{and} \qquad [OH^-] = 2\,S$$

In a saturated solution:

$$K_{sp} = [Ni^{2+}][OH^-]^2 = 1.6 \times 10^{-16}$$

which gives after substitution:

$$K_{sp} = 4\,S^3 = 1.6 \times 10^{-16}$$

$$S = 3.4 \times 10^{-6} \text{ mol L}^{-1}$$

where the two imaginary roots to the cubic equation have been rejected.

b) The presence in the solution of $OH^-\,(aq)$ from the freely soluble NaOH depresses the solubility of the $Ni(OH)_2$ by the common ion effect. Although the solubility of the $Ni(OH)_2$ goes down, its K_{sp} is unchanged and:

$$K_{sp} = [Ni^{2+}][OH^-]^2 = 1.6 \times 10^{-16}$$

If S mol L^{-1} of $Ni(OH)_2$ dissolves then $[Ni^{2+}] = S$ and $[OH^-] = 0.100 + S$. The OH^- *(aq)* ion comes from two sources, and its concentration is therefore *not* equal to the concentration of Ni^{2+} *(aq)*:

$$1.6 \times 10^{-16} = [Ni^{2+}][OH^-] = S(0.100 + S)^2$$

The above expression gives a cubic equation in S which can of course be solved. Mathematical labor is avoided if it is remembered that S was 3.4×10^{-6} mol L^{-1} in part a) and is expected to be *less* now. Because S is so small compared to 0.100:

$$0.100 + S \simeq 0.100$$

hence:

$$1.6 \times 10^{-16} \simeq S(0.100)^2 \quad \textit{and} \quad S = 1.6 \times 10^{-14} \text{ mol L}^{-1}$$

---

29. As long as the solution is in equilibrium with $Mg(OH)_2$ *(s)*, then the solubility product constant expression is satisfied:

$$K_{sp} = 1.2 \times 10^{-11} = [Mg^{2+}][OH^-]^2$$

Let S represent the solubility of the $Mg(OH)_2$ *(s) before* any NaOH is added. At this stage:

$$[Mg^{2+}] = S \quad \textit{and} \quad [OH^-] = 2S$$

$$4S^3 = 1.2 \times 10^{-11}$$

$$S = 1.44 \times 10^{-4} \text{ mol L}^{-1}$$

NaOH dissociates completely in water. Adding it to the solution is the same as the addition of Na^+ *(aq)* and OH^- *(aq)* ions in a 1:1 ratio. $[OH^-]$ goes up with the addition of NaOH, and $[Mg^{2+}]$ must diminish to maintain the product $[Mg^{2+}][OH^-]^2$ at a constant value. This exemplifies the common ion effect. Additional magnesium hydroxide precipitates. The problem states that the solubility of $Mg(OH)_2$ is reduced to 0.001 of its original value. This means that after the addition:

$$[Mg^{2+}] = 0.001(1.44 \times 10^{-4}) = 1.44 \times 10^{-7} \text{ M}$$

$$[OH^-]^2 = K_{sp} / (1.44 \times 10^{-7})$$

$$[OH^-] = 9.1 \times 10^{-3} \text{ M}$$

---

31. The first solution contains 5.00×10^{-3} mol of $AgNO_3$ and the second 1.80×10^{-3} mol of Na_2CrO_4. When they are mixed the reaction follows the equation:

$$2\,Ag^+ (aq) + CrO_4^{2-} (aq) \rightarrow Ag_2CrO_4 (s)$$

Assume for the moment that the reaction goes *completely* to the right. Then all of the CrO_4^{2-} *(aq)*, the limiting reagent, is consumed and $5.00 \times 10^{-3} - 2(1.80 \times 10^{-3}) = 1.40 \times 10^{-3}$ mol of Ag^+ *(aq)* remains. Because the total volume is 0.080 L, this would give the answers:

$$[CrO_4^{2-}] = 0 \text{ M} \quad \textit{and} \quad [Ag^+] = 1.75 \times 10^{-2} \text{ M}$$

However *some* of the Ag_2CrO_4 *(s)* must dissolve at equilibrium so that the solubility product constant expression:

$$K_{sp} = 1.9 \times 10^{-12} = [Ag^+]^2[CrO_4^{2-}]$$

is satisfied. Let $x = [CrO_4^{2-}]$. Then $[Ag^+] = 1.75 \times 10^{-2} + 2x$ because the passage into solution of one CrO_4^{2-} ion releases 2 Ag^+ ions. The K_{sp} expression in terms of x is:

$$1.9 \times 10^{-12} = (1.75 \times 10^{-2} + 2x)^2 \, x$$

Solving, $x = 6.2 \times 10^{-9}$, hence $[CrO_4^{2-}] = 6.2 \times 10^{-9}$ M. Now substitute to get the silver ion concentration:

$$[Ag^+] = 1.75 \times 10^{-2} + 2(6.2 \times 10^{-9}) = 1.75 \times 10^{-2} \text{ M}$$

The answer must be rounded off to 2 significant digits. The result is $[Ag^+] = 1.8 \times 10^{-2}$ M. The concentration of Ag^+ *(aq)* ion that comes from the dissolution of the silver chromate is quite insignificant compared to the concentration already present. It is small compared even to the *change* caused by rounding-off.

---

33. ***a)*** The K_{sp}'s of MgC_2O_4 *(s)* and PbC_2O_4 *(s)* are 8.6×10^{-5} and 2.7×10^{-11} respectively. When $C_2O_4^{2-}$ *(aq)* ion is added to the solution of Mg^{2+} *(aq)* and Pb^{2+} *(aq)* the *more insoluble* oxalate, PbC_2O_4, starts to precipitate first, as its K_{sp} is exceeded. Therefore Pb^{2+} is in the solid. As more $C_2O_4^{2-}$ *(aq)* comes in, $[C_2O_4^{2-}]$ continues to rise. When MgC_2O_4 *(s)* is just about to precipitate the following equation holds:

$$8.6 \times 10^{-5} = [Mg^{2+}][C_2O_4^{2-}]$$

Assume that the addition of $C_2O_4^{2-}$ *(aq)* has not diluted the solution. Then $[Mg^{2+}] = 0.10$ M and, substituting in the K_{sp} expression, $[C_2O_4^{2-}] = 8.6 \times 10^{-4}$ M.

b) Some lead(II) ion always remains in solution in equilibrium with the solid PbC_2O_4 and the oxalate ion. Its concentration is dictated by the K_{sp} expression:

$$2.7 \times 10^{-11} = [Pb^{2+}][C_2O_4^{2-}]$$

Because $[C_2O_4^{2-}] = 8.6 \times 10^{-4}$ M when the MgC_2O_4 *(s)* just begins to come down $[Pb^{2+}] = 3.1 \times 10^{-8}$ M. This is 3.1×10^{-7} of the original 0.10 M Pb^{2+} *(aq)* or about 0.31 ppm. The separation of Mg^{2+} *(aq)* and Pb^{2+} *(aq)* is quite effective.

---

35. The I_2 partitions itself between the CCl_4 and H_2O which, being immiscible liquids, form separate layers at equilibrium. Originally $[I_2(aq)] = 1.00 \times 10^{-2}$ M. After the CCl_4 treatment $[I_2(aq)] = 1.30 \times 10^{-4}$ M. Evidently 9.87×10^{-3} mol I_2 is removed from every liter of water that is shaken with the CCl_4. This I_2 transfers into an equal volume of CCl_4 so:

$$[I_2(CCl_4)] = 9.87 \times 10^{-3}\ \text{M}$$

The equilibrium concentrations of I_2 are now known in both phases. The partition coefficient is nothing more than the equilibrium constant for the process:

$$I_2(aq) \rightleftarrows I_2(CCl_4)$$

$$K = [I_2(CCl_4)]\,/\,[I_2(aq)] \qquad \textit{ergo} \qquad K = 75.9$$

---

37. For the chemical equation given in the problem:

$$K = \frac{P_{CH_3OH}}{P_{CO}\,P_{H_2}^{\,2}} = 6.08 \times 10^{-3}$$

The problem also states:

$$2\,P_{CO} = P_{H_2} \qquad \textit{and} \qquad P_{CH_3OH} = 0.50\ \text{atm}$$

Let $P_{CO} = y$. Then:

$$6.08 \times 10^{-3} = 0.50\,/\,4y^3$$

Solving for y gives $P_{CO} = 2.74$ atm and $P_{H_2} = 5.48$ atm.

---

39. ***a)*** The reaction quotient for this reaction has the form:

$$Q = \frac{P_{Cl_2}\,P_{PCl_3}}{P_{PCl_5}}$$

All of the partial pressures are given. By substitution, $Q = 120$.

b) As originally prepared, the system has $Q = 120$ and $K = 11.5$. It will tend to react from right to left, adjusting partial pressures until $Q = K$. Let y equal the partial pressure of $Cl_2(g)$ that is removed by this reaction. An equal pressure of $PCl_3(g)$ is removed, and an equal pressure of $PCl_5(g)$ is produced. This follows from the stoichiometry of the equation. Therefore:

$$K = 11.5 = (6.0 - y)(2.0 - y)/(0.10 + y)$$

Rearrangement of this equation gives a quadratic in y:

$$y^2 - 19.5y + 10.85 = 0$$

The roots of this equation are 0.573 and 18.93. Only the first makes physical sense. The second corresponds to negative partial pressures. Complete the computation by finding the various partial pressures:

$$P_{PCl_3} = 2.0 - 0.573 = 1.43 \text{ atm}$$

$$P_{PCl_2} = 6.0 - 0.573 = 5.43 \text{ atm}$$

$$P_{PCl_5} = 0.10 + 0.573 = 0.673 \text{ atm.}$$

c) By LeChatelier's principle an increase in volume causes the reaction to shift to the side having *more* moles of gas. The amount of PCl_5 *(g)* will decrease.

---

41. ***a)*** The only material originally placed in the flask is NOBr*(g)*. At 350 K it decomposes:

$$NOBr(g) \rightleftarrows NO(g) + \tfrac{1}{2} Br_2(g)$$

The reaction increases the chemical amount of gases inside the flask and in that way makes the total flask pressure *higher* than what it would have been if NOBr behaved ideally. If NOBr did *not* decompose then:

$$P_{NOBr} = (n/V)RT$$

where n/V is the molar density of NOBr The molecular weight of NOBr is 109.915. The molar density of the hypothetically undecomposed NOBr at 350 K is:

$$2.219 \text{ g L}^{-1} / 109.915 \text{ g mol}^{-1} = 0.02019 \text{ mol L}^{-1}$$

so that the pressure of NOBr would, absent the reaction, be:

$$P_{NOBr} = (0.02019 \text{ mol L}^{-1})(0.08206 \text{ L atm K}^{-1}\text{mol}^{-1})(350 \text{ K}) = 0.5798 \text{ atm}$$

In the actual set-up the total pressure exceeds this. It is 0.675 atm:

$$P_{tot} = P_{NOBr} + P_{NO} + P_{Br_2} = 0.675 \text{ atm.}$$

The partial pressure of NOBr is less than 0.5798 atm and P_{NO} and P_{Br_2} both are greater than zero. Let α be the fraction of the NOBr*(g)* that decomposes at equilibrium. Then:

$$P_{NOBr} = 0.5798(1 - \alpha) \text{ atm}$$

$$P_{NO} = 0.5798\,\alpha \text{ atm}$$

$$P_{Br_2} = 0.5798\,(\alpha / 2) \text{ atm}$$

Adding these three equations:

$$P_{tot} = P_{NOBr} + P_{NO} + P_{Br_2} = 0.5798(1 + \tfrac{1}{2}\alpha) \text{ atm}$$

The total pressure is 0.675 atm. Therefore:

$$0.675 = 0.5798(1 + \tfrac{1}{2}\alpha) \quad \textit{and} \quad \alpha = 0.3284$$

Substitute this value in the expressions for the various partial pressures:

$$P_{NOBr} = 0.3894 \text{ atm}; \quad P_{NO} = 0.1904 \text{ atm}; \quad P_{Br_2} = 0.0952 \text{ atm}$$

b) The equilibrium constant expression is:

$$K = \frac{P_{NO}\, P_{Br_2}^{1/2}}{P_{NOBr}}$$

All of the partial pressures are known from part a). K equals 0.1509.

---

43. The reaction is the oxidation of sulfur dioxide to sulfur trioxide:

$$SO_2\,(g) + \tfrac{1}{2}\, O_2\,(g) \rightleftarrows SO_3\,(g)$$

The equilibrium constant expression is:

$$K = 0.587 = \frac{P_{SO_3}}{P_{SO_2}\, P_{O_2}^{1/2}}$$

The quartz vessel is charged with the product $SO_3\,(g)$ so equilibrium is approached from the right. The direction by which it is approached makes no difference in the nature of the final equilibrium state.

The 0.800 g of SO_3 (molar weight 80.06 g mol^{-1}) amounts to 9.99×10^{-3} mol SO_3. The pressure of $SO_3\,(g)$ in the vessel before any breakdown to $SO_2\,(g)$ and $O_2\,(g)$ is:

$$P = (n/V)\,RT = (9.99 \times 10^{-3} \text{ mol}/0.100 \text{ L})(0.08206 \text{ L atm K}^{-1}\text{mol}^{-1})(900 \text{ K})$$

$$P_{SO_3} = 7.379 \text{ atm} \quad \text{(no decomposition)}$$

At equilibrium, equal pressures of $SO_2\,(g)$ and $O_2\,(g)$ are present. Let the pressure of $SO_2\,(g)$ = y atm. Then $P_{O_2} = \tfrac{1}{2}\, y$ and $P_{SO_3} = 7.379 - y$. At equilibrium:

$$K = 0.587 = \frac{7.379 - y}{y\,(y/2)^{1/2}}$$

The easiest way to determine y in the equation is by successive approximation. Prepare a table. The first entry for y is a frank guess somewhere in the range 0 to 7.379. Q is the reaction quotient calculated on the basis of that guess:

y	$7.379 - y$	$y\,(y/2)^{1/2}$	Q
4.00	3.379	5.6568	0.597
4.10	3.279	5.8703	0.559
4.03	3.349	5.7206	0.585
4.02	3.359	5.6993	0.589

From the table it is clear that a y value between 4.02 and 4.03 will make Q equal to K (which is 0.587). The equilibrium partial pressure of O_2 is $y / 2$ which is, to three significant digits, 2.01 atm.

---

45. The reaction, at 25° C, is:

$$KOH(s) + CO_2(g) \rightleftarrows KHCO_3(s)$$

which has the equilibrium constant expression:

$$K = 6 \times 10^{15} = 1/P_{CO_2}$$

As long as both solid KOH and $KHCO_3$ are present the pressure of $CO_2\,(g)$ is fixed by the operation of the equilibrium. This pressure is 1.7×10^{-16} atm.

If the closed container is huge, then P_{CO_2} could in principle be *less* than 1.7×10^{-16} atm. In this scenario all 9.41 g of $KHCO_3\,(s)$ would be used up so that the equilibrium no longer existed. Such complete conversion to the left (reactant) side would give a maximum of about 0.09 mol of $CO_2\,(g)$ in the system. The container would have to be bigger than about 10^{16} L. If spherical, this container would be about 30 km in diameter.

---

47. The equilibrium involves two forms of an organic compound: the chair and boat forms:

$$\text{chair} \rightleftarrows \text{boat}$$

$$K = [\text{boat}] / [\text{chair}]$$

At 580 K, 6.42 percent of the molecules are in the chair form. Since no other forms are mentioned, conclude that the other 93.58 percent of the molecules are in the boat form. Whatever the volume of the system the *ratio* of the concentrations of the two forms will be 93.58 / 6.42. Therefore $K = 14.6$.

---

49. Magnesium ammonium phosphate dissolves according to the equation:

$$MgNH_4PO_4\,(s) \rightleftarrows Mg^{2+}(aq) + NH_4^{+}(aq) + PO_4^{3-}(aq)$$

Assume that the 0.0173 g of $MgNH_4PO_4\,(s)$ that dissolves in 2.0 L of water saturates the solution. The molar weight of $MgNH_4PO_4$ is 137.31 g mol^{-1} so there is $1{,}26 \times 10^{-4}$ mol dissolved in 2.0 L or 6.30×10^{-5} mol L^{-1} of $MgNH_4PO_4$. If the only source of $Mg^{2+}(aq)$, $NH_4^{+}(aq)$ and $PO_4^{3-}(aq)$ is the solid and if none of the three ions is reduced in concentration by further reactions:

$$K_{sp} = [Mg^{2+}][NH_4^{+}][PO_4^{3-}] = S^3$$

Since $S = 6.30 \times 10^{-5}$ mol L^{-1}, $K_{sp} = 2.50 \times 10^{-13}$.

---

51. In using the Mohr technique there are two K_{sp} expressions:

$$[Ag^+][Cl^-] = 1.6 \times 10^{-10}$$

$$[Ag^+]^2[CrO_4^{2-}] = 1.9 \times 10^{-12}$$

The K_{sp} values are taken from text Table 5-2. If $[Cl^-] = 0.100$ M then the concentration of Ag^+ *(aq)* that just causes precipitation of $AgCl$*(s)* is 1.6×10^{-11} M. If $[CrO_4^{2-}] = 0.00250$ M, then the $[Ag^+]$ that just causes precipitation of Ag_2CrO_4 *(s)* is $(7.6 \times 10^{-10})^{1/2} = 2.76 \times 10^{-5}$ M.

As $AgNO_3$ *(aq)* solution is added dropwise $[Ag^+]$ increases from zero. When it reaches the threshold value of 1.6×10^{-11} M, white $AgCl$*(s)* starts to precipitate, removing Ag^+ *(aq)* from solution and keeping its concentration near the threshold value. Only after nearly all the Cl^- *(aq)* ion is removed can $[Ag^+]$ rise to 2.76×10^{-5} M where red Ag_2CrO_4 *(s)* starts to precipitate.

---

53. ***a)*** This is the same distribution equilibrium considered in problem 33. Note that *K* is somewhat different here than in the answer to that problem. The temperatures differ slightly in the two cases which helps account for the difference. However, *K* values for the same reaction under the same conditions do sometimes vary slightly from reference to reference.

The equilibrium constant is:

$$K = 85 = [I_2](CCl_4)/[I_2](aq)$$

The 0.500 L of water originally contains 1.50×10^{-2} mol of dissolved I_2. Suppose the fraction of I_2 remaining in the H_2O at equilibrium is *f*. Then the equilibrium molarity of I_2 in the water is:

$$[I_2](aq) = 1.50 \times 10^{-2}(1 - f) / 0.150$$

and the equilibrium molarity of I_2 in the CCl_4 is:

$$[I_2]_{CCl_4} = (1.50 \times 10^{-2})\, f / 0.500$$

Substitution of these concentration values into the *K* expression gives:

$$85 = \frac{1.50 \times 10^{-2}(1 - f)\,(0.500)}{1.50 \times 10^{-2}(f)\,(0.150)}$$

Notice that the 1.50×10^{-2} cancels away. It was *not* necessary to know the original concentration of the I_2 *(aq)*. Proceed now to solve for *f*:

$$85\,(0.150) / 0.500 = (1 - f) / f$$

$$f = 0.0380$$

b) The same analysis will apply to the second treatment with CCl_4. After the two washes, the fraction of I_2 left in the aqueous phase will be $f \times f = 0.038^2 = 1.4 \times 10^{-3}$ of the original.

c) Using one big (0.300 L) wash with CCl_4 will give:

$$85(0.300) / 0.500 = (1 - f) / f$$

by analogy to the equation in part a). Here f comes out to 0.0192. Two 0.150 L washes are 13 times more effective than one 0.300 L wash.

---

55. The solubility of I_2 in water and in benzene are represented by two quite similar equilibria:

$$I_2\,(s) \rightleftarrows I_2\,(aq) \quad \text{and} \quad I_2\,(s) \rightleftarrows I_2\,(benzene)$$

The equilibrium constant expressions for these two processes are:

$$K_1 = [I_2](aq) \quad \text{and} \quad K_2 = [I_2](benzene)$$

The two equilibrium constants are just the solubilities of I_2 in the two solvents. A new equilibirum represents the partition of I_2 between water and benzene:

$$I_2\,(aq) \rightleftarrows I_2\,(benzene) \qquad K = [I_2](benzene) / [I_2](aq)$$

The K for the partition is $K_1 \ / \ K_2$ because the partition is equal to the first process above minus the second. $K = 164.6 / 0.300 = 549$.

Starting with saturated $I_2\,(aq)$ in V L of water the first extraction leaves a fraction f of the I_2 behind and transfers $1 - f$ of the I_2 into the benzene. At equilibrium after this first extraction:

$$[I_2](aq) = f / V \quad \text{and} \quad [I_2](benzene) = (1 - f) / V$$

Substitution in the K expressions gives:

$$K = 549 = (1 - f) / f \quad \text{and} \quad f = 1.8 \times 10^{-3}$$

The second extraction leaves behind only 1.8×10^{-3} of the small fraction left by the first. Hence, two extractions leave only $(1.8 \times 10^{-3})^2$ of the original I_2. If the original I_2 in the water was 0.300 g L^{-1}, then only 1×10^{-6} g L^{-1} is left.

Chapter 6

ACIDS AND BASES

6.1 *PROPERTIES OF ACIDS AND BASES*

According to the Brønsted-Lowry definition, an acid is a proton-donor and a base is a proton-acceptor. The equation:

$$\text{acid} \rightleftarrows \text{base} + H^+$$

summarizes the definition. The acid and base in this equation make a *conjugate pair.* That is, the acid is the conjugate acid of the base and the base is the conjugate base of the acid. The definition is easy to memorize. A common error is to reverse acid and base. To avoid this, become familiar with a few important acids and bases and use them as touchstones in recollection.

There are just six common *strong acids* in water: $HClO_4$, HNO_3, HCl, HBr, HI, and H_2SO_4. Each contains H^+ and is very effective at donating it. When HCl acts as an acid:

$$HCl \rightleftarrows Cl^- + H^+ \qquad \text{HCl as acid}$$

the well-known chloride ion forms as the conjugate base. If it served as a base, HCl would gain H^+:

$$HCl + H^+ \rightleftarrows H_2Cl^+ \qquad \text{HCl as base}$$

The rare H_2Cl^+ ion is the conjugate acid of HCl.

The common strong bases in water are NaOH, KOH, RbOH, CsOH, $Ca(OH)_2$, $Sr(OH)_2$ and $Ba(OH)_2$. All dissolve to give OH^- *(aq)*, an excellent proton acceptor.

Many substances are weak acids in water. Important weak acids are acetic acid, $HC_2H_3O_2$, and formic acid, HCOOH. Some important weak bases are NH_3, ammonia, and $C_2H_3O_2^-$, the acetate ion. The oxide ion, O^{2-}, is a strong base that contains no H.

A standard exercise is to write the formulas of the conjugate acids or conjugate bases of a list of chemical species. **See problems 35 and 36.** A variation asks for *both* the conjugate acid and conjugate base of a single species. One writes the conjugate acid by adding H^+ to the formula of the species. One writes the conjugate base by subtracting H^+ from the formula. If there is no H in a compound's formula, then that compound has no Brønsted-Lowry conjugate base.

Under the Brønsted-Lowry definition, NaOH is a base in water. It dissolves by dissociation to Na^+ *(aq)* and OH^- *(aq)* ions. The OH^- *(aq)* ions accept protons from water molecules. Sodium acetate ($Na^+CH_3COO^-$) behaves similarly. It dissolves in water to give $Na^{+}(aq)$ and $CH_3COO^{-}(aq)$ ions. The $CH_3COO^{-}(aq)$ ions accept protons from water to form their conjugate acid, $CH_3COOH(aq)$:

$$CH_3COO^{-}(aq) + H_2O(l) \rightleftarrows CH_3COOH(aq) + OH^{-}(aq)$$

Sodium acetate does the same kind of thing in water as sodium hydroxide. It just does it to a lesser extent. **See problem 5c** for detailed computations involving exactly this equilibrium.

A substance like thiamine hydrochloride, $C_{12}H_{17}ON_4SCl{\cdot}HCl$, is an acid in water. It dissolves by dissociation into thiamineH^{+} ions and Cl^{-} ions. The thiamineH^{+} ion, which is analogous to the NH_3H^{+} ion, donates protons which water molecules accept. Note how the formula of thiamine hydrochloride has been written to emphasize the presence of one acidic hydrogen. **See problem 17.**

There is no qualitative difference between the donation of a proton by a positively charged species (*e.g.* NH_4^{+}) and a neutral (*e.g.* HCN) or negatively charged species (*e.g.* HCO_3^{-}). Each of these species is an acid. The three do differ substantially in their acid strength. Similarly, bases are negatively charged (like CN^{-}), neutral (like NH_3) and positively charged (like $NH_2NH_3^{+}$).

Amphoterism

A species that can act as both acid and base is *amphoteric.* Under the Brønsted-Lowry definition, *every* species containing an H is in principle amphoteric because formulas of both a conjugate acid and base can be written. The idea is that some super-strong base could force even a strong base to act as an acid by wresting away a proton and that even a strong acid in the presence of a super-strong acid can be forced to accept protons (function as a base).

Common amphoteric species are H_2O, HCO_3^{-}, HS^{-}, HSO_4^{-}. The way in which an amphoteric species reacts depends on its surroundings. In the presence of strong acids an amphoteric molecule or ion *accepts* protons. In the presence of strong bases the same species *donates* protons. **See problems 33 and 51.**

Autoionization of Water

The conjugate acid and base of H_2O are of great importance. H_3O^{+}, the conjugate acid, is the *hydronium* ion. OH^{-}, the *hydroxide* ion, is water's conjugate base. Water is amphoteric and establishes an acid-base equilibrium with itself:

$$2\,H_2O(l) \rightleftarrows H_3O^{+}(aq) + OH^{-}(aq)$$

In pure water, $[H_3O^{+}] = [OH^{-}]$, that is, there is one H_3O^{+} ion for every OH^{-} ion. When a solute that donates or accepts protons interacts with water this equality is destroyed. Nevertheless, it is always true in aqueous solutions at equilibrium that:

$$[H_3O^{+}][OH^{-}] = K_w$$

The product of the hydronium ion and the hydroxide ion concentrations is constant. The two quantities have a seesaw relationship. As one pivots up the other pivots down.

The strongest acid that may exist in water is the H_3O^+ ion; the strongest base is the OH^- ion. An acid stronger than hydronium ion immediately donates protons to H_2O molecules when placed in water and hydronium ion results. Bases stronger than hydroxide ion accept protons from H_2O when placed in water and OH^- results.

The autoionization constant K_w is equal to 1.0×10^{-14} *at 25° C.* Like all equilibrium constants it depends on the temperature. Many calculations emphasize other aspects of acid-base equilibria, assuming tacitly that the temperature is 25° C. **See problem 37** for a treatment of the temperature dependence of K_w.

A good understanding of autoionization allows generalization to other solvent systems. The case of liquid ammonia often occurs in problems. $NH_3(l)$ is amphoteric and autoionizes, just like water:

$$2\ NH_3(l) \rightleftarrows NH_4^+(amm) + NH_2^-(amm)$$

Ammonia has a K_{amm} that differs in magnitude but is otherwise quite analogous to K_w. In liquid ammonia NH_4^+ is the strongest acid and NH_2^- is the strongest base.

The pH Function

The pH of an aqueous solution is defined as:

$$pH = -\log_{10} [H_3O^+]$$

Logarithms of numbers less than 1 are negative. The minus sign in the definition makes the pH of commonly-encountered solutions, in which $[H_3O^+]$ is less than one, positive. Unfortunately, it also makes the pH go *down* as the concentration of hydronium ion goes *up*, a source of confusion for beginners.

Negative pH's are uncommon but not impossible. A standard pH problem asks the hydronium ion concentration in an aqueous solution at 25° C with a pH of 0. (Ans.: 1 M).

The "p" operator (it says: "take the negative logarithm of what follows") is applied to quantities other than $[H_3O^+]$. Thus, pOH is the negative logarithm of the concentration of hydroxide ion and pK_w is the negative logarithm of the autoionization constant of water. Because:

$$[H_3O^+] \times [OH^-] = K_w$$

$$pH + pOH = pK_w$$

At 25° C, pK_w is 14. If the pH of an exceedingly acidic, aqueous solution at 25° C is for instance -1, then its pOH is 15. If the pH of an aqueous solution at 25° C is 7, then the pOH is also 7 and the solution is *neutral.*

With an electronic calculator the computation of pH is as easy as pushing the "log" button. Getting from pH to $[H_3O^+]$ requires taking an antilogarithm (changing the sign of the pH and pushing "inverse log"). There are two kinds of logarithm in ordinary use. The pH function uses "$\log_{10}$", logarithm to the base 10, *not* "ln", logarithm to the base *e.*

Acid and Base Strength

The donation of protons does not go on in a vacuum. To be donated a proton must also be accepted. Free protons do not exist in solution, and it is *impossible* to determine the intrinsic tendencies of acids to donate them (or bases to accept them). Instead, in determining acid and base strength, the ions of the solvent (H_3O^+ and OH^- in water) serve as the reference acid and base. An observed equilibrium constant (symbolized by K_a, for *acid equilibrium constant*) of some acid's ionization in water in reality measures the outcome of a *competition* between two acids in donating a proton. The value of K_a speaks as much about the second as the first.

Consider the following equilibrium, set up in a solution of BH^+ ion in water:

$$BH^+(aq) + H_2O(l) \rightleftarrows B(aq) + H_3O^+(aq)$$

The competitive donors are BH^+ and H_3O^+. If H_3O^+ wins, the proton sticks with B. H_3O^+ is a stronger acid than BH^+, and the above equilibrium lies to the left. Any equilibrium that lies to the left has an equilibrium constant less than 1, so K_a is less than 1.

• • By definition, a weak acid has K_a less than 1.

Therefore BH^+ is weak, but not in any absolute sense. It is weak relative to H_3O^+ ion.

What about the base strength of B, the conjugate base of BH^+? In aqueous solution the reference for base strength is the OH^- ion. The strength of B relative to OH^- is judged in the equilibrium:

$$B(aq) + H_2O(l) \rightleftarrows BH^+(aq) + OH^-(aq)$$

This equilibrium is neither the acid ionization nor the reverse of the acid ionization of BH^+. Instead, it shows B acting as a base to seize protons from H_2O and generate OH^-. Its equilibrium constant is, quite logically, symbolized K_b.

- K_b's refer to equilibria that form OH^- from a base plus water.

- K_a's refer to equilibria that form H_3O^+ from an acid plus water.

- A weak acid by definition has a K_a less than 1. A weak base, by similar definition, has a K_b less than 1.

Juxtapose the K_b equilibrium of $B(aq)$ and the K_a equilibrium of BH^+ and compare them:

$$B(aq) + H_2O(l) \rightleftarrows BH^+(aq) + OH^-(aq) \qquad K_b$$

$$BH^+(aq) + H_2O(l) \rightleftarrows B(aq) + H_3O^+(aq) \qquad K_a$$

Now add these two reactions. The result is:

$$2\,H_2O(l) \rightleftarrows OH^-(aq) + H_3O^+(aq)$$

which is the water autoionization reaction with the equilibrium constant K_w. Adding any two reactions entails multiplication of their equilibrium constants. Therefore (at 25° C):

$$K_a K_b = K_w = 1.0 \times 10^{-14}$$

This relationship is useful **(problem 5c)** and should be memorized. It shows that weaker acids (smaller K_a's) have stronger conjugate bases (larger K_b's) and vice-versa. Examples help:

$$HNO_2(aq) + H_2O(l) \rightleftarrows NO_2^-(aq) + H_3O^+(aq)$$

This reaction has K_a equal 4.6×10^{-4}, a value smaller than 1, so HNO_2 is a weak acid relative to H_3O^+. NO_2^- is the conjugate base of HNO_2. Its K_b refers to the reaction:

$$NO_2^-(aq) + H_2O(l) \rightleftarrows HNO_2(aq) + OH^-(aq)$$

K_b for nitrite ion is equal to K_w divided by K_a for HNO_2. The answer is 2.2×10^{-11}. The nitrite ion is a weak base relative to OH^-. If HNO_2 is placed in liquid ammonia, then the equilibrium:

$$HNO_2(amm) + NH_3 \rightleftarrows NO_2^-(amm) + NH_4^+(amm)$$

would be established. K_a for this equilibrium is much larger than the K_a of HNO_2 in water.

For the reaction:

$$HClO_4(aq) + H_2O(l) \rightleftarrows ClO_4^-(aq) + H_3O^+(aq)$$

the value of K_a is huge (10^7), so perchloric acid is a strong acid relative to hydronium ion. K_b for the ClO_4^- ion is 10^{-21}. It is an exceedingly weak base relative to hydroxide ion.

H_2O and its ions do not always have to be among the competitors in acid-base interactions. **Problem 39** covers several competitive reactions that do not involve H_3O^+ or OH^-. These examples also show, once again, that acid strength and base strength are relative.

A strong acid or base is not necessarily either concentrated or nasty and difficult to handle. The term "strong" rates the relative equilibrium ability of a substance to donate or accept protons.

6.2 *ACID AND BASE IONIZATION*

Calculating the pH of Solutions of Acid and Base

Suppose C mol L^{-1} of the acid HA is placed in water. It interacts with H_2O, which serves as a base and accepts protons, and the following equilibrium is quickly established:

$$HA(aq) + H_2O(l) \rightleftarrows A^-(aq) + H_3O^+(aq)$$

For every $A^-(aq)$ that forms one H_3O^+ also forms, and one $HA(aq)$ is consumed. If the reaction goes all the way to the right, then the final concentration of HA is zero, and the final concentrations of A^- and H_3O^+ are both C. A large value of K_a corresponds to the acid ionization lying, at equilibrium, *almost* all the way to the right. A large K_a makes HA a strong acid. Therefore if HA is strong the $[H_3O^+]$ of the solution equals C. **See problem 1.**

If K_a is less than 1, then HA is scarely ionized at equilibrium. It is a weak acid. If y mol L^{-1} of it actually does ionize, then the equilibrium concentrations of A^- and H_3O^+ are both y, and the concentration of the HA*(aq)* that is left unreacted is $C - y$. The equilibrium constant expression becomes:

$$K_a = y^2 / (C - y)$$

Many acid-base calculations require nothing more than setting up and solving equations just like this one. It is always possible to solve for y, either by using the quadratic formula or by successive approximations. **See problem 5a.** The analysis for aqueous solutions of bases is exactly similar, except that K_b replaces K_a and y represents the concentration of OH^-. **See problem 7.**

The big problem with restricting attention to a single equilibrium as a source of H_3O^+ (or OH^-) is that *every* aqueous solution has a second, simultaneous equilibrium going on. This is the autoionization of water. Most of the time, the concentration of H_3O^+ (or OH^-) coming from autoionization is negligible compared to the concentration coming from the weak acid (or base). But sometimes it is not. **Problem 5a** shows how to make sure that neglecting water's autoionization is justified, as does **Problem 17.**

In Section 6-4 the text derives an *exact* equation for the H_3O^+ concentration of an aqueous solution of a monoprotic weak acid of concentration C:

$$[H_3O^+]^3 + K_a[H_3O^+]^2 - (K_w + C\,K_a)[H_3O^+] - K_aK_w = 0$$

The derivation includes no approximations. In principle, all problems involving aqueous solutions of a single weak acid could be solved by applying this equation. Further, the equation could be recast, by replacing hydronium ion concentrations with hydroxide ion concentrations and K_a with K_b, to apply to solutions of a single weak base.

Memorizing the equation is *not* recommended. For one thing, relying on a memorized equation leaves one powerless to deal with obvious variations, such as a mixture of two different weak acids. What counts is chemical insight. Therefore, study the equation for insight into acid-base equilibria. Start a series of what-if questions.

- *What if K_a gets smaller and smaller?* This corresponds to making the acid weaker and weaker. Clearly all of the terms in which K_a multiplies another quantity go toward zero. The cubic equation becomes:

$$[H_3O^+]^2 = K_w$$

The hydronium ion concentration becomes equal to $K_w^{1/2}$, its concentration in pure water. An infinitely weak acid has no effect on pH.

- *What if K_a in the cubic equation gets very large?* This corresponds to making the weak acid strong. The terms containing K_a then dwarf the terms that do not contain K_a. Neglect the small terms. The cubic equation becomes:

$$K_a[H_3O^+]^2 - C\,K_a[H_3O^+] - K_aK_w = 0 \qquad K_a \text{ large}$$

But K_a appears in each term and can be divided out:

$$[H_3O^+]^2 - C[H_3O^+] - K_w = 0 \qquad K_a \text{ large}$$

The chemical insight here is that once K_a is big *it does not matter how big.* When an acid is already effectively at 100 percent efficiency in donating protons, further increases in K_a achieve no more. Two very strong acids cannot be distinguished in strength. The stronger is said to be *leveled* to the strength of the weaker.

K_w equals 1.0×10^{-14} at room temperature. Unless C is very small the third term is much smaller than the second and can be neglected:

$$[H_3O^+]^2 - C[H_3O^+] = 0 \qquad K_a \text{ large, } C \text{ not tiny}$$

This means $[H_3O^+]$ *equals* C *when* K_a *is large, and* C *is anything but tiny.* In this context, "tiny" is less than about 10^{-5} M. In other words, the hydronium ion concentration of a solution of a strong acid in water is (nearly always) equal to the concentration of the acid.

• *What if* C *becomes very small?* Chemically, this corresponds to the removal of the weak acid from the solution. If C goes to zero the cubic equation becomes:

$$[H_3O^+]^3 + K_a[H_3O^+]^2 - K_w[H_3O^+] - K_aK_w = 0$$

This equation has only one physically meaningful root *regardless* of the value of K_a. That root is $K_w^{1/2} = 10^{-7}$. The solution becomes neutral.

• *What if* K_a *and* C *are typical practical values?* The relative importance of the terms changes with dilution. Unless the solution is quite dilute (C tiny) the fourth term, K_aK_w, is much smaller than the third, $(K_w + CK_a)[H_3O^+]$, in which K_w is much smaller than CK_a. The result is the tractable quadratic:

$$[H_3O^+]^2 + K_a[H_3O^+] - CK_a = 0$$

The absence of K_w in this equation means that it takes no account of the effect of the autoionization of water on the hydrogen ion concentration. It is equivalent to the quadratic (in y) that worked out when attention was restricted to the acid ionization reaction. Because K_w is so small it is usually (but not always) safe to ignore it (set it to zero) in practical cases. Remember to consider K_w in:

- problems in which the acid is very weak (K_a less than about 10^{-10}).
- problems in which the acid is very dilute (C less than about 10^{-5}).

Example. Calculate the pH of a 1×10^{-7} M solution of HCl at 25° C.

Solution. The naive answer is 7. But making pure water (pH 7) even a little bit acidic will inevitably lower the pH. The correct answer is 6.79. Although HCl is a strong acid, this solution of it is so dilute that the autoionization of water makes a non-negligible contribution to the H_3O^+ concentration. **See problems 5d and 17** for further discussion.

Hydrolysis

In general, *hydrolysis* refers to the reaction of a substance with water. Every acid-base equilibrium in water has H_2O among the reactants and is in the general sense an hydrolysis. In a more restricted sense hydrolysis refers to those acid-base reactions that occur when a salt dissolves in water to give ions with large enough K_a or K_b to change the pH of the solution. For example, sodium acetate dissolves in water to give sodium ions and acetate ions. Acetate ion, the conjugate base of the weak acid, acetic acid, has a K_b of about 10^{-10}. Placing sodium acetate in water raises the pH because the reaction:

$$C_2H_3O_2^-(aq) + H_2O(l) \rightleftarrows C_2H_3O_2H(aq) + OH^-(aq)$$

provides OH^- ion to a measurable extent. **Problem 5c** is a calculation treating an acetate hydrolysis. In contrast to the acetate ion, the chloride ion arising from dissolving NaCl in water has a K_b so small that hydrolysis is effectively non-existent.

If NH_4Cl is dissolved in water, the NH_4^+ ion hydrolyzes to give H_3O^+ and NH_3:

$$NH_4^+(aq) + H_2O(l) \rightleftarrows NH_3(aq) + H_3O^+(aq)$$

The equilibrium constant is the K_a of the NH_4^+ ion. The hydrolysis lowers the pH.

If some ammonium acetate, $NH_4^+C_2H_3O_2^-$, is dissolved in water, then two ions hydrolyze simultaneously and the net effect on the pH is a compromise between the separate effects of the two.

6.3 *BUFFER SOLUTIONS*

Buffer solutions resist changes in pH when either a strong acid or a strong base is added. A typical buffer solution is made by mixing a weak acid with its conjugate base. **See problem 13.** A solution of a weak base and its conjugate acid also is a buffer solution. Compare problem 13 to **problem 14.** The weak base tends to neutralize any acid added to the buffer, and the weak acid tends to neutralize any added base. Buffers are not magic. Enough acid or base will always overwhelm the *buffer capacity* of a buffer solution and lower or raise the pH substantially.

Consider a typical buffer. Supppose that a weak acid HA is mixed in water with a source of its conjugate base such as the salt NaA, which dissociates completely when it dissolves. Let the original concentration of HA be C_A and the original concentration of the conjugate base be C_B. In Section 6-4, it is shown that:

$$[H_3O^+] = K_a \frac{C_A - [H_3O^+] + [OH^-]}{C_B + [H_3O^+] - [OH^-]}$$

This equation is actually a rearrangement of equation **c'** in text Section 6-4. Again, do not memorize but rather study for chemical insight. Suppose K_a is neither very big nor very small but is a number like 1×10^{-5}:

- As long as the concentrations of the weak acid and its conjugate base are big they drown out the terms added to and subtracted from them. The equation becomes:

$$[H_3O^+] \simeq K_a (C_A / C_B)$$

Under these circumstances, the concentration *ratio*, not the actual values, determines the hydrogen ion concentration. Adding a little acid from an outside source slightly reduces C_B and slightly increases C_A. The ratio, C_A / C_B, hardly changes. Adding a little base from outside also causes only a slight change in the ratio. This is the crux of the solution's buffer action.

• Start with the above approximate equation:

$$[H_3O^+] \simeq K_a (C_A / C_B)$$

Taking the negative logarithm of both sides of this approximate equation gives:

$$pH \simeq pK_a - \log (C_A / C_B)$$

which is the *Henderson-Hasselbalch equation.* This equation suffices to solve a vast majority of practical buffer problems.

• If either C_A or C_B goes to zero, their ratio goes either to zero or infinity. Either way the ratio no longer has any meaning in determining $[H_3O^+]$. Buffer action no longer occurs. Both acid and conjugate base must be present for buffer action; the buffer is most effective when C_A equals C_B. Although the (approximate) Henderson-Hasselbalch equation gives impossible answers, the original (exact) equation transforms to describe a weak base by itself or a weak acid by itself in water.

• Rewriting the equation with A for B (and B for A), OH^- for H_3O^+ (and the reverse) and K_b for K_a, describes a buffer solution of a weak base and its conjugate acid.

• The equation has $[H_3O^+]$ on both the left and right hand sides. To solve for $[H_3O^+]$ without neglecting any terms use the method of successive approximations.

Example. An aqueous solution is 0.01 M in monochloroacetic acid and 0.01 M in sodium monochloroacetate. K_a for the monochloroacetic acid is 1.4×10^{-3}. Compute the pH.

Solution. Substitution of $C_A = C_B = 0.01$ in the Henderson-Hasselbalch equation gives $[H_3O^+] = 1.4 \times 10^{-3}$ M, the numerical value of K_a. But this answer is more than 10 percent of 0.01. The assumption that $[H_3O^+]$ is negligible in comparison to C_A and C_B is not justified. Take 1.4×10^{-3} M as a preliminary value and substitute it in on the right hand side of the exact equation. (The concentration of OH^- is 7.14×10^{-12} M and still quite negligible.) Subtraction on the top and addition on the bottom give $[H_3O^+] = 1.08 \times 10^{-3}$ M. This second approximation is closer to the correct value. Another cycle of substitution gives $[H_3O^+]$ equal to 1.12×10^{-3} M. Further cycles do not improve on this answer.

• Taking K_a to zero transforms the equation to one describing a solution of a strong base with concentration C_B. The equation becomes $[OH^-] = C_B$. The chemical insight is that as K_a gets arbitrarily small the acid A becomes no real acid at all and the conjugate base B automatically becomes a strong base.

• Making K_a very large transforms the equation to one describing a solution of a strong acid with concentration C_A. As A becomes a strong acid its conjugate base B becomes so weak that it no longer exerts any effect on the pH of the solution. Thus, $[H_3O^+] = C_A$.

Effects of pH on Solubility

The pH of a solution can strongly influence its ability to dissolve salts. The effect occurs when the ions that comprise the salt take part in acid-base equilibria. For example, in **problem 3** magnesium oxide dissolves in water:

$$MgO(s) + H_2O(l) \rightleftarrows Mg^{2+}(aq) + 2\ OH^-(aq)$$

Increasing the hydronium ion concentration drives this equilibrium to the right. The extra H_3O^+ reacts with the OH^- on the right, thus removing it from the equilibrium. LeChatelier's principle predicts the shift to the right.

6.4 *EXACT TREATMENT OF ACID-BASE EQUILIBRIA*

Principle of Electrical Neutrality

Every solution must be electrically neutral. When a solution contains ions the total amount of positive charge equals the total amount of negative charge. This fact furnishes an important mathematical relationship among the concentrations of the ions in a solution. For example, in an aqueous solution of NaCl:

$$[Na^+] + [H_3O^+] = [Cl^-] + [OH^-]$$

The principle of electrical neutrality applies to electrical charges and not to ions. In a solution of sulfuric acid in water:

$$[H_3O^+] = [OH^-] + [HSO_4^-] + 2\ [SO_4^{2-}]$$

where the coefficient of 2 for the sulfate concentration reflects the contribution by each sulfate of two negative charges to the total negative charge in the solution. **See problem 41** for the use of this principle in analyzing a problem.

Material Balance

Substances often, when dissolved in water, end up partially converted to a new form. A substance may distribute itself in any number of ways among different forms. **See problem 17.**

• *The sum of the concentrations of the derivative forms and the unconverted original form always equals the number of moles per liter originally added.*

For example, if 0.1 mol of H_2SO_4 is placed in a liter of solution it interacts with the solvent to produce HSO_4^- and SO_4^{2-}. The material balance condition is:

$$0.1 = [H_2SO_4] + [HSO_4^-] + [SO_4^{2-}]$$

The Exact Treatment

The exact treatment of acid-base equilibria involves first identifying all the species in the equilibrium solution and then writing equations to relate their concentrations. Specifically:

1. List all of the species that can be present in solution at equilibrium. In aqueous solutions this includes H_3O^+, OH^- and H_2O.

3. Employ the principle of electrical neutrality to write a single mathematical equation relating the concentrations of ions in the solution.

4. Apply the principle of material balance to each substance that was placed in solution. If five weak acids are mixed in the solution there are five material balance equations, one for each of the five.

5. Write a mass action expression for every equilibrium taking place in the solution. Do not omit the autoionization of water.

6. Look up the equilibrium constants (at the proper temperature) for all of the equilibria that have been identified.

7. Verify that the number of mathematical equations equals the number of different chemical species in solution. Follow the procedure properly and it always will.

8. Solve the system of independent equations.

An exact solution to such a set of equations usually involves considerable algebraic tedium. The best course is to use chemical insight to simplify the equations. This is done by neglecting terms in the equations which make only tiny contributions when added to or subtracted from other terms. Do *not* neglect terms that are multipliers or divisors even if they *are* small. **See problems 41 and 51.**

6.5 *ACID-BASE TITRATIONS*

A acid-base titration is an analytical procedure for determining the concentration of an acid or base in a solution. During a titration, acid and base *neutralize* each other. Titrating an acid with a base means adding the base to the acid until some indication shows that the neutralization is just complete. At this point the chemical analyst notes the volume of base added. This point is the *end-point* of the titration. If HCl for example is titrated with NaOH then at the end-point:

$$V_{HCl}\, M_{HCl} = V_{NaOH}\, M_{NaOH}$$

where the *V*'s and M's stand for the volume and molarity of the two solutions. Note that the units on both sides of the above equation are *moles.* Even if the volume of the HCl solution is not known, the chemical amount of HCl (in mol) is given by the titration. **See problem 23.**

Titrating a base with an acid reverses the order of addition but otherwise is fundamentally similar. Everything said about the titration of an acid with a base applies to the titration of a base with an acid, with pOH replacing pH and K_b replacing K_a. Thus **problem 22** closely resembles **problem 21.**

To titrate an acid solution of unknown concentration, the chemical analyst carefully adds measured portions of a base of known (*standardized*) concentration to a quantity of the acid.

Sometimes the progress of the neutralization is followed by measuring the pH of the solution as a function of the volume of titrant added. A *titration curve* is a plot of pH versus volume of added titrant. Figures 6-3 through 6-5 in the text are typical titration curves.

Methods of Monitoring the pH

The most direct method of measuring the pH during a titration is with a *pH meter.* A pH meter uses the solution under test as the electrolyte in an electrochemical cell (Chapter 9). Specially constructed electrodes generate a voltage which depends on the hydrogen ion concentration in the solution. With proper calibration and scaling the voltage may be read directly as a pH.

A second method uses indicators. An *indicator* is a conjugate acid-base pair in which the acid and base forms have intense and different colors. It is added to a solution in such small molar amount that its effect on the pH of the solution is negligible. Its effect on the color of the solution is on the other hand profound. The acid ionization equilibrium of the indicator and the (slightly rearranged) mass action expression for it are:

$$HIn(aq) + H_2O(l) \rightleftarrows H_3O^+(aq) + In^-(aq)$$

$$[H_3O^+] = K_a(\text{indicator}) \times [HIn] / [In^-]$$

At high hydronium ion concentrations (low pH's), most of the indicator is in the HIn form. This is the only way the right hand side of the above equation can get as large as the left. The solution takes on the color of the predominant acid form of the indicator. As the hydronium ion concentration decreases (at higher pH) the $[HIn] / [In^-]$ ratio gets small, in order to maintain the equation. Now, the base form of the indicator predominates and the solution has a different color.

If $[H_3O^+]$ is about equal to the indicator's K_a then the acid and base forms of the indicator have about the same concentrations. The color of the solution is a mixture of the colors of the two forms. As a rule of thumb this mixed, intermediate range of color exists when the the $[HIn]$ to $[In^-]$ ratio is between $^1/_{10}$ and 10. In other words a complete color change requires a hundred-fold change in $[H_3O^+]$, or a change of two pH units.

There are numerous structurally complex organic acids and bases having different colors in the acid and base forms. These indicators have different K_a (or K_b) values and different colors. See text Table 6-3.

One can get a estimate of the pH of an unknown solution by adding a different indicator to each of several small portions and noting the colors. Finding that a solution is acidic to one indicator and basic to another brackets the solution's pH. **See problem 43.**

Calculation of Points on a Titration Curve

During the course of a titration two effects occur to change the pH of the solution:

- The acid and base react. In water, the reaction is the neutralization reaction:

$$H_3O^+(aq) + OH^-(aq) \rightarrow 2\,H_2O(l)$$

Naturally the pH goes up when an acid is titrated with a base and down when a base is titrated with acid.

- The two solutions dilute each other. Simple dilution affects pH. For example, 10 mL of 0.1 M HCl has a pH of 1. Diluting it to 100 mL changes the concentration to 0.01 M HCl and raises the pH to 2.

In calculations concerning titrations consider the two effects separately. A general procedure is:

1. Imagine that no reaction occurs until mixing is complete. Compute the effects of dilution on the concentrations of all species of importance.

2. Then, mentally, allow the neutralization reaction to proceed 100 percent to the right until the limiting reactant, whether acid or base is entirely consumed.

3. Identify the important equilibria and use them with appropriate *K*'s to compute $[H_3O^+]$.

Suppose a strong base is being added to a strong acid. The course of the titration has four regions.

1. The starting point. No titrant has been added. The $[H_3O^+]$ is just the molar concentration of the acid.

2. Approach to Equivalence. Some base has been added. The reaction:

$$H_3O^+(aq) + OH^-(aq) \rightarrow 2\,H_2O(l)$$

may be assumed to go to completion. The *amount* (not concentration) of H_3O^+ that remains in solution equals the amount originally present minus the amount of OH^- added. The concentration of H_3O^+ is this answer divided by the volume of the solution. The volume of the solution is the sum of the original volume of acid solution and the volume of titrant added.

3. Equivalence Point. At the equivalence point the amount of base that has been added equals the amount of H_3O^+ originally present. The $[H_3O^+]$ in the solution comes entirely from the autoionization of water. The pH is 7.

4. Beyond the Equivalence Point. The excess strong base builds up in solution. The concentration of OH^- is the molar amount of this excess divided by the total volume of the solution. **Problem 21** shows the calculation of points on the titration curve when a weak base is titrated with a strong acid.

6.6 *POLYPROTIC ACIDS*

Acids that contain two or more ionizable H atoms are *polyprotic* acids. Sulfuric acid (H_2SO_4) is a diprotic acid; phosphoric acid (H_3PO_4) is a triprotic acid. Although acetic acid (CH_3COOH) contains four hydrogen atoms per molecule, it is only a monoprotic acid. The three H atoms bonded to the C are not ionizable.

Polyprotic acids ionize in two or more stages. Typically the equilibrium constant for the first stage (K_{a1}) is about 10^5 times larger than for the second stage (K_{a2}).

A key in working with equilibria involving polyprotic acids (and their conjugate bases) is the realization that when two or more ionization reactions go on simultaneously each contributes some $H_3O^+(aq)$, but the $[H_3O^+]$ used in all the equilibrium constant expressions is one and the same. The different stages of ionization of a polyprotic acid interact with each other.

Exact calculations therefore require solving sets of several simultaneous equations and are hard. In practical problems, K_{a1}, K_{a2} and other K_a's almost always differ so much in magnitude that the equilibria can be treated as if they were separate. Thus, in **problem 33** the second ionization of maleic acid adds only negligibly, at the start of the titration, to the hydronium ion concentration coming from the first.

Amphoteric Equilibria

One class of problem does require consideration of simultaneous equations involving K_{a1} and K_{a2}. The two-step ionization of a diprotic acid gives an amphoteric intermediate. This species is ready to donate H^+ to form its conjugate base (according to K_{a2}) and also ready to accept H^+ to give back the original diprotic acid (according to $K_{b2} = K_w/K_{a1}$). When such an amphoteric species is placed in water the hydrogen ion concentration is related to the K's as follows:

$$[H_3O^+]^2 \simeq \frac{K_{a1}K_{a2}\,C_o + K_{a1}K_w}{K_{a1} + C_o}$$

where C_o is the original concentration of the amphoteric species. If K_{a1} is negligible compared to C_o, then this equation becomes:

$$[H_3O^+]^2 \simeq K_{a1}\,K_{a2}$$

These equations are used in **problem 33** and discussed further both there and in **problem 51.**

DETAILED SOLUTIONS TO ODD-NUMBERED PROBLEMS

1. The dilute aqueous nitric acid has pH 2.32. From the definition of pH, $[H_3O^+]$ equals the antilogarithm of 2.32 or 4.79×10^{-3} M. An 8-fold dilution reduces $[H_3O^+]$ to $1/8$ of this value: 5.98×10^{-4} M. Taking the negative logarithm of this number gives pH 3.22. At 25° C the following equation holds for all aqueous solutions:

$$K_w = 1.00 \times 10^{-14} = [H_3O^+][OH^-]$$

If $[H_3O^+] = 5.98 \times 10^{-4}$ M, then $[OH^-] = 1.67 \times 10^{-11}$ M.

---

3. Magnesia dissolves in water and raises the pH. This means it generates OH^- *(aq)* upon solution:

$$MgO(s) + H_2O(l) \rightleftarrows Mg^{2+}(aq) + 2\ OH^-(aq)$$

The pH of the solution at 25° C is 10.16. At 25° C the sum of the pH and pOH of an aqueous solution is 14.0, making the pOH of the saturated solution of magnesia 3.84. By the definition of pOH, $[OH^-]$ equals antilog (−3.84) which is 1.44×10^{-4} M. Assume that the dissolution of the magnesia is the predominant source of OH^- in the solution, far outweighing the autoionization of water. Then:

$$2\,[Mg^{2+}] = [OH^-]$$

The equilibrium constant expression for the dissolution reaction is:

$$K = [Mg^{2+}][OH^-]^2$$

$MgO(s)$ and $H_2O(l)$ do not appear because they are a pure solid and liquid with activities equal to 1. Substituting on the right-hand side of this equation gives the value of *K*:

$$K = [OH^-]^3 / 2 = (1.44 \times 10^{-4})^3 / 2 = 1.5 \times 10^{-12}$$

The solubility, *S*, of the MgO, equals the final concentration of Mg^{2+} *(aq)*. This is:

$$S = [Mg^{2+}] = \tfrac{1}{2}[OH^-] = 7.2 \times 10^{-5}\ M$$

---

5. ***a)*** Acetic acid is a weak acid, reacting to a slight extent with water:

$$HOAc(aq) + H_2O(l) \rightleftarrows H_3O^+(aq) + OAc^-(aq)$$

At equilibrium:

$$K = \frac{[H_3O^+][OAc^-]}{[HOAc]} = 1.76 \times 10^{-5}$$

where HOAc stands for CH_3CH_2COOH.

Assuming that HOAc is the *only* source of H_3O^+ in solution then, $[H_3O^+] = [OAc^-]$ and $[HOAc] = 0.100 - [H_3O^+]$. Let $[H_3O^+] = x$ and substitute in the expression for K:

$$1.76 \times 10^{-5} = x^2 / (0.100 - x)$$

which gives the quadratic equation:

$$x^2 + (1.76 \times 10^{-5})x - (1.76 \times 10^{-6}) = 0$$

Solving by use of the quadratic formula and discarding the negative root:

$$x = 1.32 \times 10^{-3}$$

$$\text{pH} = -\log[H_3O^+] = -\log x = 2.88$$

Can one justify the assumption that the acetic acid was the sole source of H_3O^+ *(aq)*? In this solution it is *exactly* true that:

$$[H_3O^+] = [OAc^-] + [OH^-]$$

because hydronium ion in fact comes from *two* sources: HOAc ionization, which creates OAc^- *(aq)*, and H_2O autoionization, which creates OH^- *(aq)*. But the concentration of hydroxide ion is 7.59×10^{-12} M (computed from $K_w = 1.0 \times 10^{-14} = [H_3O^+][OH^-]$). This justifies setting $[H_3O^+]$ equal to $[OAc^-]$. The two numbers differ by only about one part in a billion.

b) Represent diethylamine as Et_2NH. It reacts as a base with water:

$$Et_2NH(aq) + H_2O(l) \rightleftarrows Et_2NH_2^+(aq) + OH^-(aq)$$

and the equilibrium constant expression is:

$$K = \frac{[OH^-][Et_2NH_2^+]}{[Et_2NH]} = 9.6 \times 10^{-4}$$

In this computation let $x = [OH^-]$. If $[OH^-] = [Et_2NH_2^+]$ then $[Et_2NH] = 0.100 - x$ and:

$$9.6 \times 10^{-4} = x^2 / (0.100 - x)$$

which rearranges to give the quadratic equation:

$$x^2 + (9.6 \times 10^{-4})\, x - (9.6 \times 10^{-5}) = 0$$

Solution of this equation gives two roots. The positive root is $x = 9.32 \times 10^{-3}$ and leads to:

$$\text{pOH} = -\log[OH^-] = -\log x = 2.03$$

Since pH + pOH = 14.00 at 25° C, pH = 11.97.

The entire computation follows the pattern of part a). The difference is that the equilibrium makes the solution *basic* and not *acidic.* Finally, the relationship:

$$[OH^-] = [Et_2NH_2^+] + [H_3O^+]$$

holds exactly at equilibrium, but dropping $[H_3O^+]$ is justified because it is small compared to $[Et_2NH_2^+]$. Indeed, $[H_3O^+]$ is 1.1×10^{-12} M, nearly 10 orders of magnitude smaller than $[Et_2NH_2^+]$, which is 9.3×10^{-3} M.

c) Sodium acetate dissolves in water by dissociating into Na^+ *(aq)* and OAc^- *(aq)* ions. The Na^+ *(aq)* ions are so weakly acidic that they effectively do not react with H_2O to raise the concentration of H_3O^+ *(aq)*. The OAc^- *(aq)* ions on the other hand are measurably basic. Their reaction with water:

$$OAc^-(aq) + H_2O(l) \rightleftarrows HOAc(aq) + OH^-(aq)$$

raises the pH of the solution.

The material balance for the two forms of acetate provides that: [HOAc] + $[OAc^-]$ = 0.100 M.

Let x equal the equilibrium concentration of OH^- in the solution. Assuming that *all* of the OH^- in the solution comes from the reaction of OAc^- with water:

$$[HOAc] = x \qquad \textit{and} \qquad [OAc^-] = 0.100 - x$$

K_b, the equilibrium constant for the above reaction, is expressed as:

$$K_b = [OH^-][HOAc] / [OAc^-] = x^2 / 0.100 - x$$

The value of K_b for acetate ion is K_w divided by K_a of acetic acid, and works out to 5.68×10^{-10}.

Substitution in and rearrangement of the above equation give a quadratic equation in x:

$$x^2 + (5.68 \times 10^{-10})x - 5.68 \times 10^{-11} = 0$$

It is better to *think* about hard-looking equations like this than mechanically to try to solve them. Any x in this equation that makes the first term comparable in magnitude to the third makes the second much smaller than either. Neglecting the second term gives:

$$x^2 = 5.68 \times 10^{-11} \qquad \textit{hence} \qquad x = \pm 7.54 \times 10^{-6}$$

Discarding the negative root gives $[OH^-] = +7.54 \times 10^{-6}$ M. The pOH of the solution is 5.123 and the pH is 14.000 − 5.123 = 8.877, or 8.88.

d) The set-up for the computation exactly parallels that used in part c). Because the original sodium acetate is now one hundred times more dilute, the equilibrium concentration of hydroxide will be substantially lower in this case.

The quadratic equation of part c) becomes:

$$x^2 + 5.68 \times 10^{-10}x - 5.68 \times 10^{-13} = 0$$

where x again represents the concentration of OH^- *(aq)*. The second term of the revised quadratic is still negligible compared to the third. Therefore:

$$x = \pm 7.53 \times 10^{-7}$$

This gives a pOH of 6.12 and pH of 14.00 − 6.12 = 7.88.

In such a dilute solution the question of another source of OH^- ion must be considered. The water itself contributes to the OH^- *(aq)* concentration, thanks to its autoionization. If this second source of OH^- *(aq)* is considered then the answer to this problem is a trifle different. The $[OH^-]$ is 7.60×10^{-7} M instead of 7.53×10^{-7} M. The 0.01 M sodium acetate verges on being so dilute that water autoionization is a respectable source of OH^- *(aq)*.

The autoionization of water does *not* contribute by adding a flat 1×10^{-7} M to the concentration of OH^- *(aq)* from the hydrolysis of the acetate ion. Rather, OH^- *(aq)* from the acetate hydrolysis suppresses the autoionization substantially, in accord with LeChatelier's principle, so water's contribution is much less than 1×10^{-7} M.

- -

7. Let the abbreviation *morph* stand for morphine. This base reacts with water to *increase* the concentration of OH^-:

$$\text{morph}(aq) + H_2O(l) \rightleftarrows \text{morphH}^+(aq) + OH^-(aq)$$

$$K = \frac{[\text{morphH}^+][OH^-]}{[\text{morph}]} = 8 \times 10^{-7}$$

Before any reaction occurs, the concentration of morphine is (0.0400 mol / 0.600 L) = 0.0667 M. Assume that H_2O by itself is a negligible source of OH^- *(aq)*. Then:

$$[OH^-] = [\text{morphH}^+]$$

because in the absence of other sources of OH^-, one morphH^+ ion must appear for every OH^- *(aq)*. Furthermore:

$$[\text{morph}] = 0.0667 - [OH^-]$$

Letting $x = [OH^-]$, substituting in the equilibrium constant expression and rearranging:

$$x^2 + (8 \times 10^{-7})x - (5.34 \times 10^{-8}) = 0$$

Solving for x and discarding the negative root:

$$x = 2.31 \times 10^{-4} \qquad \textit{hence} \qquad [OH^-] = 2.31 \times 10^{-4}\ \text{M}$$

Applying the definition of pOH and pH gives pOH = 3.64 and pH = 10.36.

- -

9. Hydrofluoric acid is a weak acid in water:

$$HF(aq) + H_2O(l) \rightleftarrows H_3O^+(aq) + F^-(aq)$$

Since the pH of the HF solution at 25° C is 2.13, $[H_3O^+]$ equals the antilogarithm of −2.13 or 7.41×10^{-3} M. If the hydrofluoric acid is the only source of $H_3O^+(aq)$ then, $[F^-]$ is also 7.41×10^{-3} M.

The existence of the equilibrium guarantees that:

$$K_a = \frac{[H_3O^+][F^-]}{[HF]}$$

In this expression all of the quantities except [HF] are known:

$$[HF] = (7.41 \times 10^{-3})^2 / 3.5 \times 10^{-4} = 0.16 \text{ M}$$

---

11. As the 0.100 M sodium formate is added to 100 mL of 0.200 M formic acid *two* major factors affect the concentration of the various species: the dilution and the formic acid-formate equilibrium. Let x equal the volume in *liters* of Na^+HCOO^- solution added. The final volume of the solution is $(0.100 + x)$ L. When the pH reaches 3.50, $0.200x$ mol of Na^+HCOO^- has been added to the 0.0200 mol of formic acid originally present. Meanwhile the equilibrium acts to interconvert formate and formic acid:

$$HCOOH(aq) + H_2O(l) \rightleftarrows HCOO^-(aq) + H_3O^+(aq)$$

Because of this equilibrium the final concentrations of $HCOO^-(aq)$ and $HCOOH(aq)$ are *not* exactly $0.100x / (0.100 + x)$ M and $0.0200 / (0.100 + x)$ M respectively. Rather:

$$[HCOO^-] = [0.100x / (0.100 + x)] + [H_3O^+]$$

$$[HCOOH] = [0.200 / (0.100 + x)] - [H_3O^+]$$

The concentration of hydronium ion is given in terms of the pH:

$$[H_3O^+] = \text{antilog } -3.50 = 3.162 \times 10^{-4} \text{ M}$$

Now write the equilibrium constant expression for the formic acid:

$$K = 1.77 \times 10^{-4} = [H_3O^+]\,[HCOO^-] / [HCOOH]$$

All the quantities on the right are known either outright or in terms of x:

$$\frac{1.77 \times 10^{-4}}{3.162 \times 10^{-4}} = \frac{[0.100x / (0.100 + x] + [3.162 \times 10^{-4}]}{[0.0200 / (0.100 + x)] - [3.162 \times 10^{-4}]}$$

Suppose that 3.162×10^{-4} is small compared to the terms involving x. The equation simplifies greatly:

$$0.560 \simeq (0.100x / 0.0200)$$

$$x \simeq 0.112 \text{ L} = 112 \text{ mL}$$

If the equation is solved without any short-cuts then $x = 0.111$ L. The similarity of these answers means that the action of the equilibrium does not change $[HCOO^-]$ and $[HCOOH]$ very much from the values which the dilution alone gives them.

---

13. If there were no interaction between the HOAc (acetic acid) and OAc^- (acetate ion, from sodium acetate) then the concentrations would be:

$$[HOAc] = 0.0500 \text{ mol} / 0.500 \text{ L} = 0.0100 \text{ M}$$

$$[OAc^-] = 0.0200 / 0.500 \text{ L} = 0.0400 \text{ M}$$

But the equilibrium:

$$HOAc(aq) + H_2O(l) \rightleftarrows H_3O^+(aq) + OAc^-(aq)$$

acts to change the concentrations. Suppose that this equilibrium is the principal source of $H_3O^+(aq)$ in the solution. Letting $[H_3O^+]$ equal x then gives:

$$[OAc^-] = (0.0400 + x) \quad \textit{and} \quad [HOAc] = (0.0100 - x)$$

After equilibrium is attained:

$$1.76 \times 10^{-5} = [H_3O^+][OAc^-]/[HOAc] = x(0.0400 + x) / (0.100 - x)$$

Suppose x is negligible in comparison to 0.100 and 0.0400. Then:

$$1.76 \times 10^{-5} = 0.400\, x$$

$$x = 4.40 \times 10^{-5} \quad \textit{hence} \quad [H_3O^+] = 4.40 \times 10^{-5} \text{ M}$$

The neglect of x relative to 0.0400 is justified because x comes out to be only about 0.1 percent of 0.0400.

---

15. In this problem, 500 mL of 0.100 M formic acid is titrated with 0.0500 M NaOH. Before any NaOH is added the solution contains 0.500 L × 0.100 mol L^{-1} or 0.0500 mol of formate-containing species. Most of this is $HCOOH(aq)$, un-ionized formic acid, although there is a small equilibrium quantity of $HCOO^-(aq)$, the formate ion. As the strong base is added to this solution nothing happens to change the total *amount* of formate-containing species. What does happen is that $HCOOH(aq)$ is converted to $HCOO^-(aq)$ according to the reaction:

$$HCOOH(aq) + NaOH(aq) \rightarrow Na^+(aq) + HCOO^-(aq) + H_2O(l)$$

and the pH is raised.

Suppose that exactly one mole of strong base is added for every mole of formic acid. To reach this point, the *equivalence* point, requires 0.0500 mol of

NaOH which is furnished by 1000 mL of 0.0500 M NaOH. At the equivalence point the solution is dilute sodium formate. Sodium formate is the salt of a weak acid and strong base, so the solution is basic. The pH of the solution *exceeds 7 at the equivalent point.*

To raise the pH only to 4.00 clearly requires *less* than 1000 mL of 0.0500 M NaOH. At pH = 4.00 $[H_3O^+] = 1.0 \times 10^{-4}$ M and:

$$1.77 \times 10^{-4} = [H_3O^+][HCOO^-]/[HCOOH]$$

$$[HCOO^-]/[HCOOH] = 1.77$$

Suppose that V L of the 0.0500 M NaOH brings the 0.100 M formic acid up to the pH of 4.00. The total volume of the titration mixture when the pH hits 4.00 is $(0.500 + V)$ L. Assume that each mole of NaOH converts one mole of $HCOOH$*(aq)* to $HCOO^-$*(aq)* and that this is the only source of $HCOO^-$*(aq)*. This assumption neglects the fact that at equilibrium additional $HCOOH$*(aq)* indeed does ionize to give H_3O^+*(aq)* and some more $HCOO^-$*(aq)*. However, if only a small amount of $HCOOH$*(aq)* undergoes the ionization, then the assumption will be *nearly* correct. The question of *how* near must be checked later. Meanwhile, because $0.0500V$ mol of NaOH has been added:

$$[HCOO^-] = 0.0500\,V/(0.500 + V)$$

The numerator of this expression is the chemical amount of $HCOO^-$*(aq)* produced by the acid-base reaction. The denominator is the total volume of the solution in liters.

Dilution alone reduces the total concentration of formate-containing species from 0.100 M to $(0.500/0.500 + V) \times 0.100$ M. There are only two formate-containing species, $HCOOH$*(aq)* and $HCOO^-$*(aq)*. Therefore:

$$[HCOOH] = \frac{(0.500)(0.100) - 0.0500\,V}{(0.500 + V)}$$

Inserting the equations for [HCOOH] and $[HCOO^-]$ into the expression for their ratio and cancelling the denominators gives:

$$1.77 = 0.0500V/(0.0500 - 0.0500V)$$

It is easy to solve this equation for V which equals 0.639 L or 639 mL. The volume of the solution is 1.139 L and the final concentrations of $HCOOH$*(aq)* and $HCOO^-$*(aq)* are 0.0158 and 0.0280 M respectively. Both are large in comparison to the hydronium ion concentration, 1.0×10^{-4} M.

---

17. The molar weight of thiamine hydrochloride is 337.27 g mol^{-1}. Placing 3.0×10^{-4} g of this substance in 1.00 L of water makes a solution that is 8.89×10^{-7} M in $ThiH^+Cl^-$, (analogous to $NH_4^+Cl^-$).

The $ThiH^+$*(aq)* cation is a weak acid:

$$ThiH^+(aq) + H_2O(l) \rightleftarrows Thi(aq) + H_3O^+(aq) \quad K_a = 3.4 \times 10^{-7}$$

This equilibrium produces only small amounts of H_3O^+ *(aq)* because K_a is small and the original concentration of $ThiH^+$ *(aq)* is quite small. The simultaneous equilibrium:

$$2\,H_2O(l) \rightleftarrows H_3O^+(aq) + OH^-(aq)$$

must be considered as a source of H_3O^+ *(aq)*.

The following four mathematical relationships always hold in this solution:

$$3.4\times 10^{-7} = [Thi][H_3O^+]/[ThiH^+] = K_a$$

$$1.0\times 10^{-14} = [H_3O^+][OH^-] = K_w$$

$$8.89\times 10^{-7} = [Thi] + [ThiH^+] = C$$

$$[H_3O^+] + [ThiH^+] = [OH^-] + [Cl^-]$$

The last equation follows from the principle of electrical neutrality. For every positive charge in the solution there must be a negative charge. The second-to-last equation represents a material balance. Whatever the distribution between its two possible forms, the *total* concentration of thiamine-material is known. The first two equations are the usual mass-action expressions.

The $[Cl^-]$ is 8.89×10^{-7} because Cl^- *(aq)* does not react to any extent with other species. The set of four equations therefore involves four unknowns. It is a question of careful algebra to solve for $[H_3O^+]$. The details of the algebra are given in the text for a similar case (See Section 6-4). The result is a cubic equation in $[H_3O^+]$:

$$[H_3O^+]^3 + K_a[H_3O^+]^2 - (K_w + CK_a)[H_3O^+] - K_aK_w = 0$$

When the numbers specific to this case are substituted:

$$[H_3O^+]^3 + (3.4\times 10^{-7})[H_3O^+]^2 - (3.12\times 10^{-13})[H_3O^+] - 3.4\times 10^{-21} = 0$$

The $[H_3O^+]$ is certain to be at least a bit larger than 10^{-7} because the solution is after all acidic (pH less than 7). For this reason the last term is likely to be small compared to the second-to-last and:

$$[H_3O^+]^2 + (3.4\times 10^{-7})[H_3O^+] - (3.12\times 10^{-13}) \simeq 0$$

This quadratic is readily solved for $[H_3O^+]$. The $[H_3O^+]$ is 4.14×10^{-7} M and the pH is 6.38.

If the autoionization of water is *neglected* then, using the method of **problems 5c and 5d,** the quadratic equation:

$$[H_3O^+]^2 + (3.4\times 10^{-7})[H_3O^+] - 3.02\times 10^{-13} = 0$$

results. Solving *this* quadratic, which differs from the previous one only in the magnitude of the last term, gives $[H_3O^+] = 4.05\times 10^{-7}$ M, about 1 percent less than previously. The pH is then 6.39. This pH is so close to the pH previously computed that, contrary to the statement in the problem, the autoionization of water *can* be neglected.

---

19. Aspirin (AspH) is a weak acid. Its molar weight is 180.16 g mol^{-1} so 5.0×10^{-3} g of it is 2.775×10^{-5} mol. This chemical amount of AspH would require 27.75 mL of 0.0010 M NaOH to titrate it exactly to the equivalence point. Note that 0.02775 L NaOH × 0.0010 mol L^{-1} = 2.775×10^{-5} mol NaOH). At the equivalence point, the solution has a volume of 127.75 mL and contains 2.775×10^{-5} mol of Na^+Asp^- for a concentration of "sodium aspirinate" of 2.1722×10^{-4} M. What is its pH? The Asp^- *(aq)* ion reacts with water:

$$Asp^-(aq) + H_2O(l) \rightleftarrows AspH(aq) + OH^-(aq)$$

Thanks to this equilibrium, the solution is *basic* at the equivalence point. K_b for this equilibrium is K_w / K_a. But the Na^+Asp^- solution is quite dilute, and K_b is small (2.9×10^{-11}). The pH at the equivalence point should exceed 7 only slightly. Under such circumstances, water itself makes an important contribution to the OH^- *(aq)* concentration. Since the K_w equilibrium cannot be neglected, the following cubic equation (See p. 106 of this Guide) applies where C is the concentration of the Na^+Asp^-, and K_b is the equilibrium constant:

$$[OH^-]^3 + K_b[OH^-]^2 - (K_w + CK_b)[OH^-] - K_bK_w = 0$$

(Refer to problem 6-17 and text p. 190 for derivation.) Substitution into this equation and neglect of the fourth term result in:

$$[OH^-]^2 + (2.9412 \times 10^{-11})[OH^-] - 1.6389 \times 10^{-14} = 0$$

The concentration of OH^- *(aq)* comes out to be 1.28×10^{-7} M. The pH at the equivalence point is therefore 7.11. As expected, the pH is just slightly on the basic side of neutrality.

Titration to a phenolphthalein end-point raises the pH to a value between 8.2 and 10.0 because phenolphthalein changes color in that pH range. Suppose the color change is detected at pH 9.1, which corresponds to $[OH^-] = 1.26 \times 10^{-5}$ M. Since this concentration is a lot bigger than the OH^- *(aq)* concentration at the equivalence point, it is clear that going to the phenolphthalein end-point *overshoots* the equivalence point. To reach the phenolphthalein end-point requires *more* than 27.75 mL of 0.0010 M base. Let the additional volume of NaOH solution be V. It adds V L to the total volume of the solution and adds $0.0010V$ mol to the quantity of OH^- *(aq)* present. Hence, the added concentration of OH^- *(aq)* is $0.0010\ V / (0.12775 + V)$. To raise the OH^- *(aq)* concentration from 1.28×10^{-7} M to 1.26×10^{-5} M requires approximately 1.25×10^{-5} moles per liter of OH^- *(aq)*. This assumes that the Asp^- to AspH equilibrium furnishes only negligible quantities of OH^- *(aq)*. Thus:

$$1.25 \times 10^{-5} = 0.0010V / (0.12775 + V)$$

Solving for V gives 0.00162 L or 1.62 mL. The titration to the phenolphthalein end-point requires 27.75 + 1.62 or 29.37 mL.

The exact answer to this problem depends on the exact pH selected as the detectable phenolphthalein end-point. More importantly however, the original mass of aspirin is quoted to 2 significant figures (5.0 mg). The answer must also be given to 2 significant figures. It is 29 mL.

---

21. The problem describes the titration of a weak base, ethylamine, with a strong acid, HCl. The titration falls into four ranges: *before* the addition of acid; *between* the first addition of acid and the equivalence point; *at* the equivalence point; *beyond* the equivalence point.

The pH of the starting 40.00 mL of 0.1000 M ethylamine substantially exceeds 7 because ethylamine is a base. As 0.1000 M HCl is added to this solution the pH falls.

¤ *Before Addition of Acid.*

Ethylamine raises the pH of pure water by the equilibrium reaction:

$$C_2H_5NH_2\,(aq) + H_2O(l) \rightleftarrows C_2H_5NH_3^+\,(aq) + OH^-\,(aq)$$

The corresponding mass-action expression is:

$$K_b = 6.41 \times 10^{-4} = [C_2H_5NH_3^+][OH^-] \,/\, [C_2H_5NH_2]$$

Let $[OH^-] = y$. Assume that the concentration of hydroxide ion from the autoionization of water is small. Then, because no HCl has been added:

$$[OH^-] = [C_2H_5NH_3^+] = y \quad and \quad [C_2H_5NH_2] = 0.100 - y$$

$$6.41 \times 10^{-4} = y^2 \,/\, 0.1000 - y$$

This expression is identical in form to the one developed in the text for the titration of acetic acid with NaOH. The underlying difference is that y is now $[OH^-]$. It is quadratic. Rearranging it and solving gives $y = [OH^-] = 0.00769$ M. The pOH is 2.11, and the pH is 14 − 2.114 = 11.89.

Formulas exactly analogous to the ones developed in the text can be used to compute the pH in the other three ranges. The only complication is that, as just shown, the natural choice for an unknown is $[OH^-]$ rather than $[H_3O^+]$.

¤ *After First Addition of Acid, Before Equivalence Point.*

In the range of the titration between the first addition of acid and the equivalence point:

$$[C_2H_5NH_3^+] = \frac{C_t\,V}{V_o + V} + y$$

where C_t is the concentration of the titrant (0.1000 M), V_o is 0.0400 L, the original volume of ethylamine solution, V is the volume of titrant added and y is the concentration of OH^-. The quantity C_tV is in mol and is the chemical amount of $C_2H_5NH_3^+$ generated by the 1 to 1 reaction between the titrant and the $C_2H_5NH_2\,(aq)$. $V_o + V$ (in L) is the total volume of the solution. The first of these divided by the second is the concentration of $C_2H_5NH_3^+\,(aq)$ from the neutralization reaction alone. The equilibrium that produces $OH^-\,(aq)$ is an additional source of $C_2H_5NH_3^+\,(aq)$. The addition of y on the right-hand side of the above equation takes it into account. Similarly:

$$[C_2H_5NH_2] = \frac{C_o\,V_o - C_tV}{V_o + V} - y$$

where the new symbol, C_0 stands for the original ethylamine concentration (0.1000 M).

After 5.00 mL of HCl has been added, V = 0.00500 L so:

$$[C_2H_5NH_3^+] = 0.01111 + y \quad and \quad [C_2H_5NH_2] = 0.07777 - y$$

At this point in the titration:

$$6.41\times 10^{-4} = y(0.01111 + y) / (0.07777 - y)$$

This rearranges to a quadratic equation in y. Solution for y shows that $[OH^-] = 3.31\times 10^{-3}$ M, pH = 11.52.

In this solution y is *not* negligible compared to 0.01111 or 0.07777. Omitting it from the two terms on the right-hand side gives y = 0.00487 which is 43 percent of 0.01111.

At 20.00 mL, similar substitution and solution give $y = [OH^-] = 6.18\times 10^{-4}$, and a pH of 10.79. At 39.90 mL, the same sequence of operations gives pH = 8.20.

¤ *At the Equivalence Point.*

At equivalence, the titration has produced what is nothing other than 80.00 mL of 0.05 M ethylammonium chloride, $C_2H_5NH_3^+Cl^-$. The cation of this salt is a weak acid:

$$C_2H_5NH_3^+(aq) + H_2O(l) \rightleftarrows C_2H_5NH_2(aq) + H_3O^+(aq)$$

The equilibrium constant for its acid ionization is: $K_a = K_w / K_b$. Let $x = [H_3O^+]$. Then:

$$K_a = 1.56\times 10^{-11} = x^2 / (0.05000 - x)$$

Solution for x gives 8.83×10^{-7} so the pH is $-\log(8.83\times 10^{-7})$ or 6.05.

Using the formula that works in the range *before* the equivalence point now gives the deceptive result:

$$[C_2H_5NH_2] = 0 - [OH^-]$$

which cannot be right since $[OH^-]$ and $[C_2H_5NH_2]$ both are positive.

¤ *Beyond the Equivalence Point.*

In this range the solution behaves like a simple solution of HCl. In comparison to the strong acid HCl, the weakly acidic ethylammonium ion contributes essentially nothing to the $H_3O^+(aq)$ concentration.

When 40.10 mL of HCl has been added, the first 40.00 mL is consumed in producing $C_2H_5NH_3^+(aq)$ ion by neutralization of all the $C_2H_5NH_2(aq)$. The remaining 0.10 mL then is free to act as a strong acid. Thus, 0.10 mL of 0.1000 M HCl is diluted to 80.10 mL. Every HCl gives 1 H_3O^+ so the $[H_3O^+]$ is $(0.10 / 80.10)\times 0.1000 = 1.248\times 10^{-4}$ M. The pH is 3.90.

At 50.00 mL, $[H_3O^+]$ is (10.00 / 90.00) × 0.1000 = 1.111×10^{-2} M; pH = 1.95.

---

23. If it requires 15.90 mL of 0.0750 M HCl to titrate the diethylamine to the equivalence point then, *at* the equivalence point:

$$(0.0750 \text{ mol L}^{-1}) \times (0.01590 \text{ L}) = 1.190 \times 10^{-3} \text{ mol HCl}$$

has been added. HCl neutralizes $(C_2H_5)_2NH$ in a 1:1 molar ratio:

$$HCl(aq) + (C_2H_5)_2NH(aq) \rightarrow (C_2H_5)_2NH_2^+(aq) + Cl^-(aq)$$

Therefore, the original solution contained 1.190×10^{-3} mol of $(C_2H_5)_2NH$. The molar weight of this base is 73.139 g mol^{-1}, so the weight of $(C_2H_5)_2NH$ was:

$$(1.190 \times 10^{-3} \text{ mol}) \times (73.139 \text{ g mol}^{-1}) = 0.0870 \text{ g}$$

At the equivalence point the volume of the solution is 115.90 mL. The solution contains 1.190×10^{-3} mol of $(C_2H_5)_2NH_2Cl$, the product of the neutralization of diethylamine with HCl. The concentration of this diethylammonium chloride is 1.027×10^{-2} M. The diethylammonium ion is a weak acid:

$$(C_2H_5)_2NH_2^+(aq) + H_2O(l) \rightleftarrows (C_2H_5)_2NH(aq) + H_3O^+(aq)$$

Let x equal the equilibrium concentration of H_3O^+ from this reaction. Then:

$$[(C_2H_5)_2NH] = x \quad \textit{and} \quad [(C_2H_5)_2NH_2^+] = 0.01027 - x$$

The equilibrium constant for the reaction is $K_a = K_w / K_b$ since diethylammonium ion is the conjugate acid of diethylamine. K_b is given in the problem, and K_w is 1.0×10^{-14} (at 25° C). K_a equals 3.29×10^{-11}. Substitution in the K_a equilibrium constant expression gives:

$$3.29 \times 10^{-11} = x^2 / (0.01027 - x)$$

Solving for x gives $[H_3O^+] = 5.81 \times 10^{-7}$ M and a pH of 6.24. The autoionization of water is a negligible source of hydronium ion.

The pH at equivalence is slightly acidic. Bromthymol blue (which changes color between pH 6.0 and 7.6) should work as an indicator. Note incidentally how diethylamine's formula derives from that of ammonia, NH_3.

---

25. The magnitudes of the two acid ionization constants for H_2A show that it is a weak acid. Moreover the second ionization occurs to less than $^1/_{1000}$ the extent of the first, most likely making it a negligible source of $H_3O^+(aq)$ ion compared to the first. Assume then that all the $H_3O^+(aq)$ in the solution comes from the first equilibrium:

$$H_2A(aq) + H_2O(l) \rightleftarrows HA^-(aq) + H_3O^+(aq)$$

$$K_{a1} = 4.00 \times 10^{-4} = [HA^-][H_3O^+] / [H_2A]$$

If $[H_3O^+] = x$, then $[HA^-] = x$ and $[H_2A] = 0.100 - x$. At equilibrium:

$$4.00 \times 10^{-4} = x^2 / (0.100 - x)$$

$$x^2 + (4.00 \times 10^{-4})x - 4.00 \times 10^{-5} = 0$$

Applying the quadratic formula $x = 6.13 \times 10^{-3}$; $[H_3O^+] = [HA^-] = 6.13 \times 10^{-3}$ M. The concentration of $[H_2A]$ is:

$$[H_2A] = 0.100 - 0.00613 \times 10^{-3} = 0.094 \text{ M}$$

The H_2A is about 6 percent ionized. If x had been neglected relative to 0.100 in the term $(0.100 - x)$, an unacceptable error would have resulted.

The second stage of the ionization acts to reduce $[HA^-]$ slightly from the value just calculated. Will it be significant? From the chemical equilibrium:

$$HA^-(aq) + H_2O(l) \rightleftarrows H_3O^+(aq) + A^{2-}(aq)$$

the following mathematical equation must hold:

$$K_{a2} = 2.00 \times 10^{-7} = [H_3O^+][A^{2-}] / [HA^-]$$

Substituting the values of $[H_3O^+]$ and $[HA^-]$ just calculated gives $[A^{2-}] = 2.00 \times 10^{-7}$ M. Such a small answer means *first* that the concentration of $HA^-(aq)$ is only negligibly reduced from 6.13×10^{-3} by the second step ionization and *second* that the concentration of $H_3O^+(aq)$ is *not* meaningfully raised by the second step. Finally, the autoionization of water is a negligible source of $H_3O^+(aq)$.

---

27. Sodium oxalate dissolves in water to give two ions. $Na^+(aq)$ is a *very* weak acid and has no effect on the pH of the solution. Oxalate ion is a weak base and reacts with water, *hydrolyzes*, in two stages:

$$C_2O_4^{2-}(aq) + H_2O(l) \rightleftarrows HC_2O_4^-(aq) + OH^-(aq) \quad K_{b1}$$

$$HC_2O_4^-(aq) + H_2O(l) \rightleftarrows H_2C_2O_4(aq) + OH^-(aq) \quad K_{b2}$$

These two equilibria give rise to the two mass action expressions:

$$K_{b1} = [HC_2O_4^-][OH^-] / [C_2O_4^{2-}]$$

$$K_{b2} = [H_2C_2O_4][OH^-] / [HC_2O_4^-]$$

Since oxalic acid is a diprotic weak acid its acid ionization has two steps and two *other* equilibrium constant expressions result:

$$K_{a1} = [HC_2O_4^-][H_3O^+] / [H_2C_2O_4]$$

$$K_{a2} = [H_3O^+][C_2O_4^{2-}] / [HC_2O_4^-]$$

It is known that:

$$K_w = [H_3O^+][OH^-]$$

Dividing the K_w equation by the K_{a2} equation gives:

$$K_w / K_{a2} = [HC_2O_4^-][OH^-] / [C_2O_4^{2-}]$$

The right-hand side of this equation is the mass action expression for the first hydrolysis reaction. K_{b1} is therefore equal to $K_w / K_{a2} = (1.00 \times 10^{-14}) / (6.4 \times 10^{-5}) = 1.56 \times 10^{-10}$. Similarly, $K_{b2} = K_w / K_{a1}$.

Let $x = [OH^-]$ at equilibrium and assume that the second stage of the hydrolysis, the K_{b2} reaction, proceeds to only a slight extent. Then $[HC_2O_4^-] = x$ and $[C_2O_4^{2-}] = 0.1 - x$. Substituting in the K_{b1} expression:

$$1.56 \times 10^{-10} = x^2 / (0.1 - x)$$

In this quadratic equation $x = 4.0 \times 10^{-6}$ i.e. $[OH^-] = [HC_2O_4^-] = 4.0 \times 10^{-6}$ M. Also:

$$[C_2O_4^{2-}] = 0.100 - (4.0 \times 10^{-6}) = 0.100 \text{ M}$$

$$[H_3O^+] = 1.0 \times 10^{-14} / [OH^-] = 2.5 \times 10^{-9} \text{ M}$$

The equilibrium constant expression for K_{a1} allows computation of the oxalic acid concentration:

$$K_{a1} = \frac{[H_3O^+][HC_2O_4^-]}{[H_2C_2O_4]} = 5.9 \times 10^{-2}$$

All but one of the quantities are known. By substitution $[H_2C_2O_4] = 1.7 \times 10^{-13}$ M.

---

29. Some of the phosphate ion will hydrolyze to give OH^- *(aq)* and HPO_4^{2-} *(aq)*:

$$PO_4^{3-}(aq) + H_2O(l) \rightleftarrows HPO_4^{2-}(aq) + OH^-(aq)$$

The product HPO_4^{2-} *(aq)* will then, to lesser extent, hydrolyze to $H_2PO_4^{2-}$ *(aq)* and, finally, some of the $H_2PO_4^{2-}$ *(aq)* will hydrolyze to give H_3PO_4 *(aq)*. Thus the original phosphate supply spreads itself out over three additional species and:

$$0.05 = [PO_4^{3-}] + [HPO_4^{2-}] + [H_2PO_4^-] + [H_3PO_4]$$

The equilibrium constants for the three hydrolysis reactions are readily computed from the three acid ionization constants:

$$K_{b1} = K_w / K_{a1} = [OH^-][HPO_4^{2-}] / [PO_4^{3-}] = 4.55 \times 10^{-2}$$

$$K_{b2} = K_w / K_{a2} = [OH^-][H_2PO_4^-] / [HPO_4^{2-}] = 1.61 \times 10^{-7}$$

$$K_{b3} = K_w / K_{a1} = [OH^-][H_3PO_4] / [H_2PO_4^-] = 1.33 \times 10^{-12}$$

The first hydrolysis, of PO_4^{3-} *(aq)* to HPO_4^{2-} *(aq)*, is *the* major producer of OH^- *(aq)* in the solution. Disregard all other equilibria, and let $x = [OH^-]$:

$$4.55 \times 10^{-2} = x^2 / (0.050 - x)$$

$$x^2 + 4.55 \times 10^{-2} x - 2.275 \times 10^{-3} = 0$$

This quadratic equation is routinely solved giving $x = 3.0 \times 10^{-2}$. Therefore $[OH^-] = [HPO_4^{2-}] = 0.030$ M and $[PO_4^{3-}] = 0.020$ M. Substitution of these values in the second hydrolysis equilibrium expression gives $[H_2PO_4^-] = 1.61 \times 10^{-7}$. Finally, going on to the last equilibrium expression:

$$1.33 \times 10^{-12} = [OH^-][H_3PO_4] / [H_2PO_4^-]$$

$$1.33 \times 10^{-12} = (3.00 \times 10^{-2})[H_3PO_4] / (1.61 \times 10^{-7})$$

$$[H_3PO_4] = 7.1 \times 10^{-18} \text{ M}$$

Since $[OH^-] = 0.030$ M, $[H_3O^+] = 3.3 \times 10^{-13}$ M. The product of these two concentrations always equals K_w.

---

31. Henry's Law states that the solubility of a gas is directly proportional to its partial pressure above the solvent:

$$P(\text{gas}) = kX$$

where X is the mole fraction of the gas and k is a constant.

The first task is to determine k for carbon dioxide in water. The problem states that 1 g of $H_2O(l)$ dissolves 0.759 cm³ of CO_2 *(g)* at 25° C and 1 atm pressure. Assuming that the CO_2 *(g)* is an ideal gas, this is 3.10×10^{-5} mol of CO_2 *(g)*. The mole fraction of CO_2 in the solution is the number of moles of CO_2 divided by the total number of moles present. Since the molar weight of water is 18.015 g mol^{-1}, there is 55.51×10^{-3} mol of $H_2O(l)$ in 1 cm³ of water. Hence, the mole fraction of CO_2 is 3.102×10^{-5} mol divided by 55.54×10^{-3} mol or 5.58×10^{-4}. Substituting into the Henry's law expression gives:

$$1.000 \text{ atm} = k(5.58 \times 10^{-4}) \quad \textit{ergo} \quad k = 1.79 \times 10^2 \text{ atm}$$

At Denver, the partial pressure of CO_2 is only 0.833 atm, but k is of course the same. Applying Henry's law, the mole fraction of CO_2 in saturated water at Denver is 0.833 atm / 1.79×10^2 atm = 4.65×10^{-4}. In a liter of saturated water (with density 1.00 g cm^{-3}) at Denver there is 55.51 mol of water. There must be 2.58×10^{-2} mol of CO_2 in this liter of water to make the mole fraction of CO_2 come out to 4.65×10^{-4}. Thus, the *molarity* of CO_2 in the water at Denver is 2.58×10^{-2} M.

Now go on to find the pH of the solution. The CO_2 decreases the pH of the water by serving as a weak acid in solution:

$$CO_2(g) + 2\,H_2O(l) \rightleftarrows HCO_3^-(aq) + H_3O^+(aq)$$

$$HCO_3^-(aq) + H_2O(l) \rightleftarrows CO_3^{2-}(aq) + H_3O^+(aq)$$

K_{a1} is 4.3×10^{-7} and K_{a2} is about four orders of magnitude less. This fact justifies ignoring the second equilibrium as a source of $H_3O^+(aq)$. Working from the first equilibrium and letting $x = [H_3O^+]$:

$$4.3 \times 10^{-7} = x^2 / (2.58 \times 10^{-2} - x)$$

$$x = 1.05 \times 10^{-4}$$

If $[H_3O^+] = 1.05 \times 10^{-4}$ M, then the pH is 3.98.

---

33. Maleic acid is a diprotic acid. Its two ionization constants differ by about four orders of magnitude. The two ionization steps can therefore be treated separately when the 0.1000 M solution of maleic acid is titrated. Let y equal $[H_3O^+]$. Before any 0.1000 M NaOH is added the predominant source of $H_3O^+(aq)$ in the solution is the equilibrium:

$$H_2mal(aq) + H_2O(l) \rightleftarrows Hmal^-(aq) + H_3O^+(aq)$$

for which:

$$K_{a1} = 1.42 \times 10^{-2} = [Hmal^-][H_3O^+] / [H_2mal] = y^2 / (0.1000 - y)$$

Solving this equation for y gives 0.0313. This is the concentration of $[H_3O^+]$ in mol L^{-1}. The pH equals 1.51.

As the first 5.00 mL of 0.1000 M NaOH is added to the aqueous maleic acid, it converts $H_2mal(aq)$ to $Hmal^-(aq)$. The 5.00×10^{-3} L of the 0.1000 M NaOH contributes 5.00×10^{-4} mol of NaOH. There is 50.00×10^{-4} mol of H_2mal. The base is the limiting reagent, and the greatest possible yield of $Hmal^-(aq)$ is 5.00×10^{-4} mol. The reaction leaves 45.00×10^{-4} mol of $H_2mal(aq)$ untouched. Adding the 5.00 mL of liquid dilutes the solution to a volume of 55.00 mL so, still assuming a complete acid-base reaction:

$$[Hmal^-] = 9.091 \times 10^{-3} \text{ M} \quad \textit{and} \quad [H_2mal] = 8.182 \times 10^{-2} \text{ M}$$

The equilibrium now acts to alter these concentrations slightly. For every $H_3O^+(aq)$ that the equilibrium produces it uses up one $H_2mal(aq)$ and makes one additional $Hmal^-(aq)$. Thus:

$$1.42 \times 10^{-2} = y[9.091 \times 10^{-3} + y] / [8.182 \times 10^{-2} - y]$$

Rearranging and solving the quadratic equation for y gives 0.0244 M. This is the concentration of $H_3O^+(aq)$. The pH equals 1.61.

At the next point in the titration 25.00 mL of titrant has been added and the total volume is 75.00 mL. A similar equation develops:

$$1.42 \times 10^{-2} = y[0.0333 + y) / [0.0333 - y]$$

Solving gives $y = 8.45 \times 10^{-3}$ and pH = 2.07.

When 50.00 mL of 0.1000 M NaOH has been added the titration exactly attains the *first* equivalence point. The solution consists of 100.00 mL of 0.0500 M Na^+Hmal^- (sodium hydrogen maleate). The $Hmal^-(aq)$ ion is amphoteric. It behaves as an acid:

$$Hmal^-(aq) + H_2O(l) \rightleftarrows H_3O^+(aq) + mal^{2-}(aq) \qquad K_{a2} = 8.57 \times 10^{-7}$$

and it behaves as a base:

$$Hmal^-(aq) + H_2O(l) \rightleftarrows OH^-(aq) + H_2mal(aq) \quad K_b = K_w / K_{a1} = 7.04 \times 10^{-13}$$

$Hmal^-(aq)$ is just like $HCO_3^-(aq)$ in this respect. By an analysis just like the one presented for $HCO_3^-(aq)$ in Section 6-6 (text p. 205):

$$[H_3O^+]^2 \simeq \frac{K_{a1}K_{a2}[Hmal^-]_0 + K_{a1}K_w}{K_{a1} + [Hmal^-]_0}$$

in which $[Hmal^-]_0$ is the "original" concentration of $Hmal^-(aq)$, 0.0500 M. Solving gives $[H_3O^+] = 9.73 \times 10^{-5}$ M for a pH of 4.01.

Notice that the approximate formula:

$$[H_3O^+]^2 \simeq K_{a1} K_{a2}$$

gives a wrong pH of 3.96. It fails because it depends on the assumption that K_{a1} is negligible compared to $[Hmal^-]$. K_{a1} is in fact equal to 0.0142, 28 percent of 0.050, and obviously *not* negligible.

After 75.00 mL of 0.1000 M NaOH has been added, the titration is halfway to the *second* equivalence point. In this range, the main source of $H_3O^+(aq)$ is the second ionization:

$$Hmal^-(aq) + H_2O(l) \rightleftarrows mal^{2-}(aq) + H_3O^+(aq) \qquad K_{a2} = 8.57 \times 10^{-7}$$

The concentration of $H_2mal(aq)$ is quite close to zero because so much base has been added. For this reason the first ionization now has a negligible effect on the pH. The titration can now be viewed as the addition of 25.00 mL of 0.1000 NaOH to 100.00 mL of 0.0500 NaHmal, making a total volume of 125.00 mL. Assuming complete neutralization, this much NaOH chemically converts exactly half of the $Hmal^-(aq)$ to $mal^{2-}(aq)$. Dilution then reduces the concentrations of both species by the factor $^{100}/_{125}$. Thus *after* neutralization but *before* the action of any equilibrium:

$$[Hmal^-] = 0.025 \times {}^{100}/_{125} = 0.020$$
$$[mal^{2-}] = 0.025 \times {}^{100}/_{125} = 0.020$$

The second ionization equilibrium now acts to change these concentrations slightly. It adds y to the concentration of $mal^{2-}(aq)$ and removes y from the concentration of $Hmal^-(aq)$, where y is the concentration of hydronium ion that it produces. The mass action expression for the second ionization becomes:

$$8.57 \times 10^{-7} = y[0.020 - y]/[0.020 + y]$$

y is 8.57×10^{-7} so the pH is 6.07.

By similar computations, the pH after 99.90 mL of 0.1000 M NaOH has been added is 8.76.

At the *second* equivalence point the solution is really 150.00 mL of 0.0333 M Na_2mal. Consider the *hydrolysis* of $mal^{2-}(aq)$ as the principal equilibrium at this point:

$$mal^{2-}(aq) + H_2O(l) \rightleftarrows Hmal^-(aq) + OH^-(aq)$$

The K_b for this equilibrium is $K_w / 8.57\times 10^{-7}$ or 1.17×10^{-8}. Let x be the concentration of $OH^-(aq)$. Then, based on the mass-action expression for the hydrolysis equilibrium:

$$1.17\times 10^{-8} = x^2 / 0.0333 - x$$

Solving for x gives 1.97×10^{-5} so pOH = 4.71 and pH = 9.29.

After 105 mL of 0.1000 M NaOH has been added, the above hydrolysis is completely overshadowed as a source of $OH^-(aq)$ by the excess strong base NaOH. The first 100.00 mL of NaOH was used up neutralizing the acid. The remaining 5.00 mL of 0.100 NaOH makes a solution that is (5.00 / 155.00) × 0.100 or 3.23×10^{-3} M in $OH^-(aq)$. This corresponds to pH = 11.51. The calculation requires no use of equilibrium constants. NaOH is a strong base in water. This means that its K_b is large.

---

35. Simply add H^+ to the formula of each base.

a) NH_4^+ *b)* CH_3COOH *c)* NH_3 *d)* HPO_4^{2-} *e)* H_2CO_3 *f)* HS^-

---

37. At 60° C K_w is *larger* (9.61×10^{-14}) than at room temperature. If the pH of a solution at 60° C is 7 then $[H_3O^+] = 1.00\times 10^{-7}$ and $[OH^-] = 9.61\times 10^{-7}$. Because $[OH^-]$ exceeds $[H_3O^+]$ the solution is *basic*.

---

39. *a)* The equation:

$$H_2SO_3(aq) + CH_3COO^-(aq) \rightleftarrows CH_3COOH(aq) + HSO_3^-(aq)$$

can be constructed as the sum of the three equilibria:

$$H_2SO_3(aq) + H_2O(l) \rightleftarrows H_3O^+(aq) + HSO_3^-(aq)$$
$$CH_3COO^-(aq) + H_2O(l) \rightleftarrows CH_3COOH(aq) + OH^-(aq)$$
$$H_3O^+(aq) + OH^-(aq) \rightleftarrows 2\,H_2O(l)$$

The overall K is the *product* of K_a for the first reaction, K_b for the second reaction and $1 / K_w$ (for the third reaction):

$$K_{overall} = 1.54\times 10^{-2} \times (K_w / (1.76\times 10^{-5}) \times 1 / K_w$$

$$K_{overall} = 8.75\times 10^{2}$$

$CH_3COO^-(aq)$ is a stronger base than $HSO_3^-(aq)$, and $H_2SO_3(aq)$ is a stronger acid than $CH_3COOH(aq)$.

b) Here HF donates H^+ ions and $ClCH_2COO^-$ serves as the base to accept them. This is the acid ionization of hydrofluoric acid running forward combined with the acid ionization of chloroacetic acid running backward:

$$K_{overall} = 3.5 \times 10^{-4} / 1.4 \times 10^{-3} = 0.25$$

$K_{overall}$ is less than one because chloroacetic acid is a stronger acid than HF. F^- is a stronger base than $ClCH_2COO^-$.

c) $K_{overall} = 4.93 \times 10^{-10} / 4.6 \times 10^{-4} = 1.1 \times 10^{-6}$. Hydrocyanic acid is much weaker than nitrous acid so at equilibrium the reaction:

$$HCN(aq) + NO_2^-(aq) \rightleftarrows HNO_2(aq) + CN^-(aq)$$

lies *far* to the left. Cyanide ion is a stronger base than nitrite ion.

---

41. The solution is prepared by mixing 0.10 mol of $CuCl_2$ and 0.10 mol of NaCN in 1.0 L of water. The $Cu^{2+}(aq)$ ion and $CN^-(aq)$ react to form a *complex ion*:

$$Cu^{2+}(aq) + 4\ CN^-(aq) \rightleftarrows Cu(CN)_4^{2-}(aq)$$

The K for this reaction is huge. It is 2×10^{30}, the reciprocal of the K of its *reverse*, which is given in the problem. In addition, $CN^-(aq)$ reacts with $H_2O(l)$ to give $HCN(aq)$:

$$CN^-(aq) + H_2O(l) \rightleftarrows HCN(aq) + OH^-(aq)$$

This hydrolysis reaction has K_b equal to K_w / K_a where K_a is the ionization constant of HCN. K_b comes out to 2.02×10^{-5} (using the K_a from Table 6-2, text p. 174).

Here is a list of the species in the solution: $Na^+(aq)$, $Cu^{2+}(aq)$, $H_3O^+(aq)$, $OH^-(aq)$, $CN^-(aq)$, $Cu(CN)_4^{2-}(aq)$, $HCN(aq)$ (and, of course, $H_2O(l)$). The principle of electrical neutrality allows the following equation to be written:

$$2\,[Cu^{2+}] + [H_3O^+] + [Na^+] = [CN^-] + [OH^-] + 2\,[Cu(CN)_4^{2-}] + [Cl^-]$$

All of the $Cu^{2+}(aq)$ is present either as unreacted $Cu^{2+}(aq)$ or in the form of the complex:

$$[Cu^{2+}] + [Cu(CN)_4^{2-}] = 0.10$$

Also, $[Na^+] = 0.10$ and $[Cl^-] = 0.20$. Algebraic combination of all these relations gives:

$$4\,[Cu^{2+}] + [H_3O^+] = [CN^-] + [OH^-] + 0.30$$

The concentration of $H_3O^+(aq)$ is certainly quite small compared to 0.30 because the only source of $H_3O^+(aq)$ ion is the feeble autoionization of water. Neglect it:

$$4\,[Cu^{2+}] \simeq [CN^-] + [OH^-] + 0.30$$

Now suppose that both $[CN^-]$ and $[OH^-]$ are *also* small compared to 0.30. Then $[Cu^{2+}]$ = 0.075 M. This would imply that $[Cu(CN)_4{}^{2-}]$ = 0.025. Substitute these figures back into the K expression for the formation of the complex ion. The $[CN^-]$ comes out to 2.0×10^{-8} M, much less than 0.30 M. The hydrolysis of even a thousand times this concentration of CN^- *(aq)* produces OH^- *(aq)* in concentrations that are *still* negligible compared to 0.30 M. This justifies the omission of $[OH^-]$ and $[CN^-]$ in comparison to 0.30 M and confirms that the concentration of Cu^{2+} *(aq)* in the solution is 0.075 M.

---

43. According to text Table 6-3, a yellow indication with methyl orange means the pH of the rain-water exceeds 4.4. A red color (in a separate experiment) with the second indicator, methyl red, means the pH is less than 4.8. The pH of the rain-water is therefore between 4.4 and 4.8.

---

45. The vinegar under analysis contains around 5 g of acetic acid per 100 g. Acetic acid is CH_3COOH (molar weight 60.053 g mol^{-1}). If the density of the vinegar is 1.0 g cm^{-3}, then Joe Dalton proposes to titrate 50.00 mL samples of about 0.83 M acetic acid. Using 1.000 M NaOH he can expect to need around 42 mL of base for each titration. If he detects the equivalence point to within $\pm$.02 mL then his precision is $(\pm.02 / 42) \times 100 \simeq 0.05$ percent.

Charlie Cannizzaro's method of analysis would give an uncertainty of $\pm$0.01 in the pH and $10^{\pm 0.01}$ in $[H_3O^+]$. 10 raised to the 0.01 power is 1.023. This means the hydrogen ion concentration would have $\pm$2.3 percent uncertainty. Joe Dalton's method is more precise. Its drawback is that it is slower and requires a little skill.

---

47. The buffer solution is to have a pH of 10.00 (pOH of 4.00). This means its $[H_3O^+] = 1.00 \times 10^{-10}$ M, and its $[OH^-] = 1.00 \times 10^{-4}$ M (assuming the temperature is 25° C). If substantial quantities of Na_2CO_3 and $NaHCO_3$ are present then the principal equilibrium controlling the pOH is:

$$CO_3{}^{-2}(aq) + H_2O(l) \rightleftharpoons HCO_3{}^-(aq) + OH^-(aq)$$

For this equilibrium:

$$K_{b1} = K_w / K_{a2} = [HCO_3{}^-][OH^-] / [CO_3{}^{2-}] = 2.08 \times 10^{-4}$$

If $[OH^-] = 1.00 \times 10^{-4}$ then the *ratio* of bicarbonate to carbonate ion concentrations must be 2.08. Because the two ions are dissolved in the same solution their chemical amounts also have this ratio.

Let x equal the mass of Na_2CO_3 originally placed in the solution and let y equal the original mass of $NaHCO_3$. Then:

$$x + y = 10.0 \text{ g}$$

The molar weights of Na_2CO_3 and $NaHCO_3$ are 106.0 g mol^{-1} and 84.01 g mol^{-1} respectively. The original chemical amounts of Na_2CO_3 and $NaHCO_3$

then are (x / 106.0) mol and (y / 84.01) mol. If only negligible amounts of $CO_3^{2-}(aq)$ are converted to $HCO_3^-(aq)$ or vice-versa then:

$$2.08\ (x / 106.0) = (y / 84.01)$$

The two simultaneous equations involving x and y can be quickly solved. x = 3.78 g and y = 6.22 g.

---

49. Suppose that all of the sulfur in the water is in the form of $SO_3(aq)$. We then are dealing with a 0.0004 M solution of sulfuric acid, H_2SO_4. Sulfuric acid ionizes in two steps:

$$H_2SO_4(aq) + H_2O(l) \rightleftarrows HSO_4^-(aq) + H_3O^+(aq)$$

$$HSO_4^-(aq) + H_2O(l) \rightleftarrows SO_4^{2-}(aq) + H_3O^+(aq)$$

The equilibrium constants for these two reactions are $K_{a1} = 10^2$ and $K_{a2} = 1.2 \times 10^{-2}$. Suppose for the moment that both reactions go 100 percent to completion. Then $[H_3O^+]$ would be 0.0008 M. This is unlikely to happen because K_{a2} is relatively small. Nevertheless the *first* equilibrium does lie quite far to the right so $[H_3O^+]$ lies between 0.0004 (from 100 percent completion of first step) and 0.0008 M.

Based on the second step:

$$1.2 \times 10^{-2} = [H_3O^+][SO_4^{2-}] / [HSO_4^-]$$

Let $x = [SO_4^{2-}]$. Then:

$$[H_3O^+] = 0.0004 + x \quad \textit{and} \quad [HSO_4^-] = 0.0004 - x$$

Substitution gives:

$$1.2 \times 10^{-2} = x(0.0004 + x) / (0.0004 - x)$$

Rearrangement of this equation produces:

$$x^2 + 1.24 \times 10^{-2} x - 4.8 \times 10^{-6} = 0$$

of which the only physically meaningful root is $x = 3.76 \times 10^{-4}$. If $[SO_4^{2-}] = 3.76 \times 10^{-4}$ M then:

$$[H_3O^+] = 0.0004 + 3.76 \times 10^{-4} = 7.8 \times 10^{-4}\ M$$

$$[HSO_4^-] = 0.0004 - 3.76 \times 10^{-4} = 2.4 \times 10^{-5}\ M$$

These numbers allow a check of the concentration of un-ionized H_2SO_4:

$$[H_2SO_4] = [H_3O^+][HSO_4^-] / K_{a1} = 1.9 \times 10^{-10}\ M$$

Indeed, little H_2SO_4 remains unreacted. The pH of the solution is $-\log(7.8 \times 10^{-4}) = 3.1$.

If the sulfur is entirely in the +IV oxidation state, then the solution is acidified by the equilibria:

$$SO_2(aq) + 2\,H_2O(l) \rightleftarrows H_3O^+(aq) + HSO_3^-(aq) \qquad K_{a1} = 1.45 \times 10^{-2}$$

$$HSO_3^-(aq) + H_2O(l) \rightleftarrows H_3O^+(aq) + SO_3^{2-}(aq) \qquad K_{a2} = 1.02 \times 10^{-7}$$

Focus on the first equilibrium. Since its K_a is thousands of times larger than K_{a2} it furnishes essentially all the $H_3O^+(aq)$ in the solution. In such a case, $[H_3O^+] = [HSO_3^-]$ and $[SO_2] = 0.0004\text{ M} - [H_3O^+]$:

$$1.45 \times 10^{-2} = [H_3O^+]^2 / (0.0004 - [H_3O^+])$$

Solving this equation for $[H_3O^+]$:

$$[H_3O^+] = 4 \times 10^{-4}\text{ M} \quad \textit{and} \quad pH = 3.4$$

---

51. Represent the amino acid glycine as glyH. The conjugate acid is then $glyH_2^+$, and the conjugate base is gly^-.

Copy the two equilibria given in the problem:

$$glyH(aq) + H_2O(l) \rightleftarrows gly^-(aq) + H_3O^+(aq) \qquad K_a = 1.7 \times 10^{-10}$$

$$glyH(aq) + H_2O(l) \rightleftarrows glyH_2^+(aq) + OH^-(aq) \qquad K_b = 2.2 \times 10^{-12}$$

The following equations convey the *same* chemical relationships:

$$glyH_2^+(aq) + H_2O(l) \rightleftarrows glyH(aq) + H_3O^+(aq) \qquad K_1 = K_w / K_b$$

$$glyH(aq) + H_2O(l) \rightleftarrows gly^-(aq) + H_3O^+(aq) \qquad K_2 = K_a$$

In this latter pair of equilibria the second equation given in the problem has been *reversed* and *added* to the water autoionization equation. Now, the emphasis is on $glyH_2^+$ ($^+H_3N{-}CH_2{-}COOH$) as a diprotic acid rather than on glycine's self-neutralization.

The advantage of this presentation is that it emphasizes the similarity of a solution of glycine to solutions of other amphoteric species. The 0.10 M aqueous glycine is like 0.10 M $HCO_3^-(aq)$, for example. The case of the pH of an aqueous solution of $HCO_3^-(aq)$ is exhaustively treated in the text (Section 6-6, text p. 204). The approximate formula:

$$[H_3O^+]^2 \simeq K_1 K_2$$

is derived. According to this formula the $[H_3O^+]$ in the solution is *independent* of the glycine's concentration. Substitution in the formula gives:

$$[H_3O^+]^2 = (K_a K_w / K_b) = (1.7 \times 10^{-24} / 2.2 \times 10^{-12})$$

$$[H_3O^+] = 8.8 \times 10^{-7}\text{ M} \quad \textit{and} \quad pH = 6.06$$

A more exact analysis (also in text Section 6-6) gives:

$$[H_3O^+]^2 \simeq \frac{K_1 K_2 C_0 + K_1 K_w}{K_1 + C_0}$$

The more exact treatment leading to *this* formula includes only one approximation: that the equilibrium concentration of glycine is close to C_0, its original concentration. Substitution of $C_0 = 0.10$ M and the three constants gives $[H_3O^+] = 8.6 \times 10^{-7}$ M and pH = 6.07, not much different from 6.06.

If C_0 were 0.01 M, one-tenth of the value given in the problem, the pH of the glycine solution would be 6.14. This deviates substantially from 6.06, the first answer. Thus, as C_0 goes down it becomes *more* important in determining the pH, at least at first. As C_0 becomes exceedingly small the solution approximates pure water with the pH equal to 7.00.

Chapter 7

THERMODYNAMIC PROCESSES AND THERMOCHEMISTRY

Thermodynamics is the study of transfers of energy accompanying physical and chemical changes. The goals of thermodynamics are to predict which types of processes are possible and the conditions under which desired processes can occur.

7.1 *SYSTEMS, STATES, AND PROCESSES*

The basic terms in thermodynamics require careful definition:

• *System.* A real or imaginary portion of the universe which is confined by physical boundaries or mathematical constraints. In essence, a thermodynamic system is that part of the universe which we decide to study. In problem-solving, a shrewd choice of system clarifies difficult situations. A useful tactic is to define a system as the sum of a set of sub-systems. **See problem 5.** One then concentrates attention on each of the sub-systems in turn. In complex problems the system should be fully defined *in writing* before starting any computations. **See problem 43.**

• *Closed System.* A system which has boundaries that do not allow the transfer of matter. Closed systems are common in problems.

• *Open System.* A system that *does* allow the transfer of matter across its boundaries. The typical chemical reaction, performed in an open flask in the laboratory, takes place in an open system.

• *Surroundings.* The portion of the universe which lies outside the system.

• *Extensive Property.* Extensive properties depend on the *extent,* or size of a system. The value of an extensive property for the whole system is the sum of the values for the individual sub-systems. A good example of an extensive property is the volume. Two sub-systems of volume 3.0 and 7.0 L make a system of volume 10.0 L. The energy is a particularly important extensive property.

• *Intensive Property.* A property of a system that does not depend on how big the system is. Examples are the pressure and temperature.

• *Thermodynamic State.* A system is in a unique thermodynamic state when each of its properties (pressure, temperature, volume, energy, etc.) has a definite, time-independent value. Left to itself such a system will remain unchanged, in a state of equilibrium, indefinitely.

• *State Function.* A state function is a property of a system which depends only on the thermodynamic state of the system. The value of a state function does not depend on the way in which the state was achieved. State functions are symbolized by capital letters. A change in a state function is indicated by a Δ. Thus:

$$\Delta P = P_2 - P_1$$

means the difference in pressure between the final state (state 2) and the initial state (state 1). In many applications the *change* in a state function is much more important than the actual final and initial values of the function.

In a process, the change in a state function depends only on the initial and final states and not on the path by which the change occurred.

In problems it is always best to adhere strictly to the convention that Δ applied to a function means the *final* value minus the *initial* value. Feckless reversal of this convention leads to tiresome sign errors.

• *Thermodynamic Process.* Such processes lead to changes in the thermodynamic states of systems. The *path* of a process is the sequence of intermediate conditions occurring in the system during a change.

An *irreversible* process passes through intermediate conditions that are not thermodynamic states. If the outside pressure on a sample of gas is suddenly relaxed, then the gas expands irreversibly.

A *reversible* process proceeds through a continuous series of equilibrium states. At any moment during a reversible change an infinitesimal change in external conditions is enough to reverse the direction of a reversible process. Reversible compression of a sample of gas would require an infinite series of infinitely small reductions in volume. A reversible decrease in temperature would require a similar arrangement of temperature reductions. **See problem 35.**

7.2 *ENERGY, WORK, AND HEAT*

For convenience two types of energy are distinguished. The kinetic energy of a particle is its energy of motion:

$$K.E. = \frac{1}{2} M v^2$$

where M is its mass and v its velocity. A system has kinetic energy based on its own motion as a whole and internal kinetic energy based on the relative motions of the atoms that compose it. The second type of energy is that of position, or potential energy. An object has for instance gravitational potential energy relative to the earth's center. If its height h near the surface of the earth is changed, then the change in its potential energy is:

$$\Delta(P.E.) = Mg\Delta h$$

where g is the acceleration of gravity (9.80 m s^{-2}) and M is its mass. **See problem 1.** One kind of internal potential energy is *chemical potential energy* which is the energy of position of a set of chemical reactants relative to their conversion to a set of products. A pound of TNT for instance has much chemical potential energy relative to its conversion to the gas mixture it forms when it explodes.

The total internal energy, E, of a system is the sum of its internal kinetic energy and internal potential energies:

$$E = K.E. + P.E.$$

The total energy of a system cannot be determined because of the impossibility of setting a meaningful zero. Should one take the zero of the gravitational potential energy with reference to the center of the earth or the center of the sun? It is an impossible question. The entire emphasis in thermodynamics is therefore upon *changes* in energy, ΔE.

Work (w) and heat (q) are the two different means by which energy is transferred into or out of a system. The sign of w and q tells the direction of the transfer:

- Positive q means the system gains heat.
- Negative q means the system loses heat
- Positive w means the system gains work (is worked upon).
- Negative w means the system loses work (performs work).

Strict adherence to this sign convention reduces the incidence of errors in problem solving. See the solutions to **problems 7 and 33.**

Work

Work involves an oriented, non-random transfer of energy. It is measured in joules. Of major concern in chemistry are electrical work (see Chapter 9) and mechanical work. Mechanical work occurs when the application of a force moves an object a certain distance. The work to move a block of mass M over the distance $(r_f - r_i)$ applying a constant force F is:

$$w = F \cdot (r_f - r_i)$$

where the subscripts tell the final and initial locations. If the force is expressed in newtons (N) and the distance in meters (m), the work is in newton meters (N m). A newton is a kg m s^{-2} and a newton meter is a kg m^2 s^{-2} which is a joule.

- *A newton meter is equal to a joule (J).*

The transfer of mechanical work requires mechanical contact between a system and its surroundings. In the absence of levers, wheels and other mechanical contrivances the only kind of mechanical work possible in chemical processes is *pressure-volume work.* Pressure-volume work results from the change in volume (compression or expansion) of a system against a resisting pressure. If a system expands against an external pressure then work flows:

$$w = -P_{ext}\,\Delta V$$

The negative sign preserves the sign convention (that positive work is work gained by the system). For an expansion ΔV is positive because V_2 is larger than V_1. The system pushes back its surroundings. It does work on the surroundings. Under the sign convention this is stated in reverse: the system absorbs negative work. If the pressure is in atmospheres and the volume is in liters then the unit of pressure-volume work is the L atm. A L atm is equal to 101.325 J.

Heat

Heat, like work, is a way in which energy is exchanged between a system and its surroundings. It is a manifestation of energy and has the same units (joules). The transfer of heat requires thermal contact (rather than mechanical contact) across a system boundary and occurs because of a temperature difference. Heat involves random, non-directed motion instead of motion that has overall coherence and direction. Heat lacks the macroscopic organization of work.

Calorimetry and Heat Capacity

Calorimetry is the measurement of quantities of heat. The name of an alternate unit of heat, the *calorie*, reflects this. (One calorie is 4.184 J.) The fundamental calorimetric method is to let some heat flow into a system, checking the temperature of the system before and after. As the system gains heat its temperature increases. Its *heat capacity* is the quantity of heat necessary to raise its temperature 1 Kelvin. Heat capacity has the units J K^{-1}. If the heat capacity of the system is C, then the quantity of heat it absorbs in the temperature change ΔT is:

$$q = C\Delta T$$

This relationship occurs extensively in practical problems. **See problems 5 and 19.**

The heat capacity of a system is clearly an extensive property. A massive nugget of copper can absorb much more heat before its temperature goes up one K than a small copper coin. For this reason the *molar heat capacity*, the amount of heat necessary to raise the temperature of one mole of a substance by one K, is defined. Molar heat capacities have the unit J $mol^{-1}K^{-1}$. If a molar heat capacity is used then:

$$q = n\,C\Delta T \qquad (C = \text{molar heat capacity})$$

The *specific heat* of a substance is similar to the molar heat capacity but on a per gram basis. The units for specific heat are J $g^{-1}K^{-1}$. If a specific heat is used:

$$q = m\,C\Delta T \qquad (C = \text{specific heat})$$

where m is the mass of the substance in grams.

- Always check a heat capacity's units. Mixing up the heat capacity, an extensive variable, and the molar heat capacity and specific heat, intensive variables, is a common error.

Although the absolute temperature scale has a zero different from the zero of the Celsius scale, 1 K is exactly the same size as 1° C.

- Values of heat capacities and specific heats quoted in J $(°C)^{-1}$ or J $g^{-1}(°C)^{-1}$ are for this reason numerically equal to values quoted in J K^{-1} or J $g^{-1}K^{-1}$, respectively.

The observed heat capacity of a system depends on whether the calorimetry experiment is carried out at constant volume or constant pressure. The constant pressure molar heat capacity is symbolized C_p. Constant volume molar heat capacity is C_v. The difference exists because a system at constant pressure changes volume a bit as it absorbs heat. While changing volume, the system exchanges some work with its surroundings. This affects its internal energy and final temperature. At constant volume there can be no such pressure-volume work. At constant volume:

$$q_v = n\ C_v \Delta T$$

and at constant pressure:

$$q_p = n\ C_p \Delta T$$

The difference between C_v and C_p is small for liquids and solids and is often ignored. Thus **problem 21** gives a specific heat for water, a liquid, without stating whether volume or pressure is constant. **Problems 5 and 19** do the same. The distinction between C_v and C_p becomes important with gases and figures strongly in many problems (see below).

7.3 *THE FIRST LAW OF THERMODYNAMICS*

The first law of thermodynamics is a process-oriented re-statement of the principle of conservation of energy. Any change in the internal energy of a system during a process must equal the work *absorbed* by the system plus the heat *absorbed* by the system:

$$\Delta E = w + q$$

In a general process both heat and work cross the boundaries of the system. The two are positive when they flow into the system. This explains the emphasis on *absorption* in the statement of the first law. Although the terms "negative work" and "negative heat" seem peculiar and troublesome, negative work is really just work done *by* a system on its surroundings, and negative heat is heat flowing from a system out to the surroundings. The point is particularly clear in **problem 25.** There is a financial analogy. A positive money flow is income; a negative money flow is expense. Transferring an expense to somebody else (the surroundings) is equivalent to receiving income.

What if a system is carefully insulated so that heat cannot flow across its boundaries? Heat is neither absorbed nor lost so q is 0. If the system now does some work, it does it entirely at the expense of its internal energy. When the internal energy diminishes E_2, after the process, is smaller than E_1, before the process. $\Delta E = E_2 - E_1$ is negative. According to the first law:

$$\Delta E = w + q$$

When q is zero and ΔE is negative w is obviously negative. **See problem 27.**

Although both q and w depend on the path along which a change occurs, their sum, ΔE, is a state function and does *not* depend on the path.

Heat Capacities of Ideal Gases

At constant volume the molar heat capacity of a monatomic ideal gas is $^3/_2$ R. R is equal to 8.314 J mol^{-1}K^{-1} so this C_v is 12.471 J mol^{-1}K^{-1}. At constant pressure the same monatomic ideal gas has a molar heat capacity of $^5/_2$ R, 20.785 J mol^{-1}K^{-1}, 1.666 times larger. **Problem 9b** requires these facts as do **Problems 7, 11, 25 and 27.**

In the first two of these problems the solver must recall that argon and neon are monatomic. The qualification "monatomic" appears because diatomic and polyatomic gases have larger values of C_v and C_p than monatomic gases. C_p and C_v increase with the number of atoms per molecule. Gaseous C_8H_{18} has a C_p of 327 J mol^{-1}K^{-1} compared to only 29.8 J mol^{-1}K^{-1} for N_2 *(g)*. **See problem 43.** For ideal gases, whether monatomic or polyatomic:

$$C_p = C_v + R$$

7.4 *THERMOCHEMISTRY*

Enthalpy

The enthalpy is an additional, very useful state function. It is defined in terms of other state functions:

$$H = E + PV$$

The definition means that the enthalpy of a system is computed by determining its internal energy and adding the product of its pressure and volume. Clearly H has the same units as E (joules). Although PV ordinarily comes out in L atm, an easy conversion (1 L atm = 101.325 J) would allow adding it to E. One never actually performs such a computation. Absolute values of E are unattainable because of the impossibility of defining a zero for energy. The emphasis is on *changes* in E and H instead. For a process (change) in a system:

$$\Delta H = \Delta E + \Delta(PV)$$

The definition of enthalpy was deliberately chosen so that:

$$q_p = \Delta H$$

which means that the change in a system's enthalpy during a process equals the amount of heat that the system would absorb if the process were carried out at constant pressure.

The enthalpy function is useful because many processes occur under constant (atmospheric) pressure making q_p relatively easy to measure. The similarity of the relationships between q_p and ΔH and q_v and ΔE is not accidental:

$$q_v = \Delta E \qquad \textit{and} \qquad q_p = \Delta H$$

These are key relationships in thermochemistry.

Enthalpies of Reaction

Energy and enthalpy changes occur in chemical reactions because chemical reactions involve the making and breaking of chemical bonds, events that release and consume energy. A negative enthalpy change in a reaction means that H_2, the enthalpy of the products, is *less* than H_1, the enthalpy of the reactants:

$$\Delta H = H_2 - H_1 = q_p < 0$$

Because q_p is negative the reaction gives off heat. The reaction is *exothermic.* Reactions in which heat is absorbed are *endothermic* and have a positive ΔH.

The enthalpy change of a reaction is often written immediately after the reaction on the same line. In this position, ΔH refers to the enthalpy change of the balanced chemical reaction with which it is associated. For example, in **problem 39** the enthalpy change for the combustion of octane is computed:

$$C_8H_{18}(l) + 12\tfrac{1}{2}\, O_2(g) \rightarrow 8\, CO_2(g) + 9\, H_2O(l) \qquad \Delta H = -5531 \text{ kJ}$$

If all of the coefficients in the reaction are doubled, then ΔH is also doubled:

$$2\, C_8H_{18}(l) + 25\, O_2(g) \rightarrow 16\, CO_2(g) + 18\, H_2O(l) \qquad \Delta H = -11062 \text{ kJ}$$

An enthalpy change is an extensive property, and the coefficients in the balanced equations give the size of the system. **Problem 19** shows how the quantity of reactant must be considered in figuring out how much heat a reaction evolves.

If the direction of a reaction is reversed the roles of products and reactants are interchanged. Such an interchange makes H_2 into H_1 and H_1 into H_2. The upshot is to change the sign of ΔH, but not its magnitude. Thus, for the reverse of the combustion of octane:

$$8\, CO_2(g) + 9\, H_2O(l) \rightarrow C_8H_{18}(l) + 12\tfrac{1}{2}\, O_2(g) \qquad \Delta H = +5531 \text{ kJ}$$

H is a state function since ΔH is independent of the path followed during the conversion of reactants to products. Consequently, if two or more chemical equations are added to give some overall equation their enthalpy changes add up to the enthalpy change of the overall equation. The overall reaction can be imagined to proceed by stages consisting of the first reaction, then the second, etc. *Hess's law* is a statement of this fact:

- *The ΔH for the conversion of a given set of reactants into a given set of products is the same regardless of whether the conversion occurs in one step or a series of steps.*

The practical importance of Hess's law is that whenever a new chemical reaction can be constructed as a combination of other reactions with known ΔH's then the ΔH of the new reaction need not be measured. It can instead be calculated. **Problem 35** is an example showing the use of known ΔH's to get a ΔH for a new reaction.

To compute ΔE for a reaction given its ΔH, use the relationship:

$$\Delta E = \Delta H - \Delta(PV)$$

For solids and liquids the term Δ(PV) is small, and for reactions involving exclusively solids and liquids the difference between ΔE and ΔH can be neglected. For reactions involving gases, it is nearly always safe to assume that the ideal gas equation is followed. If it is then, at constant temperature:

$$\Delta E = \Delta H - \Delta n_g RT \qquad \text{(at constant temperature)}$$

where Δn_g is the number of moles of gas among the products minus the number of moles of gas among the reactants. As **problem 15** shows, the difference between ΔE and ΔH is usually only a small fraction of ΔH. In **problem 17,** ΔE is the experimental value and computation gives ΔH. The variation presents no special difficulty. A common mistake in applications is to compute Δn_g RT in joules and then to subtract it from a ΔH value that is in *kilo*joules. Always check the units of these quantities before adding or subtracting.

Standard State Enthalpies

The standard state of a substance is its stable form at a temperature of 298.15 K and a pressure of 1 atm. The standard state of dissolved species is unit molarity (1 M). This last definition figures in **problem 21.** A superscript of zero following a thermodynamic symbol refers to standard conditions.

Every chemical compound can, in principle, form directly from the elements that make it up. Reactions of this type are *formation* reactions. Equations representing formation reactions always have some assortment of pure elements on the left-hand side and a single compound, the compound being formed, on the right-hand side.

- *The standard enthalpy of formation*, ΔH°_f, of a substance is the enthalpy change of the reaction that produces one mole of the substance from its constituent elements in their standard states.

The units of standard enthalpies of formation are kJ mol^{-1}. For example, the formation of ammonium iodide from its constituent elements is represented:

$$\tfrac{1}{2} N_2(g) + 2 H_2(g) + \tfrac{1}{2} I_2(s) \rightarrow NH_4I(s) \qquad \Delta H_f = 201.4 \text{ kJ}$$

To emphasize that this reaction is a formation reaction its ΔH is subscripted with an f (for formation). As written the reaction produces 1 mol of $NH_4I(s)$. Hence the ΔH_f of ammonium iodide is 201.4 kJ mol^{-1}. This particular ΔH was measured at 298.15 K and 1 atm which are defined as the standard conditions. It is a standard enthalpy change, ΔH°_f.

The standard enthalpy of formation of pure elements is zero. For example, the equation:

$$O_2(g) \rightarrow O_2(g)$$

represents the formation of the element oxygen "from its constituent elements in their standard states." The two sides of the equation are the same, so there is no change in the enthalpy. ΔH°_f of $O_2(g)$ is zero.

Standard enthalpies of formation of compounds are tabulated in long lists. (See text Appendix D.) Such tables are useful because a reaction, no matter how complicated, can be imagined to go by the decomposition of all of the reactants into the constituent elements and then the formation of all of the products directly from these elements. As a consequence of Hess's law then:

$$\Delta H^\circ = \Sigma \Delta H_f^\circ \text{ (products)} - \Sigma \Delta H_f^\circ \text{ (reactants)}$$

This equation is heavily used in solving practical problems. **Problem 15** is an example. Confusion sometimes arises concerning the units. ΔH_f°'s are given in kJ mol^{-1}. To apply the above equation:

1. Multiply each compound's ΔH_f° by the number of moles of that compound shown in the balanced chemical equation.

2. Add these numbers together for all of the products and do the same for all of the reactants.

3. Subtract the latter from the former.

4. The result is ΔH° of the reaction and has the units of kJ. In those cases where ΔH of a reaction is given in kJ mol^{-1} (or J mol^{-1}), take it to mean *per mole of reaction as written.* If the reaction:

$$H_2(g) + \tfrac{1}{2} O_2(g) \rightarrow H_2O(l)$$

is said to have a ΔH° of -285.83 kJ mol^{-1}, then it follows that the new reaction:

$$2\,H_2(g) + O_2(g) \rightarrow 2\,H_2O(l)$$

(derived by doubling all of the coefficients) has a ΔH° of -571.66 kJ mol^{-1}.

Explosions and Flames

The theme in the use of Hess's law and tables of ΔH_f°'s is the substitution of imaginary pathways which make computations easy for actual pathways. An extension of this theme allows ready estimates of the temperature rise in explosions and flames. The trick is to imagine first the completion of the chemical reaction and then the use of all the heat to raise the temperature of the products. In an explosion (at constant volume):

$$-\Delta E \text{ (at } T_1) = C_v\text{(products)} (T_2 - T_1)$$

whereas, in a flame (at constant pressure):

$$-\Delta H \text{ (at } T_1) = C_p\text{(products)} (T_2 - T_1)$$

Problem 23b shows the use of the second of these equations. The minus signs on the left side in the equations signify that, as far as the system containing the products is concerned, the heat from the reaction comes from the surroundings.

7.5 *REVERSIBLE GAS PROCESSES*

Changes in state functions depend only on the initial and final state and not at all on the path along which a process or reaction takes place. Sometimes, however, such knowledge is available and can help in computing ΔE, ΔH, etc. To compute values for q and w, which *do* depend on the path a process follows, specification of the path is of course essential. Several terms describe paths and must be learned. They are:

- *Isothermal.* An isothermal process goes on at constant temperature. In other terms, $\Delta T = 0$ in an isothermal change. The internal energy of an ideal gas depends only on its temperature. If T is constant $\Delta E = 0$ and:

$$q = -w \qquad \text{(isothermal process, ideal gas)}$$

In addition ΔH is zero. For non-ideal gases, liquids, and solids, ΔE and ΔH can be non-zero even if the process is isothermal.

- *Isothermal and Reversible.* An isothermal and reversible process goes on at constant temperature in a series of infinitesimally small steps. The system is always at equilibrium and at any time the direction of the process can be changed. For an ideal gas in such a process:

$$q = -w = nRT \ln (V_2 / V_1) \qquad \text{(isothermal reversible process, ideal gas)}$$

In addition:

$$P_1 V_1 = P_2 V_2 \qquad \text{(ideal gas)}$$

which, when solved for V_2 and substituted into the previous equation, gives:

$$q = -w = nRT \ln (P_1 / P_2) \qquad \text{(isothermal reversible process, ideal gas)}$$

In an isothermal *expansion* V_2 is bigger than V_1. The gas expands and absorbs heat from the surroundings as it does so. It simultaneously performs an exactly equivalent amount of work on the surroundings. In that way ΔE remains at zero.

- *Adiabatic.* In an adiabatic process there is no transfer of heat into or out of the system. This means that q equals zero in all adiabatic processes. In an adiabatic change any work that the system performs comes at the expense of its internal energy and any work the system absorbs adds to its internal energy:

$$\Delta E = w$$

- *Adiabatic and Reversible.* For an ideal gas it can be shown that:

$$P_1 V_1^{\gamma} = P_2 V_2^{\gamma} \qquad \text{(adiabatic reversible process, ideal gas)}$$

where γ is the ratio of C_p to C_v.

In addition:

$$T_1 V_1^{\gamma - 1} = T_2 V_2^{\gamma - 1} \qquad \text{(adiabatic reversible process, ideal gas)}$$

The solution to **problem 9c** points out a common pitfall with these equations. They do not apply unless an *ideal* gas undergoes an *adiabatic, reversible* change.

All three qualifications must be met. **Problems 27** and **45** show more complicated applications of these equations.

The Carnot Cycle

A *cyclic* process takes a system through a series of thermodynamic states and returns it ultimately to the exact state where it started. In a *Carnot* cycle, an ideal gas undergoes a series of four reversible changes during an excursion out from and then back to its original state. These are:

1. An isothermal reversible expansion. The gas, at its original temperature, T_h, expands from V_A to V_B. It performs work on the surroundings. It simultaneously absorbs heat from the surroundings to keep the same internal energy.

2. An adiabatic reversible expansion. The gas continues to expand, going from V_B to V_C. Now, however, q is zero. The continued expansion of the gas performs more work on its surroundings. The energy to do the work comes from the internal energy of the gas. The temperature of the gas drops to T_l.

3. An isothermal reversible compression. At T_l the surroundings compress the gas, from V_C to V_D. Work is performed upon the gas. The temperature is constant and just enough heat flows from the gas to the surroundings so that ΔE for this step stays zero.

4. An adiabatic reversible compression. The surroundings continue to do work on the gas. Now however the gas is thermally insulated so that $q = 0$. All the work adds to the internal energy of the gas. The temperature rises toward T_h. The volume diminishes toward V_A. The compression continues until the gas reaches its original state.

In a Carnot cycle the gas extracts heat from the surroundings at T_h, converts some of it into work, and dumps the rest to the surroundings at T_l. The net work is:

$$w_{net} = -nR\,(T_h - T_l)\ln(V_B / V_A)$$

The net work is negative, consistent with the convention that work performed by a system is negative. The net work depends on how much gas (n is the chemical amount) is taken around the cycle, on the two extremes of temperature and by how much the gas expands in the first step.

7.6 *HEAT ENGINES AND REFRIGERATORS*

A heat engine is a device which converts the natural heat flow from higher to lower temperature into useful work. A refrigerator does the reverse. It destroys work to produce a difference in temperature.

A heat engine operates between two temperatures. The *efficiency*, ϵ, of a heat engine is the net work that it performs divided by the heat that it absorbs. The maximum efficiency of a heat engine operating between two temperatures is obtained when the engine operates reversibly. This maximum efficiency is:

$$\epsilon = 1 - T_l / T_h$$

This equation is derived from consideration of the Carnot cycle. Why does this equation give the theoretical maximum efficiency? The proof consists of assuming the opposite and showing that an engine with greater efficiency would make it possible to transfer heat in a continuous cycle out of a low temperature reservoir into a high temperature reservoir without expending any work. In all human experience it is impossible to do this. Heat never flows uphill (against a temperature gradient). This is the second law of thermodynamics.

The theoretical operation of a refrigerator or heat pump is modelled on a Carnot cycle running in reverse. The maximum heat absorbed, *q*, from the interior of a refrigerator is:

$$q = w_{net} (T_{\ell} / T_h - T_{\ell})$$

where T_{ℓ} and T_h are the low and temperatures and w_{net} is the work during the refrigeration cycle. These equations are used in **problems 29 and 47.**

DETAILED SOLUTIONS TO ODD-NUMBERED PROBLEMS

1. An object of mass m under the influence of gravity possesses potential energy relative to the ground. This potential energy is given by mgh where h is its height and g is the acceleration of gravity. As the object falls, its potential energy is converted into kinetic energy. At impact the object stops and its kinetic energy relative to the ground goes to zero. Simultaneously its potential energy attains zero because h is zero. All of the energy appears as heat. If this heat goes to increase the temperature of an equal mass of water then:

$$m\,C\Delta\mathrm{T} = m\,g\Delta h$$

where C is the heat capacity per unit mass (specific heat) of the water. The m's will cancel out. In this problem $\Delta\mathrm{T} = 1$ Kelvin and C is 4.184×10^3 J kg^{-1}K^{-1}. The acceleration of gravity at the earth's surface is 9.80 m s^{-2}. Substituting and solving for Δh:

$$\Delta h = (4.184\times 10^3 \text{ J kg}^{-1}\text{K}^{-1})(1 \text{ K}) / 9.80 \text{ m s}^{-2}$$

$$\Delta h = 4.27\times 10^2 \text{ J kg}^{-1}\text{m}^{-1}\text{s}^2$$

By its definition a joule is equal to a kg m^{2}s^{-2}. This means that in the above cluster all units but meters cancel out. Δh is 427 m. The object must be dropped from a height of 427 meters.

---

3. The kinetic energy of the ball just before it strikes the mitt is $\frac{1}{2}mv^2$ where m is its mass and v is its velocity. *All* of this energy goes to heat up the ball at impact. Therefore:

$$\tfrac{1}{2}mv^2 = m\,C\Delta\mathrm{T}$$

where C is the specific heat of the ball. In this problem the velocity of the ball is 150 km hr^{-1} which converts to 41.7 m s^{-1}. C is 2.0 J g^{-1}K^{-1} or 2.0×10^3 J kg^{-1}K^{-1}. The m's cancel out and:

$$\Delta\mathrm{T} = \tfrac{1}{2}\, v^2 / C$$

$$\Delta\mathrm{T} = 0.434 \text{ m}^2\text{s}^{-2}\text{J}^{-1}\text{kg K}$$

Because a joule is a kg m^{2}s^{-2} all of the units except that of temperature cancel out. The temperature rise is 0.434 K.

The problem can also be solved by computing the kinetic energy numerically (using the mass of the baseball) and then determining how much this quantity of energy will raise the temperature of the ball. However, as the above solution shows, the mass of the ball is not needed.

---

5. Define the *system* under consideration to consist of two *sub-systems*, the 61.0 g of hot metal and the 800 g of cool water. If it is assumed that the system is well insulated then $q_{sys} = 0$ and:

$$q_{metal} + q_{water} = q_{sys} = 0$$

The amount of heat absorbed by a system (or subsystem) in a temperature change, ΔT is:

$$q = m\,C\,\Delta T$$

where m is the mass and C is the specific heat. For the water:

$$q_{water} = C_{water}\,m_{water}\,\Delta T_{water}$$

and for the metal:

$$q_{metal} = C_{metal}\,m_{metal}\,\Delta T_{metal}$$

The change in temperature of the water is 29.38 − 20.00 = 9.38° C which is the same as 9.38 K (a kelvin is the same size as a Celsius degree). The change in temperature of the metal is $T_2 - T_1$ or 29.38 − 600 = −570.62° C = −570.62 K. The specific heat capacity of the water is given as 4.184 J $g^{-1}K^{-1}$ and the masses of the water and metal are 800 g and 61.0 g:

$$(C_{metal} \times 61.0\text{ g} \times -570.62\text{ K}) + (4.184\text{ J g}^{-1}\text{K}^{-1} \times 800\text{ g} \times 9.38\text{ K}) = 0$$

$$C_{metal} = 0.902\text{ J g}^{-1}\text{K}^{-1}$$

The Law of DuLong and Petit states that the *molar* heat capacity of solids is about 25 J $mol^{-1}K^{-1}$. This allows an estimate of the molar weight of the metal (27.7 g mol^{-1}) and the guess that it is aluminum.

---

7. The 0.500 mol of neon expands against a constant pressure of 0.100 atm. Let the neon comprise the system. *Before* the expansion its volume is 11.207 L (calculated by applying the ideal gas equation to 0.500 mol of gas at STP). *After* the expansion the neon's volume is 43.08 L (calculated from the ideal gas equation with P = 0.200 atm, n = 0.500 mol and T = 210 K). The gas expands against a *constant* pressure (of 0.100 atm). Presumably the pressure is suddenly raised to its final value of 0.200 atm just after the gas reaches its final volume. Therefore the work done on the system is:

$$w = -P\Delta V = -0.100\text{ atm }(43.08 - 11.207)\text{L} = -3.19\text{ L atm}$$

The gas cools from 273.15 to 210 K. Since it is an ideal monoatomic gas, the change in its internal energy is directly proportional to the change in its temperature with the constant of proportionality being $n(^3/_2 R)$, the heat capacity at constant volume:

$$\Delta E = nC_v\Delta T = n(^3/_2 R)\,\Delta T$$

Substituting n = 0.500 mol, R = 0.08206 L atm $mol^{-1}K^{-1}$ and ΔT = −63.15 K gives:

$$\Delta E = -3.89\text{ L atm}$$

By the first law:

$$\Delta E = q + w \quad \textit{and} \quad q = \Delta E - w$$

This gives q, the heat absorbed by the neon, as $-3.89 - (-3.19) = -0.70$ L atm. The units of all three answers can be converted to joules (1 L atm = 101.325 J):

$$w = -323 \text{ J} \quad \Delta E = -394 \text{ J} \quad q = -71 \text{ J}$$

---

9. ***a)*** The statement of the problem gives the initial quantity (2.00 mol), pressure (3.00 atm), and temperature (350 K) of the ideal monoatomic gas. The initial *volume* of the gas is nRT / P or 19.15 L. The *final* volume is *twice* this original volume or 38.3 L. Evidently, the change in volume, ΔV, is $38.3 - 19.15 = 19.15$ L.

b) The adiabatic expansion occurs against a *constant* pressure of 1.00 atm. Under that circumstance, the work done on the gas is:

$$w = -P\Delta V = -1.00\ (19.15) \text{ L atm} = -1.94 \times 10^3 \text{ J}$$

Because the expansion is adiabatic $q = 0$, by definition, and:

$$\Delta E = q + w = 0 - 1.94 \times 10^3 \text{ J} = -1.94 \times 10^3 \text{ J}$$

Any change in the internal energy of an ideal gas is reflected in direct proportion by a change in temperature:

$$\Delta E = n\, C_v \Delta T = 2.00 \text{ mol } (^3/_2\ R)\, \Delta T$$

This gives $\Delta T = -77.8$ K. The *enthalpy* change of the gas is:

$$\Delta H = n\, C_p \Delta T = 2.00 \text{ mol } (^5/_2\ R)\, \Delta T$$

which comes out to -3.23×10^3 J. The enthalpy change is $^5/_3$ larger in magnitude than the change in internal energy; $^5/_3$ is the ratio of C_p to C_v.

c) T_2, the final temperature, is $T_1 + \Delta T$ or $350 - 77.8 = 272$ K.

The *final pressure*, P_2, of the gas can also be computed, using the ideal gas equation with n = 2.00 mol, T_2 = 272.2 K, V_2 = 38.3 L (and R = 0.08206 L atm $mol^{-1}K^{-1}$). It is 1.16 atm. It is *wrong* to apply the relationship (where $\gamma = C_p / C_v$):

$$P_1 V_1^{\gamma} = P_2 V_2^{\gamma}$$

to this problem. This attractive equation is valid only for *reversible* adiabatic processes. It predicts that P_2 will be 0.945 atm. Yet it is impossible for a gas to double its volume against an external pressure of 1.00 atm and then come to a final pressure that is *less* than 1.00 atm.

---

11. The system consists of the 6.00 mol of argon. The energy change of this monoatomic gas (assuming ideality) is:

$$\Delta E = n\, C_v \Delta T = (6.00 \text{ mol})(^3/_2\ 8.314 \text{ J mol}^{-1}\text{K}^{-1})(150 \text{ K}) = 11.2 \times 10^3 \text{ J}$$

The change is adiabatic; $q = 0$. From the first law:

$$w = \Delta E - q = 11.2\times 10^3 \text{ J} - 0 = 11.2\times 10^3 \text{ J}$$

The work done on the argon is 11.2×10^3 J, *all* of which goes to increase its internal energy.

---

13. The ice cube conveniently is 2.00 mol in extent. It is placed in 20.0 mol of 20° C water. Warming the ice cube from −10° C to 0° C would consume 38 J $mol^{-1}K^{-1} \times 2.00 \text{ mol} \times 10$ K or 760 J. Melting it would takes 6007 J mol^{-1} $\times$ 2.00 mol or 12014 J. On the other hand, cooling the 20.0 mol of 20° C water all the way to 0° C would require the extraction of 75 J $mol^{-1}K^{-1} \times 20.0 \text{ mol} \times$ 20 K = 30000 J. By warming up to its melting point and then melting, the ice cube only accommodates 12774 J. This proves that T_f, the final temperature, of the mixture, is above 0° C. No heat is lost to the surroundings, so:

$$760 \text{ J} + 12014 \text{ J} + 2.00 \text{ mol}(75 \text{ J mol}^{-1}\text{K}^{-1})(T_f - 0) = 20.0 \text{ mol}(75 \text{ J mol}^{-1}\text{K}^{-1})(20.0 - T_f)$$

$$T_f = 10.4° \text{ C}$$

---

15. ***a)*** The standard enthalpy change for the reaction:

$$N_2H_4\,(l) + 3\,O_2\,(g) \rightarrow 2\,NO_2\,(g) + 2\,H_2O(l)$$

is:

$$\Delta H° = \Sigma\,\Delta H_f° \text{ (products)} - \Sigma\,\Delta H_f° \text{ (reactants)}$$

From Appendix D, the $\Delta H_f°$ of liquid water is −285.83 kJ mol^{-1}. Since there are *two* moles of $H_2O(l)$ among the products, the $\Delta H_f°$ of *all* the water is −571.66 kJ. Similarly:

$$\Delta H_f° \text{ of 2 mol } NO_2\,(g) = 2\times 33.18 = 66.36 \text{ kJ}$$

$$\Delta H_f° \text{ of 1 mol } N_2H_4\,(l) = 1\times 50.63 = 50.63 \text{ kJ}$$

$$\Delta H_f° \text{ of 3 mol } O_2\,(g) = 3\times 0.0 = 0.0 \text{ kJ}$$

The enthalpy change of formation of an element in its standard state (such as $O_2\,(g)$) is 0.0 kJ. The standard enthalpy change of the reaction:

$$\Delta H° \text{ (reaction)} = (-571.66 + 66.36) - (50.63 + 0) = -555.93 \text{ kJ}$$

b) The standard *energy* change of the reaction is:

$$\Delta E° = \Delta H° - \Delta(PV) = \Delta H° - \Delta(nRT) = \Delta H° - \Delta n_g\,RT$$

Assume that the product and reactant gases behave ideally and that the volumes of the liquids are negligibly small. The temperature is 298.15 K, and Δn_g is $(2 - 3) = -1$ mol so:

$$\Delta E^\circ = -555.93 \text{ kJ} - (-1 \text{ mol})(8.314 \times 10^{-3} \text{ kJ mol}^{-1}\text{K}^{-1})(298.15 \text{ K})$$

$$\Delta E^\circ = -555.93 + 2.48 = -553.45 \text{ kJ}$$

The $\Delta(PV)$ term, the difference between ΔE° and ΔH°, is only about 0.44 percent of ΔH.

---

17. ***a)*** The balanced equation for the combustion of the naphthalene is:

$$C_{10}H_8\,(s) + 12\,O_2\,(g) \rightarrow 10\,CO_2\,(g) + 4\,H_2O(l)$$

b) The energy change of a reaction taking place in a system at constant volume equals the quantity of heat *absorbed* by the system; no pressure-volume work can be performed. The heat absorbed, q, is -25.79 kJ because $+25.79$ kJ is evolved. Hence, $\Delta E = -25.79$ kJ. This is ΔE° because the reaction occurs at 298.15 K, starts from reactants in their standard states, and gives products in their standard states. However, this energy change results from the combustion of only 0.6410 g of $C_{10}H_8$. 1.00 mol of naphthalene weighs 128.1753 g. The actual sample is only 0.005001 mol (about $^1/_{200}$ mol) so the molar ΔE° is almost exactly two hundred times larger than -25.79 kJ. It is -5157 kJ mol^{-1}.

c) Assuming that the gases in the equation in part a) are ideal and using the definition of ΔH°:

$$\Delta H^\circ = \Delta E^\circ + \Delta(PV) = \Delta E^\circ + \Delta n_g RT$$

The volumes of the solid and liquid are negligible. Δn_g is the number of moles of gas among the products minus the number of moles of gas among the reactants: $\Delta n_g = 10 - 12 = -2$. Also, $T = 298.15$ K and $R = 8.314 \times 10^{-3}$ kJ mol^{-1}K^{-1}:

$$\Delta H^\circ = \Delta E^\circ - 2\,RT = -5157 - 4.96 = -5162 \text{ kJ}$$

for a mole of $C_{10}H_8\,(s)$. ΔH° of this reaction is also the sum of the standard enthalpy changes of formation of the products minus the sum of the standard enthalpy changes of formation of the reactants. Using Appendix D:

$$\Delta H^\circ = -5162 \text{ kJ} = 4(-285.83 \text{ kJ}) + 10(-393.51 \text{ kJ}) - \Delta H_f^\circ(C_{10}H_8)$$

$$-5162 \text{ kJ} = -5078.42 \text{ kJ} - \Delta H_f^\circ(C_{10}H_8)$$

$$\Delta H_f^\circ(C_{10}H_8) = 83.5 \text{ kJ mol}^{-1}$$

The ΔH_f° is in kJ mol^{-1} because one mol of $C_{10}H_8\,(s)$ is consumed in the reaction.

---

19. The candy bar provides 14.3 g of glucose, $C_6H_{12}O_6$. The molar weight of glucose is 180.16 g mol^{-1} so the candy contains 0.0794 mol of glucose. As it is oxidized each mole of glucose releases 2820 kJ to the body of the woman. The glucose from the candy bar releases 0.0794 mol × 2820 kJ mol^{-1} = 223.8 kJ. If all of this energy goes to heat up the 50 kg body weight of the candy-eater, then:

$$223.8 \times 10^3 \text{ J} = (4.184 \text{ J g}^{-1}\text{K}^{-1})(50 \times 10^3 \text{ g})(\Delta T)$$

where ΔT is the temperature rise. The ΔT is 1.07 K. When rounded off to two significant digits this becomes 1.1 K. The un-rounded answer implies precision of 1 part in 107, the rounded answer implies precision of 1 part in 11. The woman's mass is known to 1 part in 50 (50±1 kg). Both answers are defensible.

---

21. ***a)*** Data from Appendix D are needed to calculate the $\Delta H°$ of the reaction:

$$CaCl_2\,(s) \rightarrow Ca^{2+}\,(aq) + 2\,Cl^-\,(aq)$$

$$\Delta H° = \underset{\text{for } Ca^{2+}}{-542.83} + \underset{\text{for 2 } Cl^-}{2(-167.16)} - \underset{\text{for 1 } CaCl_2}{(-795.8)} \text{ kJ}$$

$$\Delta H° = -81.4 \text{ kJ}$$

b) Retain an additional digit in the answer to part a) to avoid round-off errors. Then, 81.35 kJ is released as one mole of $CaCl_2\,(s)$ dissolves to give $Ca^{2+}\,(aq)$ in its standard state and $Cl^-\,(s)$ in its standard state. 20.0 g of $CaCl_2$ is (20.0 g / 110.98 g mol^{-1}) or 0.180 mol of $CaCl_2$. As it dissolves in 0.100 L of water, the $CaCl_2$ creates a solution that is approximately 1.80 M in Ca^{2+} and 3.6 M in $Cl^-\,(aq)$. Although these concentrations are not the standard states of the products, assume that they are close enough to the standard states to justify using $\Delta H°$. On that basis, 81.35 kJ mol^{-1} × 0.180 mol = 14.66 kJ of heat evolves from the reaction. This heat goes to raise the temperature of the water:

$$14.66 \times 10^3 \text{ J} = 4.184 \text{ J g}^{-1} \times 100 \text{ g} \times \Delta T$$

$$\Delta T = 35.0 \text{ K} \quad \textit{and} \quad T_{(final)} = 20.0 + 35.0 = 55.0°\text{C}$$

---

23. ***a)*** The reaction is the combustion of isobutene:

$$C_4H_8\,(g) + 6\,O_2\,(g) \rightarrow 4\,CO_2\,(g) + 4\,H_2O(g)$$

The enthalpy change of this combustion is the sum of the enthalpy changes of formation of the products less the sum of the enthalpy changes of formation of the reactants:

$$\Delta H = -2528 \text{ kJ} = 4(-393.51) + 4(-241.82) - (1 \text{ mol})\,\Delta H_f°(\text{isobutene})$$

The numerical values on the right-hand side come from Appendix D and the balanced equation. $\Delta H_f°$ is the only unknown in the equation. It equals −13 kJ mol^{-1}.

b) When one mole of isobutene burns in a flame nearly all of the enthalpy change goes to heat up the products (because the process is nearly adiabatic).

$$-\Delta H = C_p \Delta T$$

Imagine a system consisting of the 4 mol of CO_2 and 4 mol of H_2O formed by combustion of one mole of isobutene. The heat capacity of such a system is (4 mol × 37 J $mol^{-1}K^{-1}$) + (4 mol × 36 J $mol^{-1}K^{-1}$) = 292 J K^{-1}.

All the enthalpy change of the combustion feeds into this system. $-\Delta H$ is $+252.8 \times 10^3$ J. Hence:

$$2528 \times 10^3 \text{ J} = 292 \text{ J K}^{-1} \Delta T$$

and ΔT equals 8657 K. The maximum flame temperature is 8657 K plus about 298 K (assuming room temperature) or about 9000 K.

---

25. As any system undergoes an *isothermal* change there is no change of temperature ($\Delta T = 0$). Since this system consists of an ideal gas, both ΔE and ΔH equal 0; since that the internal energy of an ideal gas depends only on its temperature. Because the expansion is reversible:

$$w = -n RT \ln V_2/V_1 = -2.00 \text{ mol } (8.314 \text{ J mol}^{-1}\text{K}^{-1})(298 \text{ K}) \ln (36.00 / 9.00)$$

$$w = -6.87 \text{ kJ}$$

The first law provides that for the gas $\Delta E = q + w$. Therefore q is +6.87 kJ.

---

27. During any adiabatic process $q = 0$. During this *reversible* adiabatic expansion of an ideal gas:

$$T_1 V_1^{\gamma-1} = T_2 V_2^{\gamma-1}$$

where γ is C_p / C_v and the subscripts refer the initial and final states of the gas. In this problem, V_1 is 20.0 L, V_2 is 60.0 L, γ is $^5/_3$ and T_1 is 300 K. Solving for T_2 and substituting gives T_2 = 144.2 K. Rounding off to the correct number of significant digits gives T_2 = 144 K.

Meanwhile, the ΔE of the ideal gas depends solely upon its change in temperature:

$$\Delta E = nC_v\Delta T = (2.00 \text{ mol})(^3/_2 \times 8.314 \text{ J mol}^{-1}\text{K}^{-1})(-155.78 \text{ K}) = -3.88 \text{ kJ}$$

This number is also w, the work done on the gas, because $\Delta E = q + w$ and q is zero in this process. Finally, ΔH of the ideal gas also can depend only on the temperature change:

$$\Delta H = nC_p\Delta T = (2.00 \text{ mol})(^5/_2 \times 8.314 \text{ J mol}^{-1}\text{K}^{-1})(-155.78 \text{ K}) = -6.48 \text{ kJ}$$

Notice that $\Delta H = \gamma \Delta E$.

---

29. ***a)*** The maximum efficiency of an engine operating between two temperatures is obtained when the engine operates reversibly. This maximum efficiency is:

$$\epsilon = 1 - T_l / T_h$$

In this case T_l is 300 K and T_h is 450 K so ϵ is 0.333.

b) The efficiency of the engine is the ratio of the net work it *performs* to the heat that it *absorbs.* If 1500 J is absorbed per cycle from the 400 K reservoir and ϵ is 0.333 the engine discards 1000 J of heat per cycle. Therefore q is −1000 J

c) The engine has absorbed 1500 J of heat and must lose this number of joules by the time it completes the cycle of operation. 1000 J goes to the 300 K reservoir as heat. 500 J therefore appears as work done by the engine. That is, $w = -500$ J.

---

31. As the gas expands against the piston it does pressure-volume work. The pressure is constant so the work is $P\Delta V$ = 1 atm(13 − 5 L) = 8 L atm. This is the work *done by* the gas. Ignoring friction in the paddle mechanism, this work all goes to heating 1.00 L of water. Because 1 L atm is 101.325 J, 8 L atm is 810.6 J. The liter of water weighs 1000 g and has a specific heat of 4.184 J $g^{-1}K^{-1}$:

$$810.6 \text{ J} = 4.184 \text{ J g}^{-1}\text{K}^{-1} \times 1000 \text{ g} \times \Delta T$$

The temperature rise, ΔT, equals 0.194 K or, to one significant figure, 0.2 K.

---

33. ***a)*** The work done *on* the system is $-P\Delta V$. Since the gas is ideal and P is constant at (1 atm) $P\Delta V = n R\Delta T$ for a system defined as the argon gas. This is convenient because ΔT is known. It is 298 − 398 = −100 K. Then:

$$w(\text{on system}) = -n R\Delta T = -2.00 \text{ mol} \times 8.314 \text{ J mol}^{-1}\text{K}^{-1} (-100 \text{ K})$$

$$w = 1.66 \times 10^3 \text{ J}$$

b) The process goes on at constant pressure so the heat absorbed is q_p:

$$q_p = n C_p \Delta T = 2.00 \text{ mol} \times {}^5/_2 \times 8.314 \text{ J mol}^{-1}\text{K}^{-1} (-100 \text{ K}) = -4.16 \times 10^3 \text{ J}$$

c) The energy change of the system is the sum of the work done on the system and the heat it absorbs:

$$\Delta E = q + w = -4157 + 1663 = -2494 \text{ J}$$

This rounds off to -2.49×10^3 kJ. Note the use of un-rounded answers from parts a) and b) in the addition.

d) The ΔH of the system is q_p, which was computed in part b). $\Delta H = -4.16$ kJ.

---

35. The reaction:

$$2\ CH_4(g) + 2\ O_2(g) \rightarrow CH_2CO(g) + 3\ H_2O(g)$$

can be constructed as the sum of *twice* reaction No. 2 from the problem:

$$2\ CH_4(g) + 4\ O_2(g) \rightarrow 2\ CO_2(g) + 4\ H_2O(g)$$

and the *reverse* of reaction No. 1:

$$2\ CO_2(g) + H_2O(g) \rightarrow CH_2CO(g) + 2\ O_2(g)$$

By Hess's law, the enthalpy change for the reaction in question is:

$$\Delta H_{reaction} = 2\,\Delta H_2 + (-\Delta H_1)$$

$$\Delta H_{reaction} = 2(-802.3\ kJ\ mol^{-1}) + (981.1\ kJ\ mol^{-1})$$

$$\Delta H_{reaction} = -623.5\ kJ\ mol^{-1}$$

Thus, for every mole of $CH_2CO(g)$ formed from 2 $O_2(g)$ and 2 $CH_4(g)$, 623.5 kJ is evolved.

---

37. ***a)*** The equation for the combustion of benzene in air is:

$$C_6H_6(l) + 7\tfrac{1}{2}\ O_2(g) \rightarrow 6\ CO_2(g) + 3\ H_2O(l)$$

b) The ΔH° of the reaction is 6 times the molar ΔH°_f of $CO_2(g)$ plus 3 times the molar ΔH°_f of $H_2O(l)$ minus 1 times the molar ΔH°_f of benzene:

$$\Delta H^\circ = 6(-393.51\ kJ) + 3(-285.83\ kJ) - 1(49.03\ kJ) = -3267.58\ kJ$$

c) The ΔE° of this reaction is ΔH° corrected by subtraction of the PV term:

$$\Delta E^\circ = \Delta H^\circ - \Delta(PV) = \Delta H^\circ - \Delta(nRT) = \Delta H^\circ - \Delta n_g RT$$

where it is assumed that the gases involved in the reaction are ideal and the volumes of the liquids are negligible. The change in the number of moles of gas, Δn_g, is $6 - 7\tfrac{1}{2} = -1\tfrac{1}{2}$ mol. Therefore:

$$\Delta E^\circ = -3267.58\ kJ - (-1\tfrac{1}{2})(8.314 \times 10^{-3}\ kJ\ mol^{-1}K^{-1})(298.15\ K)$$

$$\Delta E^\circ = -3263.90\ kJ$$

---

39. ***a)*** The combustion of the liquid fuel isooctane is represented by the equation:

$$C_8H_{18}(l) + 12\tfrac{1}{2}\ O_2(g) \rightarrow 8\ CO_2(g) + 9\ H_2O(l)$$

b) The combustion of 0.542 g of isooctane must be exothermic because isooctane is a fuel. The heat released by the combustion goes to raise the temperature of the calorimeter vessel and of the water inside it:

$$q\text{(released)} = q\text{(absorbed by water)} + q\text{(absorbed by calorimeter)}$$

The calorimeter has a ΔT of 28.670° C − 20.450° C = 8.220° C. A Celsius degree is the same size as a K so ΔT is also 8.220 K. Because its heat capacity is 48 J K^{-1}, the calorimeter absorbs 8.22 K × 48 J K^{-1} = 394.6 J. The water has the same ΔT and its specific heat is 4.184 J $g^{-1}K^{-1}$; it absorbs 4.184 J $g^{-1}K^{-1}$ × 750 g × 8.22 K = 2.58×10^4 J. The heat *released* by the reaction is $2.58 \times 10^4 + 0.03946 \times 10^4 = 2.62 \times 10^4$ J. The heat *absorbed* by the reaction is -2.62×10^4 J. No work is accomplished either by or upon the reaction because it goes on at constant volume. Hence:

$$\Delta E = q_v + 0 = -2.62 \times 10^4 \text{ J}$$

c) The molar weight of C_8H_{18} is 114.23 g mol^{-1}, making the 0.542 g of isooctane that is actually burned far less than one mole. An entire mole of isooctane would *absorb* (114.23 / 0.542) × (-2.62×10^4) J = -5.52×10^6 J = −5520 kJ. ΔE is −5520 kJ mol^{-1}.

d) The ΔH of combustion is related to ΔE:

$$\Delta H = \Delta E + \Delta(PV)$$

Assuming the gases are ideal and the liquids have nearly zero volume, then $\Delta(PV) = \Delta n_g RT$, where Δn_g is the change in the number of moles of gas during the course of the reaction. In this reaction $\Delta n_g = 8 - 12.5 = -4.5$ mol. Therefore:

$$\Delta(PV) = -4.5 \text{ mol}(8.314 \text{ J mol}^{-1})(298.15 \text{ K}) = -11.15 \text{ kJ}$$

The temperature is essentially constant during the reaction and is approximately 298 K. Any error that this approximation introduces is less than the uncertainty in the calorimetric measurements. Therefore:

$$\Delta H = \Delta E - \Delta(PV) = -5520 \text{ kJ} - 11.1 \text{ kJ} = -5531 \text{ kJ.}$$

e) The standard enthalpy change of the reaction as written above is:

$$\Delta H^\circ_{\text{(reaction)}} = 8\,\Delta H^\circ_f\,(CO_2\,(g)) + 9\,\Delta H^\circ_f\,(H_2O(l)) - \Delta H^\circ_f\text{(isooctane)}$$

This follows from Hess's law. The ΔH°_f's of the products are −393.51 kJ mol^{-1} and −285.83 kJ mol^{-1} respectively. ΔH° of the reaction is −5531 kJ so:

$$-5531 \text{ kJ} = 8(-393.51) \text{ kJ} + 9(-285.83) \text{ kJ} - \Delta H^\circ_f\text{ (isooctane)}$$

$$\Delta H^\circ_f\text{(isooctane)} = -190 \text{ kJ mol}^{-1}$$

---

41. Roasting zinc sulfide produces zinc oxide. The ΔH° of the roasting reaction:

$$2\,ZnS(s) + 3\,O_2\,(g) \rightarrow 2\,ZnO(s) + 2\,SO_2\,(g)$$

is readily computed from the data in Appendix D using Hess's Law:

$$\Delta H^\circ = 2\,\Delta H^\circ_f\,(ZnO) + 2\,\Delta H^\circ_f\,(SO_2) - 2\,\Delta H^\circ_f\,(ZnS)$$

where it is recognized that ΔH°_f of O_2 *(g)* is zero. Using the Appendix:

$$\Delta H^\circ = 2(-348.28) + 2(-296.83) - 2(-205.98)\text{ kJ}$$
$$\Delta H^\circ = -878.26\text{ kJ}$$

This result means that 878.26 kJ of heat is *released* when 2 mol of sphalerite is roasted at constant pressure according to the above equation. If 3.00 metric tons (3.00×10^3 kg) is roasted, much more heat is released. 3.00 metric ton of ZnS is 3.00×10^6 g of ZnS. The molar weight of ZnS is 97.4 g mol^{-1}. Therefore, the 3.00 metric tons is 3.08×10^4 mol of ZnS. Every *two* moles releases 878.26 kJ so the total heat released is:

$$-q = 3.08 \times 10^4\text{ mol} \times (878.26\text{ kJ / 2 mol}) = 1.35 \times 10^7\text{ kJ}$$

---

43. Define the system, which is a closed system, as the contents of the auto engine cylinder. Before the explosive combustion of the *n*-octane the temperature is 600 K, the volume is 0.150 L and the pressure is 12.0 atm. Also, the cylinder holds *n*-octane and air in a 1 to 80 molar ratio, i.e.: $80\, n_{octane} = n_{air}$. By applying the ideal gas equation to the contents of the cylinder before the combustion:

$$n_{octane} + n_{air} = 0.03656\text{ mol}$$

Solving the two simultaneous equations gives:

$$n_{octane} = 4.514 \times 10^{-4}\text{ mol} \quad \textit{and} \quad n_{air} = 3.611 \times 10^{-2}\text{ mol}$$

The system does not change in volume during the combustion of the fuel, so w is zero. Furthermore, q is zero (the combustion happens so fast that there is no time for heat to be lost or gained). Since w and q both equal zero the first law assures that ΔE of the system equals zero during the process. Imagine the combustion to occur in two stages: *step a*, the reaction going on at a constant temperature of 600 K; *step b*, the product gases heating up at constant volume. The sum of these two changes is the overall change that takes place within the cylinder. Therefore:

$$\Delta E_{sys} = 0 = \Delta E_a + \Delta E_b$$

$$\Delta E_a = -\Delta E_b$$

The problem offers data pertaining to enthalpy changes, not energy changes, in the two steps. We therefore substitute for the ΔE's in terms of ΔH's:

$$\Delta H_a - \Delta(PV)_a = -\Delta H_b + \Delta(PV)_b$$

Step a involves ideal gases, takes place at a constant temperature, and changes the chemical amount of gas. Therefore $\Delta(PV)_a$ equals Δn_g RT, where Δn_g is the change in the chemical amount of gases during the reaction. Step b is the heating of the ideal gases inside the cylinder after the first step. $\Delta(PV)_b$ therefore equals nR ΔT where n is the chemical amount of gases present *after* the reaction. Also, for any change in temperature, ΔH is equal to $nC_p\Delta T$, as long as the heat capacity C_p is independent of temperature. Substitution of these relations gives:

$$\Delta H_a - \Delta n_g RT = -n\, C_p\Delta T + n\, R\Delta T$$

In this equation T is 600 K, and ΔT is the temperature change during the heating. The approach will be to compute all the other quantities and then use this equation to get ΔT and thence the final temperature.

The fuel burns according to the balanced equation:

$$C_8H_{18}\,(g) + 12\tfrac{1}{2}\,O_2\,(g) \rightarrow 8\,CO_2\,(g) + 9\,H_2O(g)$$

Air is 80 percent N_2 and 20 percent O_2 on a molar basis. From this fact, one calculates chemical amounts of O_2, N_2 and *n*-octane within the cylinder before reaction. From the stoichiometry one computes the chemical amounts of all the different gases in the cylinder after the reaction:

$$n_{CO_2} = 8\times(4.514\times10^{-4}\text{ mol}) = 3.611\times10^{-3}\text{ mol}$$

$$n_{H_2O} = 9\times(4.514\times10^{-4}\text{ mol}) = 4.063\times10^{-3}\text{ mol}$$

$$n_{N_2} = 0.80\times(3.6109\times10^{-2}) = 2.889\times10^{-2}\text{ mol}$$

$$n_{O_2} = 0.20\times(3.6109\times10^{-2}) - 12.5\times(4.514\times10^{-4}) = 1.579\times10^{-3}\text{ mol}$$

These calculations establish that n, the chemical amount of gases in the cylinder after the reaction, is 0.03814 mol, the sum of all the chemical amounts of all the different gases. The original chemical amount of gases in the system is 0.03656 mol, so Δn_g is equal to 1.583×10^{-3} mol.

n-Octane's enthalpy change of *combustion* is 9 (ΔH°_f of $H_2O(g)$) + 8 (ΔH°_f $CO_2\,(g)$) − ΔH°_f(*n*-octane) which is 9(−241.8 kJ mol^{-1}) + 8(−393.5 kJ mol^{-1}) − (−57.4 kJ mol^{-1}) = −5266.8 kJ. This is *not* the enthalpy change of the reaction inside the cylinder because only 4.514×10^{-4} mol of *n*-octane is confined there. Instead, ΔH_a is −2.377 kJ, the product of the enthalpy change per mole and the number of moles of *n*-octane.

For the mixture in the cylinder after the reaction the overall nC_p is the sum of the individual nC_p values for the four product gases: $nC_p = (1.579\times10^{-3})35.2 + (2.889\times10^{-2})29.8 + (4.063\times10^{-3})38.9 + (3.611\times10^{-3})45.5 = 1.24\text{ J K}^{-1}$

Now, there is enough information to calculate ΔT. Substitute in the equation derived previously:

$$-2377 - (1.583\times10^{-3})(8.314)(600) = -1.24\,\Delta T + (0.03814)(8.314)\Delta T$$

All energies are in joules, temperatures in K and chemical amounts in moles. Completion of the arithmetic gives ΔT equal to 2610 K. The maximum temperature inside the cylinder is therefore 600 + 2610 or 3210 K. This is 2940° C.

---

45. *a)* The gases trapped inside the cylinder of the motorcycle's engine have volume V_1 when the piston is fully withdrawn and a smaller volume V_2 when the piston reaches the farthest limit of its compression stroke. The compression ratio is 8:1 so $V_1 = 8\ V_2$.

The area of the base of the engine's cylinder is πr^2, where r is the radius of the base. The volume of a cylinder is the area of its base times its height, h:

$$V_1 = A\,h \qquad V_2 = A\,(h - 12\text{ cm})$$

which employs the fact that full compression shortens h by 12.00 cm. Because r is 5.00 cm A is 78.54 cm². Substituting for V_1 and V_2 in terms of A and h:

$$A\,h = 8\ A(h - 12\text{ cm})$$

gives $h = 13.714$ cm; $V_1 = 1.077$ L, and $V_2 = 0.1347$ L.

The temperature and pressure of the fuel mixture are 353 K (80° C) and 1.00 atm when it enters the cylinder with fully withdrawn piston (V_1). Assuming the fuel mixture is an ideal gas:

$$n_{\text{mixture}} = (1.00\text{ atm})(1.077\text{ L}) \,/\, (0.08206\text{ L atm mol}^{-1}\text{K}^{-1})(353\text{ K}) = 0.0372\text{ mol}$$

The molar ratio of air to fuel (C_8H_{18}) is 62.5 to 1. If n_{fuel} is the chemical amount of the fuel and n_{air} is the chemical amount of air then:

$$n_{\text{fuel}} + n_{\text{air}} = 0.0372\text{ mol} \qquad \textit{and} \qquad n_{\text{air}} = 62.5\ n_{\text{fuel}}$$

Solving these two simultaneous equations establishes that at the start the cylinder contains 0.0366 mol of air and 5.86×10^{-4} mol of *n*-octane fuel.

During the compression stroke this system undergoes an irreversible adiabatic compression to one-eighth of its initial volume. None of the relationships that govern *reversible* adiabatic processes strictly applies here. Assume however that the compression is near to reversible. If it is, then:

$$T_1 V_1^{\gamma - 1} \simeq T_2 V_2^{\gamma - 1}$$

where γ equals 35 J mol⁻¹K⁻¹ divided by 26.7 J mol⁻¹K⁻¹ or 1.31. Assuming reversibility, the temperature after the compression stroke is 673 K.

b) The compressed gases occupy a volume of 0.135 L just before they are ignited, as calculated above.

c) The pressure of the compressed fuel mixture just before ignition is P_2. It is computed applying the ideal gas equation to the system with $T_2 = 673$ K, $V_2 = 0.1347$ L and $n = 0.0372$ mol and comes out to be 15.3 atm. Alternatively, one can compute this pressure using the formula for a reversible adiabatic change:

$$P_1 V_1^{\gamma} = P_2 V_2^{\gamma}$$

Use of the formula emphasizes that the value 15.3 atm depends on the assumption of a *reversible* adiabatic compression stroke.

d) For the combustion of *n*-octane ΔH is -5266.8 kJ mol^{-1} (see problem 43). The combustion mixture inside the cylinder contains 5.86×10^{-4} mol of *n*-octane. Consequently, the ΔH of the combustion in this system is -3.09 kJ.

After the combustion the cylinder contains CO_2, H_2O and unreacted O_2 and N_2. Referring to the balanced chemical equation:

$$C_8H_{18}(g) + 12\tfrac{1}{2}\, O_2(g) \rightarrow 8\, CO_2(g) + 9\, H_2O(g)$$

it is clear that the combustion consumes 5.86×10^{-4} mol of octane and $12\tfrac{1}{2} \times 5.86 \times 10^{-4}$ mol of O_2 to produce $8 \times 5.86 \times 10^{-4}$ mol of CO_2 and $9 \times 5.86 \times 10^{-4}$ mol of H_2O. The effect of the reaction is to increase the chemical amount of gases in the cylinder by $3\tfrac{1}{2} \times 5.86 \times 10^{-4}$ mol. This is Δn_g for the reaction. The original quantity of gas is 0.0372 mol; after the combustion there is 0.0393 mol.

The *energy* (not enthalpy) released from the reaction all goes to heat up the gaseous contents of the cylinder. This must be true because we assume none is lost to the cylinder walls as heat and none exits the cylinder as work until the power stroke starts. Therefore:

$$\Delta H_{react} - \Delta n_g\, RT = -n\, C_p \Delta T + n\, R\Delta T$$

In this equation, which is derived in more detail in problem 43, every quantity but ΔT is known: ΔH_{react} is -3090 J, T is 673 K, n is 0.0392 mol, Δn_g is 2.051×10^{-3} mol, and C_p is approximately 35 J mol^{-1}K^{-1}. Completion of the arithmetic shows the temperature inside the cylinder rises by 2950 K to a maximum of 3620 K.

e) Assume that the expansion stroke is not only adiabatic but reversible. Then the formula:

$$T_1 V_1^{\gamma - 1} = T_2 V_2^{\gamma - 1}$$

applies. In this case, T_1 is 3616 K. The ratio V_1 / V_2 is 1 to 8 because now the initial state is the *small* volume state just before the expansion stroke of the piston. $\gamma - 1$ is 0.31, assuming that C_p is 35 J mol^{-1}K^{-1}. Substitution gives T_2 of 1900 K. This is the temperature of the exhaust gases.

---

47. A refrigerator is a machine that consumes work to create a temperature difference. This is the reverse of a heat engine, which operates across a temperature difference to create work. The maximum heat absorbed from the interior of a refrigerator, *q* is:

$$q = w_{net}\, T_l / (T_h - T_l)$$

where T_l and T_h are the low and high temperatures. In this case T_l is 263.15 K and T_h is 293.15 K. This amount of heat is actually absorbed only if the refrigerator operates at 100 percent thermodynamic efficiency, that is, reversibly. The *q* is to be 800 J per cycle in this problem and *w* is 91.2 J for each cycle. The total heat discharged per cycle into the room (at 293.15 K) is 800 + 91.2 = 891.2 J.

---

49. A refrigerator is placed in a closed room, plugged in and left with its door open. Imagine the system to be the closed room. If it is closed, then the room is probably fairly well insulated. Electrical energy is flowing into the room yet q and w across its walls are close to zero. Therefore the internal energy, E, of the room increases and the temperature of the room also must increase.

It is true that a region of the room just at the open refrigerator door will get cooler. But the refrigerator is pouring heat into the room from the coils at its rear. Once the room comes to thermal equilibrium its temperature will be higher than before the refrigerator was plugged in.

Chapter 8

SPONTANEOUS CHANGE AND EQUILIBRIUM

8.1 *ENTROPY, SPONTANEOUS CHANGE, AND THE SECOND LAW*

The *entropy*, S, of a system is a state function just as the energy, pressure, volume and temperature are. It is associated with the *disorder* or randomness of the system.

A *spontaneous* process occurs without outside intervention. The driving force for a spontaneous process is an increase in the entropy of the universe. During any process, the entropy change of the universe is the sum of the entropy changes of the system under consideration and its surroundings:

$$\Delta S_{univ} = \Delta S_{sys} + \Delta S_{surr}$$

Either of the terms on the right may be negative in a spontaneous process. The *sum* of the terms, the total entropy change of the universe is always positive in a spontaneous change. **Problem 35,** which concerns what happens when a piece of hot iron is plunged into cold water, illustrates this point. The iron spontaneously cools. Its entropy *decreases.* The increase in the entropy of the surroundings of the iron (the water) more than compensates.

To compute a system's entropy difference, ΔS_{sys}, between two states, it is necessary to imagine one *reversible* path connecting them. The path can be as idealized as necessary (or desired) because it is used only for computational purposes. As the change proceeds along the imagined reversible path, the system absorbs heat from its surroundings. ΔS is the sum of each little quantity of heat absorbed divided by the temperature at which is it absorbed:

$$\Delta S = \sum_i q_i / T_i$$

$$\Delta S = \int dq_{rev} / T$$

As the system gives off heat then it "absorbs" negative heat; q may be either positive or negative. The T in the equation is the absolute temperature and *never* is negative. This means that for a system losing heat on the reversible path from the first state to the second the entropy change is negative.

The ΔS of a system under change may be positive, negative, or, in the case that q_{rev} is zero, equal to zero. This last case is important enough to merit a special name: an *adiabatic* reversible change.

The same net change between two states may be achieved in an infinite number of ways. Every different way may have a different total q. The above calculational procedure gives correct values for the entropy change only if a reversible path is chosen.

Mapping out a reversible path may seem difficult, especially if the final temperature differs from the original. The best tactic is to break down the total change into a combination of steps that occur at constant pressure (*isobarically*), at constant volume (*isochorically*), and at constant temperature (*isothermally*). Such an approach works to solve **problem 33b.**

Calculation of Entropy Changes

The tactic of imagining the change going in a series of steps works well because simple formulas allow one to calculate entropy changes along isochoric, isobaric and isothermal path segments. Here is a summary of formulas for computing ΔS under various circumstances:

- *Reversible absorption of heat at constant temperature.* Direct application of the formula that defines entropy change is relatively easy because T is a constant:

$$\Delta S = q_{rev} / T$$

This situation arises when a system is exchanging work with its surroundings as heat flows in or out. Also, when substances freeze melt, boil or condense they stay at the same temperature while absorbing or emitting heat. For example, as water boils it absorbs heat at a constant temperature equal to its boiling point. This heat equals its ΔH of vaporization. The entropy change of vaporization therefore is:

$$\Delta S_{vap} = \Delta H_{vap} / T_b$$

This fact is used in **problem 11.**

- *Reversible heating (cooling) at constant pressure.*

$$\Delta S = n\, C_p \ln (T_2 / T_1)$$

Whenever a systems cools, its entropy diminishes along with its temperature. The entropy change of the system is negative if its temperature drops and positive if it rises. This formula is used in **problem 5.** To use it one either assumes that C_p is not a function of temperature or gets information on how C_p varies with temperature. **Problem 39** introduces a case in which C_p *does* depend on temperature.

- *Reversible heating (cooling) at constant volume.*

$$\Delta S = n\, C_v \ln (T_2 / T_1)$$

This formula is the same as the previous one except that at constant volume C_v appears instead of C_p.

- *Reversible expansion (compression) at constant temperature.* The entropy of a system always increases with its size as long as the temperature does not change. The entropy change upon expansion *of an ideal gas*: is:

$$\Delta S = n\, R \ln (V_2 / V_1)$$

The formula:

$$\Delta S = n\, R \ln (P_1 / P_2)$$

is a obvious consequence (by combination with the ideal gas law). These two formulas are valid only if the system is an ideal gas. Fortunately, systems which are solids or liquids often experience only small volume changes. If ΔV is small, then $V_2 / V_1 \simeq 1$ and $\ln (V_2 / V_1) \simeq 0$.

Even the most complex process, one involving changes in T, V and P, can be imagined to proceed along a path involving a change first with T, then with P, and finally with V constant. Coming up with such paths is a common theme in problems involving the entropy function. It is particularly well illustrated in **problem 31.** When a path is finally sketched, the above formulas allow the computation of the change in entropy. The following is another example.

Example. 10.00 L of an ideal monatomic gas at 10.00 atm and 273.15 K is expanded and cooled to a final volume of 63.93 L and a final temperature of 174.8 K. Compute the entropy change of the gas.

Solution. During the process all of the usual state variables change values. Applying the ideal gas equation to the system in its original state shows that there is 4.457 mol of gas. None of the gas escapes during the change, so there is 4.457 mol of it in the final state. The ideal gas equation also applies to the final state. Using it, the final pressure is 1.000 atm. In summary: $P_1 = 10.0$ atm, $P_2 = 1.00$ atm; $V_1 = 10.00$ L, $V_2 = 69.93$ L; $T_1 = 273.15$ K, $T_2 = 174.8$ K; n = 4.457 mol.

One reversible path connecting the two states is *first*: to cool the gas reversibly at constant pressure from its initial to its final temperature; *second*: to expand the gas reversibly at constant temperature from its intermediate volume (the volume attained after the cooling step) to its final volume.

After the isobaric cooling, the first segment of the path (step I), the volume of the gas has *dropped* to 6.393 L. The entropy change is:

$$\Delta S_I = n\, C_p \ln (T_2 / T_1)$$

$$\Delta S_I = 4.457 \text{ mol} \times (^5/_2)\ 8.314 \text{ J K}^{-1}\text{mol}^{-1} \ln (174.8 / 273.15)$$

$$\Delta S_I = -41.35 \text{ J K}^{-1}$$

In the second segment of the reversible path, the gas expands isothermally (at T = 174.8 K) from 6.393 L to 63.929 L, and the entropy change is:

$$\Delta S_{II} = n\, R \ln (V_2 / V_1)$$

$$\Delta S_{II} = 4.457 \text{ mol} \times 8.314 \text{ J K}^{-1}\text{mol}^{-1} \times \ln (63.929 / 6.393)$$

$$\Delta S_{II} = 85.32 \text{ J K}^{-1}$$

The entropy change for the process is $85.32 - 41.35 = 43.97$ J K^{-1}.

Many additional reversible paths connect the final and initial states. The problem could also have been solved by imagining first an isothermal expansion from 10.0 to 63.93 L followed by an isochoric cooling from 273.15 to 174.8 K.

Adiabatic Does Not Mean Isentropic

In changes that are *both* adiabatic *and* reversible $\Delta S = 0$. Such changes are *isentropic.* It is a common error to suppose that all adiabatic processes are isentropic. Suppose that a real adiabatic process occurs. Like all real changes, it is irreversible. After sizing up the final and initial states it will turn out to be impossible to imagine a reversible path achieving the same change unless the path has a *non*-adiabatic segment.

Irreversible Processes

Reversibility is an idealization. All real processes are more or less *irreversible.* If a comparison is made between an idealized, reversible path connecting two states and a real, irreversible path then:

$$q_{irrev} < q_{rev} \quad and \quad -w_{irrev} < -w_{rev}$$

Suppose that as a practical matter one wishes to get work out of a system. During the process the maximum work is extracted (largest $-w$) only if one proceeds reversibly. The actual work performed during the process will be less than this maximum. Similarly, the system will absorb the maximum heat only if the process goes on reversibly. The heat absorbed in a real, irreversible process will be less than this maximum. **Problem 35** shows how an irreversible process becomes more nearly reversible as the method by which it is performed is changed. The process in that problem approaches reversibility as an unattainable limit. Reversibility is always an unattainable ideal.

The Second Law

In many applications a paramount question is whether a proposed change can occur. The entropy function is pivotal in answering this question for chemical (and all other) processes. The second law of thermodynamics states:

- In a reversible process the total entropy of a system plus its surroundings is unchanged.

- In an irreversible process the total entropy of a system plus its surroundings increases.

- *In all human experience, processes for which ΔS_{univ} is negative are impossible.* That is, in real processes:

$$\Delta S_{univ} = \Delta S_{sys} + \Delta S_{surr} > 0$$

To see if a process can occur, calculate the entropy change of the system. Then calculate the entropy change of the surroundings (regarding them momentarily as a system). If the sum of these two answers exceeds zero, then the change can occur.

The text proves, by generalization from the Carnot cycle, that the entropy is a state function. ΔS for a Carnot cycle is zero. Any other cyclic process can be divided into path segments that are portions of Carnot cycles and thus modelled as a sum of many Carnot cycles. See text Figure 8-2. The ΔS of any cyclic process is, therefore, zero.

8.2 *ENTROPY AND IRREVERSIBILITY: A STATISTICAL INTERPRETATION*

The entropy is a measure of the randomness of a system. Spontaneous change occurs from states of low probability (ordered states) to states of high probability (disordered states). Consider the expansion of an ideal gas into a vacuum. Originally, the gas is confined behind a closed stopcock in the left-hand bulb (of volume V) of a two-bulb container. The right-hand bulb (of identical volume V) is empty. The stopcock is opened. The gas now has the volume $2V$. Everybody knows that the gas expands. But why? The energy of the ideal gas does not change during the expansion. The gas absorbs neither work nor heat. As far as the *first* law of thermodynamics is concerned the ideal gas could just as well stay in the left-hand bulb.

To appreciate why the gas expands, suppose for simplicity that there are only six molecules of the gas. How can the six molecules be distributed between the two sides?

Number of Molecules on Left	*Number of Molecules on Right*	*Number of Ways Attainable*
6	0	1
5	1	6
4	2	15
3	3	20
2	4	15
1	5	6
0	6	1

There are 64 ways to distribute the six molecules between the two sides. This is the sum of the numbers in the last column of the table. These give only 7 distinguishable states of the gas, corresponding to the 7 lines in the table. *The state that is most likely to occur is the one* which can be achieved in the *greatest number of ways.*

The chance of finding all 6 molecules on the left is only $^{1}/_{64}$. The chance of finding 3 on the left and 3 on the right is $^{20}/_{64}$. The latter state is 20 times more probable than the former The gas is more likely spontaneously to fill the entire volume evenly than to stay on the left. For a large number of molecules there is a huge number of ways to achieve an even distribution and far fewer ways to achieve a state with all the molecules on the left or right. **See problem 9** for a calculation.

Entropy

Nature proceeds spontaneously toward states that have the highest probability of occurring. These states have higher entropies. Entropy is closely associated with probability. It depends on the number of arrangements available to a system existing in a given state. Such arrangements are the system's *microstates.*

A microstate is a particular distribution of molecules among the positions and momenta accessible to them; each microstate is equally likely to be occupied by the system. The number of microstates of a system is symbolized Ω.

The entropy of a system depends on the number of its microstates according to the equation:

$$S = k_B \ln \Omega$$

where k_B is Boltzmann's constant. In problems, actual numerical values for Ω are rarely attainable. Fortunately, ΔS values can be calculated even if values for S are not available. The *change* in the entropy of a system between two states depends on the *ratio* of the number of microstates:

$$\Delta S = k_B \ln (\Omega_2 / \Omega_1)$$

If state 2 of the 6 molecule gas is the state with 3 molecules on the left and three on the right and state 1 is the state with all 6 molecules on the left, then Ω_2 / Ω_1 is 20 / 1 and the difference in entropy between the states is:

$$\Delta S = k_B \ln 20 = 1.381 \times 10^{-23} \text{ J K}^{-1} (2.996) = 4.137 \times 10^{-23} \text{ J K}^{-1}$$

This approach is used in solving **problems 9 and 37a.**

Entropy and Disorder

The greater the disorder of a system, then the greater its entropy. This disorder should not be conceived of in terms of a motionless array of atoms or molecules, but rather in terms of all the possible motions of the particles and the ways in which their arrangements change with time. These are the system's microstates. Their number depends on factors like the number of atoms per molecule and the strength of the bonds between atoms. As a result different substances have different entropies even if both are solids (or liquids or gases) at the same temperature. Some helpful generalizations:

- The entropy of substances increases when they melt. In solids every atom or molecule is at or near a prescribed position. In liquids the particles may move around more.

- Gases have more entropy than their corresponding liquids. There are more microstates for a gas than a liquid. Many more arrangements of atoms or molecules correspond to the same observed state.

- Gases at low pressure have greater entropy than at high pressure.

- A dilute solution of a given quantity of a substance has greater entropy than a concentrated solution.

- When one substance dissolves in another, ΔS of the system usually is positive.

- The entropy of a system always increases with increasing temperature.

8.3 ABSOLUTE ENTROPIES AND THE THIRD LAW

Unlike energy and enthalpy, for which there is no natural zero point, the entropy function *does* have a theoretical zero. The third law of thermodynamics provides this zero for the measurement of entropies. The third law states:

In any thermodynamic process involving only pure phases in their equilibrium states the entropy change, ΔS, approaches zero as T approaches 0 K.

This experimental law means that the entropy of any pure substance in its equilibrium state approaches zero as T approaches 0 K. It provides a natural reference for the tabulation of entropies. See Appendix D.

Standard-State Entropies

Appendix D gives the absolute entropies at 1 atm pressure and 298.15 K for a variety of substances. The $S°$ values (note the absence of a Δ) of the pure solids, liquids and gases are all positive because they are all referred to a zero entropy at 0 K. The values are experimental. They were obtained by measuring C_p as a function of temperature for each substance. Then the increase from 0 K of the entropy of each substance was computed using a method similar to the one in **problem 39.** The solution phase entropies in Appendix D are not third law, absolute entropies. They are instead measured relative to an arbitrary standard. Some of them are negative.

In a chemical reaction the entropy change is the sum of the standard entropies of the products minus the sum of the standard entropies of the reactants:

$$\Delta S° = \Sigma\, S°_{\text{products}} - \Sigma\, S°_{\text{reactants}}$$

The use of this equation to get $\Delta S°$ values is like the use of $\Delta H°_f$'s (tabulated in Appendix D) to get $\Delta H°$'s of reaction. Remember these points:

- The tabulated $S°$ values are *molar* entropies. If a reaction forms, say, 2 mol of NH_3 *(g)* (for which $S° = 192.3$ J $K^{-1}mol^{-1}$), the contribution of the ammonia to the entropy of the products is 384.6 J K^{-1}. If it forms 3 mol of NH_3 *(g)* then ammonia's contribution is 576.9 J K^{-1}. **See problem 17.**

- Standard entropies are tabulated in *joules* per Kelvin per mole (J $K^{-1}mol^{-1}$). In contrast, standard enthalpy changes of formation are tabulated in *kilo*joules per mole (kJ mol^{-1}). These are entirely different units. The similar numerical values in the $S°$ and $\Delta H°_f$ columns in Appendix D is nothing but an accident.

- The $\Delta S°$ of a reaction may be positive, negative or even zero. A negative $\Delta S°$ does *not* mean that a reaction is forbidden to occur by the second law of thermodynamics. For example, in **problem 17a** the $\Delta S°$ for the rusting of iron at room conditions comes out to about *minus* 0.550 J K^{-1}. Iron still rusts.

8.4 *FREE ENERGY*

So far, predicting whether a given process can occur requires the calculation of ΔS_{univ}, the sum of ΔS of the surroundings and ΔS of the system. The universe is big and complex, and getting its ΔS for every proposed process is hard. It is possible, at a price, to avoid figuring ΔS_{univ} for every process. The method requires the definition of a new function, the *Gibbs free energy*, G:

$$G = H - TS$$

G is a state function and for any change in a system:

$$\Delta G_{sys} = \Delta H_{sys} - \Delta(TS)_{sys}$$

When T does not change it may be placed in front of the Δ:

$$\Delta G = \Delta H - T\,\Delta S \qquad \text{(T constant)}$$

Under an even more restricted set of conditions, *constant temperature and pressure*, a simple criterion for spontaneity emerges:

- If ΔG_{sys} for a proposed process at constant temperature and pressure is negative then the process is spontaneous.

- If ΔG_{sys} is positive then the proposed process at constant temperature and pressure is *not* spontaneous but the reverse process is.

- If ΔG is 0 then the system is at equilibrium at constant temperature and pressure and no change occurs.

The price paid for the convenience and simplicity of this criterion is the restriction to constant temperature and pressure. Processes at constant temperature and pressure are common in chemistry, so the price is not heavy.

A classic problem asks the calculation of ΔG for a process that is obviously spontaneous and which involves a change in temperature. ΔG comes out to be positive. This upsets those who expect ΔG to be negative for all spontaneous processes. Once again, ΔG is automatically negative *only* for processes that are spontaneous at constant temperature and pressure.

ΔG is a difference between an enthalpy and an entropy term. Its value at constant T and P represents a compromise between the tendency toward minimum energy in the system and the tendency toward maximum entropy. The solution to **Problem 17** presents a table showing how this compromise works for different ΔH's and ΔS's.

8.5 FREE ENERGY AND CHEMICAL EQUILIBRIUM

Standard State Free Energies

Just as with other thermodynamic quantities a *standard* free energy change is the free energy change measured at 298.15 K and 1 atm. As with other thermodynamic functions a superscript zero to the right of the symbol, ΔG°, designates that it is a standard change. The standard free energy change of a reaction is computed in two ways. The first is from ΔH° and ΔS° values:

$$\Delta G^\circ = \Delta H^\circ - 298.15\,\Delta S^\circ$$

The other is from tabulated values of standard free energies of formation:

$$\Delta G^\circ = \Sigma\,\Delta G_f^\circ\ \text{(products)} - \Sigma\,\Delta G_f^\circ\ \text{(reactants)}$$

Problem 13 presents a typical calculation of the second type. The procedure is identical in every respect to the procedure for getting ΔH°'s of reaction. Text Appendix D contains the necessary free energy of formation data.

Many problems emphasize the link between ΔG° and ΔH° and ΔS°. Consider a simple example, the computation of ΔG° for the sublimation of lithium. The chemical equation is:

$$Li(s) \rightarrow Li(g)$$

ΔH° for this reaction is the ΔH_f° of Li*(g)* minus the ΔH_f° of Li*(s)*. The numbers are in the first column of Appendix D. The answer is 159.37 kJ − 0 kJ or 159.37 kJ. The ΔS° of the reaction is 138.66 J K^{-1} − 29.12 J K^{-1} or 109.54 J K^{-1}. The S° values are in the second column in Appendix D. The ΔG° for the vaporization of 1 mol of Li*(s)* is:

$$\Delta G^\circ = \Delta H^\circ - T\,\Delta S^\circ$$

where the superscript zero tells us that T is 298.15 K. Substitution gives:

$$\Delta G^\circ = 159.37\ \text{kJ} - 298.15\ \text{K}\ (0.10954\ \text{kJ K}^{-1}) = 126.71\ \text{kJ}$$

The answer is the same as the ΔG° tabulated as the standard free energy of formation of Li*(g)* (in the third column of Appendix D). This works out because the equation under consideration represents the formation of Li*(g)* from "elements in their standard states."

Warning: A classic error in calculations like this to subtract a $T\Delta S$ term in joules from a ΔH term which is in kilojoules. In the above example $T\Delta S$ was carefully converted to kJ before the subtraction.

Free Energy and Phase Transitions

The equation Li*(s)* → Li*(g)* represents the sublimation of lithium, a phase transition. At 298.15 K, ΔG for this process is positive. Subliming one mole of lithium metal requires 126.71 kJ of free energy at room conditions. As the temperature increases, the TΔS for this process becomes larger and larger. Since TΔS is subtracted from ΔH, the difference, ΔG, slips down toward zero as the temperature increases. When ΔG for a process is equal to zero, then there is equilibrium between the two sides. For the sublimation of lithium, ΔG = 0 when the temperature reaches 1504 K, assuming that ΔS and ΔH are not affected by the change in temperature and equal ΔS° and ΔH°. In problem-solving this is a common assumption, particularly when the temperature does not change very much. See, for example, **problem 23b.**

The Equilibrium Constant

For a general chemical reaction involving gases:

$$a\text{A} + b\text{B} \rightleftarrows c\text{C} + d\text{D}$$

the associated equilibrium constant expression is:

$$K = \frac{P_C^c P_D^d}{P_A^a P_B^b}$$

The text derives the fundamentally important expression:

$$\Delta G^\circ = -RT \ln K$$

Memorize this relationship.

Although it is not derived for reactions involving dissolved species, it is valid in such cases. Furthermore, it applies to reactions involving gases and dissolved species.

• *This key expression allows calculation of the equilibrium position of a chemical reaction from tabulated calorimetric data alone. These data are the data collected in Appendix D.*

Problems fall into these types:

• *Given a reaction at 25° C, calculate K.* After balancing the chemical equation, one gets the ΔG°_f's of all of the reactants and products from Appendix D. Combination of these values gives ΔG° for the reaction. This intermediate answer is in kJ. It is the free energy change observed every time the chemical reaction takes place as expressed in the balanced equation. Regard it, therefore, as having the units kJ per mole of the chemical reaction as written. Multiplication by 1000 J kJ^{-1} converts it to J mol^{-1}. Then, division by 8.314 J $mol^{-1}K^{-1}$ and 298.15 K gives $-\ln K$. See **problems 13 and 15.**

• *Calculate ΔG°, given K.* This is just the reverse use of the fundamental equation. Sometimes either ΔH° or ΔS° is given and the problem extends to the calculation of whichever one is missing.

• *Given a reaction, calculate K at temperatures other than 25° C.* This is the substance of **problem 21.** One derives ΔH° and ΔS° values for the reaction by combination of the data in Appendix D. It is a waste of time to calculate ΔG° because all ΔG's are strongly dependent on temperature and the value of ΔG° (when the temperature is 298.15 K) has nothing to do with K at temperatures other than 25° C. In contrast, ΔH's and ΔS's are usually only weakly dependent on temperature. In many problems it is explicitly assumed that they are independent of temperature. If so, the equilibrium constant is calculated using:

$$-RT \ln K = \Delta H^\circ - T\Delta S^\circ \qquad (\Delta H^\circ \text{ and } \Delta S^\circ \text{ independent of T})$$

• *Given K's at two different temperatures, calculate ΔH° and ΔS° for a given reaction.* The K's are related to the ΔG values:

$$-RT_1 \ln K_1 = \Delta H^\circ - T_1 \Delta S^\circ$$
$$-RT_2 \ln K_2 = \Delta H^\circ - T_2 \Delta S^\circ$$

Substitution of the T's and K's allows calculation of the two unknowns, ΔH° and ΔS°. If the two equations are combined by eliminating ΔS°:

$$\ln \frac{K_2}{K_1} = \frac{-\Delta H^\circ}{R} \left[\frac{1}{T_2} - \frac{1}{T_1} \right]$$

This is the *van't Hoff equation.* It is valid insofar as ΔH° and ΔS° do not depend on temperature. It offers an alternative way to get ΔH° of a reaction from the observation of K's at different temperatures. **See problem 27.**

When ΔH° is *positive,* the reaction is endothermic, and its equilibrium constant increases with increasing temperature. When ΔH° is *negative,* the reaction is exothermic, and its equilibrium constant decreases with increasing temperature.

• *Given values of K at two different temperatures and either ΔH° or ΔS° calculate whichever of the two is missing.* This is a reversal of the problem just discussed.

Other problems combine the thermodynamic avenue to K's with concentration and partial pressure data. Thus, an elaborate problem would give a reaction and enough equilibrium concentration data to calculate K (as in problem 5-17, for example) at two different temperatures. Then it would ask for ΔH° and ΔS° for that reaction. **Problem 20-13** is a combination-style problem.

In all problems express gas concentrations in atm and the concentrations of dissolved species in mol L^{-1}. This must be done because the standard state for all of the thermodynamic values in Appendix D is either a pressure of 1 atm or a concentration of 1 M.

DETAILED SOLUTIONS TO ODD-NUMBERED PROBLEMS

1. The internal energy of an ideal gas depends solely on its temperature, T, and, in an isothermal process, T does not change. If we assume that the system of 4.00 mol of hydrogen behaves ideally during its isothermal expansion, then ΔE equals 0. Also, ΔH equals 0. To appreciate why, write the definition of ΔH:

$$\Delta H = \Delta E + \Delta(PV)$$

and realize that for a given portion of an ideal gas $\Delta(PV) = nR\,\Delta T$. But ΔT is zero. This means that $\Delta(PV) = 0$.

The work done *on* the gas during the reversible isothermal expansion from 12.0 L to 30.0 L is:

$$w = -nRT \ln V_2/V_1 = -4.00\text{ mol}(8.314\text{ J mol}^{-1}\text{K}^{-1})(400\text{ K}) \ln (30.0 / 12.0)$$

$$w = -1.22 \times 10^4\text{ J}$$

From the First Law, if $\Delta E = 0$ then $q = -w$. This means the gas absorbs 12.2 kJ of heat during its expansion, just enough to balance off the 12.2 kJ of work it performs. That is, $q = +12.2$ kJ. Finally, $\Delta S = q_{rev} / T$ for an isothermal process. The q_{rev} is the q just computed. Hence, $\Delta S = (+12.2 \times 10^3 / 400)$ J K^{-1} = 30.5 J K^{-1}.

---

3. The ideal diatomic gas expands reversibly and adiabatically from the original volume and temperature $V_1 = 10.0$ L, $T_1 = 300$ K. Knowing there is 5.00 mol, its original pressure is easily computed using the ideal gas equation: P_1 equals 12.31 atm. For any reversible adiabatic process, $q_{rev} = 0$. Therefore $\Delta S = 0$ for this process. As it expands the gas performs a substantial amount of work on its surroundings. All of this work must come at the expense of its internal energy inasmuch as no heat flows in. For a reversible adiabatic process:

$$T_1 V_1^{\gamma-1} = T_2 V_2^{\gamma-1}$$

where γ is the ratio C_p/C_v, the heat capacity ratio. For any ideal gas $C_p = C_v + R$. C_v is 20.8 J mol^{-1}K^{-1}; C_p equals 29.1 J mol^{-1}K^{-1}, and $\gamma - 1$ equals 0.400. This means:

$$T_2 = T_1 (V_1/V_2)^{0.400} = 300\text{ K }(0.1)^{0.400} = 119\text{ K}$$

Now that T_2 is known, $\Delta T = -181$ K and:

$$\Delta E = nC_v\Delta T = (5.00\text{ mol})(20.8\text{ J mol}^{-1}\text{K}^{-1})(-181\text{ K}) = -18.8\text{ kJ}$$

The work is also -18.8 kJ because $\Delta E = w$ for any adiabatic process. Meanwhile:

$$\Delta H = nC_p\Delta T = (5.00\text{ mol})(29.1\text{ J mol}^{-1}\text{K}^{-1})(-181\text{ K}) = -26.3\text{ kJ}$$

The similarities between this problem and text Example 7-8 (text page 241) are worth study. In Example 7-8, 5.00 mol of a monatomic (*not* diatomic) gas at 300 K was expanded reversibly and adiabatically so that the *pressure* dropped to one-tenth its original value (*not* so that its volume increased ten-fold, the case in this problem). In Example 7-8, the volume change was 12.3 L to 49.0 L. In *this* problem the *pressure* change is 12.3 atm to 0.490 atm.

In *both* problems the temperature change is − 181 K. Although it is not calculated in Example 7-8, ΔE is substantially *less* (it is − 11.3 kJ) than ΔE in this problem because the heat capacities of a monatomic gas are less than those of a diatomic gas. Reducing the temperature of the monatomic gas by 181 K requires extraction of *less* energy than reducing the temperature of a diatomic gas by 181 K. The molar C_v and C_p of an ideal *diatomic* gas are $^5/_2$ R and $^7/_2$ R respectively compared to $^3/_2$ R and $^5/_2$ R for an ideal *monatomic* gas. These differences just compensate for the difference between adiabatic expansion ten-fold in volume and adiabatic expansion to one-tenth the pressure.

---

5. Break down the process into the obvious three steps and calculate ΔS_{sys} for each. Then add up the three contributions. The steps are: warming of ice; melting of ice; warming of water. As the text proves, ΔS for any temperature change at constant pressure is given by the equation $\Delta S = n\ C_p \ln (T_2 / T_1)$. This allows the calculation of ΔS of the system for the first and third steps:

$$\Delta S_I = 2.88 \text{ J K}^{-1} \quad \textit{and} \quad \Delta S_{III} = 5.30 \text{ J K}^{-1}$$

In the second step of the process T is a constant (273.15 K), and ΔS of the system is the quantity of heat absorbed reversibly (q_{rev}) divided by this temperature:

$$\Delta S_{II} = (6007 / 273.15) \text{ J K}^{-1} = 22.0 \text{ J K}^{-1}$$

The *overall* ΔS of the system is:

$$\Delta S_{sys} = \Delta S_I + \Delta S_{II} + \Delta S_{III} = 30.2 \text{ J K}^{-1}$$

The entire process is reversible so the entropy of the universe remains constant:

$$\Delta S_{univ} = \Delta S_{sys} + \Delta S_{surr} = 0 \quad \textit{hence} \quad \Delta S_{surr} = -30.2 \text{ J K}^{-1}$$

---

7. Hot iron is plunged into cool water and the final temperature is 28.9° C. This process is far from reversible. Nevertheless, the ΔS of the iron and the ΔS of the water may be computed using the equation developed for a reversible heat flow and employed in problem 5. This is legitimate because entropy is a *state* function. Its change depends only on the original and final states of the system, not on how the path along which the change occurs. For the 30.0 g of iron (molar weight 55.847 g mol^{-1}):

$$\Delta S_{Fe} = (0.537 \text{ mol})(25.1 \text{ J mol}^{-1}\text{K}^{-1}) \ln T_2 / T_1$$

T_2 and T_1 are 302.05 and 423.15 K respectively so:

$$\Delta S_{Fe} = -4.54 \text{ J K}^{-1}$$

For the 100.0 g of water, n, the chemical amount is 5.55 mol, and C_p is 75.3 J mol^{-1} K^{-1}. T_2 is 302.05 K and T_2 is 295.15 K. Hence:

$$\Delta S_{water} = n\ C_p \ln (T_2 / T_1) = 5.43\ J\ K^{-1}$$

The overall ΔS of the system is +0.89 J K^{-1}, the sum of these two numbers.

---

9. Before the stopcock is opened, the number of *microstates* available to a single H_2 (or He) is proportional to the volume of the glass bulb:

$$\Omega = cV \quad \text{(c is a constant)}$$

There are N_0 molecules of H_2 and N_0 atoms of He. The number of microstates for each is:

$$\Omega_{H_2} = (cV)^{N_0} \quad \textit{and} \quad \Omega_{He} = (cV)^{N_0}$$

The number of microstates of the entire system, still before the valve is opened, is the *product* of the Ω's:

$$\Omega_{sys} = \Omega_{H_2}\Omega_{He} = (cV)^{2N_0}$$

This is the number of microstates in which all of the H_2 occupies the first bulb and all of the He the second. By symmetry it is also the number of microstates in which all of the H_2 is in the *second* bulb and all of the He is in the first.

After the stopcock is opened, $2N_0$ molecules occupy a volume of 2V and:

$$\Omega_{sys} = (c2V)^{2N_0}$$

The probability, *p*, of the "cross-diffused" result, the state in which the H_2 and N_2 trade places, is the number of ways in which it can be constituted divided by the number of ways in which the mixed system can be constituted:

$$p = (cV)^{2N_0} / (c2V)^{2N_0} = 2^{-2N_0}$$

Taking the logarithm of both sides of this equation:

$$\log p = -2\ N_0 \log 2 = -2\ N_0\ (0.301)$$

$$\log p = -3.62 \times 10^{23} \quad \textit{hence} \quad p = 10^{-3.62 \times 10^{23}}$$

---

11. When one mole of ethanol is vaporized at its normal boiling point, ΔH° = 38.7 kJ. Since this goes on at constant pressure q_p = ΔH and q = 38.7 kJ. The vaporization occurs isothermally and reversibly, so q is also q_{rev}, and:

$$\Delta S = q_{rev} / T = 38.7\ kJ / 351.1\ K = 0.110\ kJ\ K^{-1}$$

Turn next to the calculation of ΔE. From the definition of enthalpy:

$$\Delta E = \Delta H - \Delta(PV)$$

At constant pressure $\Delta(PV) = P\Delta V = P(V_2 - V_1)$. V_2 is the volume of one mole of ethanol vapors at 351.1 K and V_1 is volume of one mole of liquid ethanol at the same temperature. Assume that the vapors behave ideally:

$$V_2 = nRT / P = 28.8 \text{ L}$$

The volume of the liquid (V_1) is less than 0.1 L, negligibly small. Therefore, $P\Delta V = 28.8$ L atm $= 2.92$ kJ (Recall that 1.000 L atm is 0.101325 kJ). Substitute these values into the expression for ΔE:

$$\Delta E = \Delta H - \Delta(PV) = 38.7 \text{ kJ} - 2.92 \text{ kJ} = 35.8 \text{ kJ}$$

Knowing $P\Delta V$ helps to get the work for the process. By expanding against a constant pressure the system performs 2.92 kJ of pressure-volume work on its surroundings. This is the only kind of work possible so the total work absorbed by the system is -2.92 kJ.

For all reversible processes at constant temperature and pressure $\Delta G = 0$. This can be verified in this case by noting that:

$$\Delta G = \Delta H - T\Delta S = 38.7 - 38.7 = 0 \text{ kJ}$$

---

13. From Appendix D, the ΔG°_f values for H_2O, NO_2 and NH_3 (as gases) are respectively -228.59, 51.29 and -16.48 kJ mol^{-1}. The ΔG° of formation of O_2 *(g)* is zero at 25° C. The ΔG° of the reaction:

$$2\ NH_3(g) + {}^7/_2\ O_2(g) \rightleftarrows 2\ NO_2(g) + 3\ H_2O(g)$$

comes from combining the ΔG°'s of formation:

$$\Delta G^\circ_{\text{reaction}} = 3(-228.59) + 2(51.29) - 2(-16.48) - {}^7/_2\ (0) \text{ kJ}$$

$$\Delta G^\circ_{\text{reaction}} = -550.23 \text{ kJ}$$

This ΔG° applies to a system which starts with 2 mol of ammonia and $^7/_2$ mol of O_2. If it were twice as large (starting with 4 mol of NH_3 and 7 mol of oxygen), then ΔG° would be twice as large, too. Another way to recognize the difference is to think of the same chemical reaction happening twice with ΔG° equal to -550.23 kJ each time. In this view the ΔG° value is really -550.23 kJ *per mole of the reaction as it is written:* -550.23 kJ mol^{-1}.

The equilibrium constant for the reaction is related to ΔG°:

$$-RT \ln K = \Delta G^\circ$$

$$\ln K = -(-550.23 \times 10^3 \text{ J mol}^{-1}) / (8.314 \text{ J mol}^{-1}\text{K}^{-1})\ (298.15 \text{ K})$$

$$\ln K = 221.97 \quad \textit{and} \quad K = 2.5 \times 10^{96}$$

---

15. Each example is similar to problem 13. *First.* Use Appendix D to find ΔG°_f values for all the compounds involved in the reaction. *Second.* Compute the

ΔG_f° of the actual reactants and products by multiplying the molar ΔG_f° of each by the number of moles assigned in the balanced equation. *Third.* Add together these answers for all the products and from the sum *subtract* the answers for all the reactants. *Last.* Use the fundamental relationship:

$$\Delta G^\circ = -RT \ln K$$

to compute *K*. Remember to express ΔG° in joules (not kJ) if the value R = 8.314 J mol^{-1} is to be used.

a) For the reaction:

$$SO_2\,(g) + \tfrac{1}{2}\,O_2\,(g) \rightleftarrows SO_3\,(g)$$

$\Delta G^\circ = -70.89$ kJ mol^{-1} and $K = 2.63 \times 10^{12}$. This *K* corresponds to the mass-action expression:

$$P_{SO_3} / P_{SO_2}\, P_{O_2}^{1/2} = 2.6 \times 10^{12}$$

Suppose all of the coefficients in the equation were doubled. Then $\Delta G^\circ = -141.78$ kJ mol^{-1} and $K = 6.92 \times 10^{24}$, the *square* of the *K* just now computed. The new equilibrium constant expression would be:

$$P^2_{SO_3} / P^2_{SO_2} P_{O_2} = 6.92 \times 10^{24}$$

b) For the reaction:

$$3\,Fe_2O_3\,(s) \rightleftarrows 2\,Fe_3O_4\,(s) + \tfrac{1}{2}\,O_2\,(g)$$

ΔG° at 298.15 K equals 195.6 kJ mol^{-1}, and *K* equals 5.38×10^{-35}:

$$5.38 \times 10^{-35} = P_{O_2}^{1/2}$$

The two pure solids have activities of 1 and do not appear in the mass-action expression.

c) For the reaction:

$$CuCl_2\,(s) \rightleftarrows Cu^{2+}\,(aq) + 2\,Cl^-\,(aq)$$

we add $2 \times (-131.23$ kJ), for the 2 Cl^- *(aq)*, to 65.49 kJ, for the Cu^{2+} *(aq)*, and subtract -175.7 kJ, for the $CuCl_2$ *(aq)*. This chemical equation represents the dissolution of copper(II) chloride in water. The large negative ΔG_f° of 2 Cl^- *(aq)* is what drives the reaction at room temperature. ΔG° of reaction equals -21.27 kJ mol^{-1} and the equilibrium constant is 5.33×10^3:

$$5.33 \times 10^3 = [Cu^{2+}][Cl^-]^2$$

---

17. For each reaction, ΔH° and ΔS° are computed using the data in Appendix D. If the reactions go on at constant temperature and pressure (and if ΔH° and ΔS° are *not* temperature dependent):

$$\Delta G = \Delta H^\circ - T\Delta S^\circ$$

ΔG becomes equal to zero only when $T = \Delta H^\circ / \Delta S^\circ$. All absolute temperatures are positive, but the quotient $\Delta H^\circ / \Delta S^\circ$ is positive only when ΔH° and ΔS° have the same sign.

If ΔH° is positive and ΔS° is negative then ΔG *exceeds* zero for all possible T's. If ΔH° is negative and ΔS° is positive then ΔG is *less than* zero for all possible T's. If *both* are positive $\Delta G < 0$ at high T. If *both* are negative $\Delta G < 0$ at low T:

ΔH°	ΔS°	$\Delta H^\circ / \Delta S^\circ$	*Outcome*
+	−	−	$\Delta G > 0$. Always non-spontaneous
−	+	−	$\Delta G < 0$. Always spontaneous
+	+	+	spontaneous at high T
−	−	+	spontaneous at low T

a) For the reaction:

$$4\,Fe(s) + 3\,O_2(g) \rightleftarrows 2\,Fe_2O_3(s)$$

$\Delta H^\circ = 2(-824.2) - 3(0) - 4(0)$ kJ and $\Delta S^\circ = 2(87.40) - 3(205.03) - 4(27.28)$ J K^{-1} Completion of the arithmetic gives:

$$\Delta H^\circ = -1648.4 \text{ kJ} \quad \textit{and} \quad \Delta S^\circ = -0.54941 \text{ kJ K}^{-1}$$

If these values are substituted in the equation $\Delta G = \Delta H^\circ - T\Delta S^\circ$ with T set at 298.15 K, then ΔG°_{298} comes out to be *twice* the tabulated (Appendix D) ΔG°_f of $Fe_2O_3(s)$. This is correct because the above reaction forms *two* moles of $Fe_2O_3(s)$ from its constituent elements.

The reaction is spontaneous at low temperatures. When $T = \Delta H^\circ / \Delta S^\circ = 3000$ K, then ΔG equals zero. At temperatures above 3000 K, ΔG is greater than zero, and the reaction is non-spontaneous.

b) The reaction is:

$$SO_2(g) + \tfrac{1}{2}\,O_2(g) \rightleftarrows SO_3(g)$$

Using the data from Appendix D, $\Delta H^\circ = -98.89$ kJ and $\Delta S^\circ = -0.09397$ kJ K^{-1}. Note the timely conversion to kJ in the units of ΔS°. ΔG equals zero at $T = \Delta H^\circ / \Delta S^\circ = 1052$ K. Below this temperature ΔG is *less* than zero, and the reaction is spontaneous.

c) The reaction is:

$$NH_4NO_3(s) \rightleftarrows N_2O(g) + 2H_2O(g)$$

From the data in Appendix D, $\Delta H^\circ = -36.03$ kJ and $\Delta S^\circ = 0.4461$ kJ K^{-1}. Because ΔH° is negative and ΔS° is positive, only a negative T makes ΔG. equal zero. But negative absolute temperatures are impossible. The reaction is spontaneous at all temperatures.

----

19. The process is the dilution of aqueous ammonia from 0.100 M to 0.020 M at 298.15 K. The free energy change for n moles of ammonia going from C_1 to C_2 is:

$$\Delta G = nRT \ln (C_2 / C_1)$$

In this case:

$$\Delta G / n = (8.314 \text{ J mol}^{-1}\text{K}^{-1})(298.15 \text{ K})(-1.6094)$$

Completing the arithmetic shows that $\Delta G / n$ is -3.99 kJ mol^{-1}. A volume of 10.0 L contains 1.00 mol of NH_3 so if 10.0 L of the solution is diluted to 50.0 L then n = 1.00 and -3.99 kJ is the free energy change.

---

21. **a)** The reaction is the same SO_2 *(g)* to SO_3 *(g)* conversion considered in problem 15a, except now it occurs at 348.15 K (75° C). The $\Delta G°$ values in Appendix D are no good at this higher temperature because $\Delta G°$ is strongly dependent on T. However, $\Delta H°$ and $\Delta S°$ are assumed to be *independent* of temperature. Once $\Delta H°$ and $\Delta S°$ are computed, the relationship:

$$\Delta H° - T\Delta S° = -RT \ln K$$

allows determination of *K*. $\Delta H°$ of the reaction is -98.89 kJ and $\Delta S°$ is -0.09397 kJ K^{-1}. (See problem 17b.) Substituting T = 348.15 K and R = 8.314×10^{-3} kJ mol^{-1}K^{-1} into the above expression gives $K = 8.5 \times 10^9$. The increase in temperature from 25 to 75° C has *decreased* the equilibrium constant 300-fold from the value calculated in problem 15a.

b) For the hematite to magnetite reaction of problem 15b, $\Delta H°$ is 235.8 kJ and $\Delta S°$ is 0.133 kJ K^{-1}. These numbers result from combining values from Appendix D in the usual way; fractional coefficients are nothing special. Then:

$$K = e^{(-\Delta H°/RT + \Delta S°/R)}$$

which is another form of the equation used in part a). Substituting T = 348.15 K and taking the usual care with units gives $K = 3.7 \times 10^{-29}$.

c) For the reaction:

$$CuCl_2\,(s) \rightleftarrows Cu^{2+}(aq) + 2\,Cl^-\,(aq)$$

$$\Delta H° = 2(-167.16) + (64.77) - (-220.1) \text{ kJ}$$

$$\Delta S° = 2(56.5) + (-99.6) - (108.07) \text{ J K}^{-1}$$

The numbers are standard enthalpy changes of formation and standard entropies from Appendix D. Completion of the arithmetic gives $\Delta H° = -49.45$ kJ and $\Delta S° = -0.09467$ kJ K^{-1}. Substitution of these values and a temperature of 348.15 K into the equation:

$$-RT \ln K = \Delta H° - T\Delta S° \quad \textit{gives} \quad K = 3.0 \times 10^2$$

---

23. The balanced chemical equation for the synthesis of ammonia is:

$$3\ H_2(g) + N_2(g) \rightleftarrows 2\ NH_3(g)$$

Knowing K at 298.15 K allows the computation of ΔG°_{298} from the relationship:

$$\Delta G^\circ_{298} = -R\,(298.15)\ \ln K$$

ΔG°_{298} comes out to be -32.92 kJ for the reaction as written above (the reaction giving 2 NH_3's). This answer is twice the molar free energy of formation of ammonia at 298.15 K, available in Appendix D. This is not a discrepancy. The chemical equation shown above forms two moles of ammonia, but the tabulated free energies are all on the basis of the formation of one mole of compound.

If the temperature and pressure are constant:

$$\Delta G^\circ = \Delta H^\circ - T\Delta S^\circ$$

From this relationship and the given ΔH° (-92.2 kJ), ΔS° of the reaction is -0.199 kJ K^{-1}. To get the equilibrium constant at 600 K, use these values of ΔH° and ΔS° along with T = 600 K in the expression:

$$-RT\ \ln K = \Delta H^\circ - T\Delta S^\circ$$

Substitution gives:

$$-(8.314 \times 10^{-3})\ 600\ \ln K = -92.2 - 600(-0.199)$$

$$\ln K = -5.45 \quad \textit{and} \quad K = 4.3 \times 10^{-3}$$

The problem can also be solved by direct substitution into the van't Hoff equation. The details are presented on text p. 685, but using 800 K instead of 600 K as the second temperature.

---

25. The problem is an application of the van't Hoff equation:

$$\ln \frac{K_2}{K_1} = \frac{-\Delta H^\circ}{R}\left[\frac{1}{T_2} - \frac{1}{T_1}\right]$$

State 2 refers to the reaction at 200° C: $K_2 = 1.21 \times 10^{-3}$ and $T_2 = 473.15$ K. State 1 refers to the reaction at 25° C: $K_1 = 6.8$ and $T_1 = 298.15$ K. It is wise to size up the probable answer. The term in brackets will be negative. The ratio of the K's is less than 1 and will have a negative logarithm. ΔH° will therefore be negative. Notice that exchanging the labels (1 and 2) makes the term in brackets and the logarithm both positive, but leaves the sign and magnitude of ΔH° unaffected. Completing the arithmetic gives $\Delta H^\circ = -5.79 \times 10^4$ J or $-$ 57.9 kJ. This is the *average* ΔH° over the range in temperature because ΔH° does vary slightly with temperature.

The ΔH° can be checked using the Appendix D data for $NO_2\,(g)$ and $N_2O_4\,(g)$. From those data, ΔH° is -57.2 kJ.

---

27. The process in question is the vaporization of ammonia:

$$NH_3\,(l) \rightleftarrows NH_3\,(g)$$

The equilibrium constant for this process is just the equilibrium vapor pressure of $NH_3\,(g)$; pure liquid ammonia has unit activity. We therefore have K_1 = 0.4034 at T_1 = 223.15 K, and K_2 = 4.2380 at T_2 = 273.15 K. Substitution of these numbers into the van't Hoff equation (see problem 25) gives ΔH° of vaporization. It is 23.84 kJ mol^{-1}.

The normal boiling temperature of a liquid is the temperature at which its vapor pressure is 1.000 atm. This is a definition. Assign 1.000 as K_1 in the van't Hoff equation. Then T_1 is the desired boiling point:

$$\ln (4.328 / 1.000) = (-23840 / 8.314) [^1/_{273.15} - 1 / T_1]$$

where $\Delta H = -23840$ J mol^{-1} comes from the first part of the problem. Solving for T_1 gives 239.7 K or -33.5° C (The measured normal boiling temperature of ammonia is -33.35° C.)

---

29. The text develops the relationship:

$$\Delta T_b = (RT_b^2 / \Delta H_{vap}) X_2$$

where X_2 is the mole fraction of the *solute*, T_b and ΔH_{vap} are the boiling temperature and enthalpy change of vaporization of the *solvent* and ΔT_b is the boiling point elevation. In this problem, 2.00 g of the solute $C_{14}H_{10}$ (molar weight 178.23 g mol^{-1}) is in 100 g of solvent C_6H_6 (benzene, molar weight 78.115 g mol^{-1}). The solute contributes 0.01122 mol to a total of 1.29142 mol in the solution: X_2 is 0.008688.

Substitute this and the other given values into the above equation, remembering that T_b is 353.35 K and that ΔH and RT_b both must both be expressed in the same units. Then ΔT_b = 0.230 K. The boiling point of the solution is 353.35 K + 0.230 K = 353.58 K, is 80.4° C.

---

31. ***a)*** The compression of the oxygen is reversible and adiabatic. Therefore, ΔS_{sys} equals 0.

b) If 2.60 mol of oxygen is compressed reversibly and adiabatically from a state (P_1, V_1) to a state (P_2, V_2) then:

$$P_1 V_1^{\gamma} = P_2 V_2^{\gamma}$$

where γ is the ratio of C_p to C_v. The exponent, γ, in this case is 29.4 J mol^{-1}K^{-1} divided by 21.09 J mol^{-1}K^{-1} or 1.394. The original volume (V_1) of the oxygen is 64.0 L, computed from the ideal gas equation with T_1 = 300 K. P_2 is 8.00 atm so:

$$(1.00 \text{ atm})(64.0 \text{ L})^{1.394} = (8.00 \text{ atm})(V_2)^{1.394}$$

Knowing P_2 and V_2 (with ideality assumed) gives T_2. In summary:

$$P_1 = 1.00 \text{ atm} \quad V_1 = 64.0 \text{ L} \quad T_1 = 300 \text{ K}$$
$$P_2 = 8.00 \text{ atm} \quad V_2 = 14.4 \text{ L} \quad T_2 = 540 \text{ K}$$

The problem traces an alternative path between state 1 and state 2: The oxygen is first heated to T_2 at constant pressure and then compressed reversibly and isothermally to P_2. Compute all of the state variables in the *intermediate* state (subscripted *i*), after the isochoric heating but before the isothermal compression:

$$P_i = 1.00 \text{ atm} \quad T_i = 540 \text{ K} \quad V_i = 115.2 \text{ L}$$

The last of these comes from the ideal gas law, remembering that n = 2.60 mol. The entropy change during the constant pressure heating is:

$$\Delta S_{1 \text{ to } i} = n\, C_p \ln(T_i / T_1) = 2.60(29.4) \ln(540 / 300) \text{ J K}^{-1}$$

$$\Delta S_{1 \text{ to } i} = 44.9 \text{ J K}^{-1}$$

The entropy change during the isothermal compression is:

$$\Delta S_{i \text{ to } 2} = nR \ln(V_2 / V_i) = 2.60(8.314) \ln(14.40 / 115.2) \text{ J K}^{-1}$$

$$\Delta S_{i \text{ to } 2} = -44.9 \text{ J K}^{-1}$$

The ΔS for the overall process is the sum of these values; it is zero.

---

33. ***a)*** If the motion of air masses through the atmosphere is adiabatic and reversible, then $q_{rev} = 0$ and $\Delta S = 0$.

b) During the upward displacement of an air mass we can expect both its temperature and pressure to drop simultaneously. Break down this overall process into two parts: a temperature change at constant pressure (step I) and a pressure change at constant temperature (step II). The original values of temperature and pressure are T_0 and P_0 and the final values are T and P. For the two steps:

$$\Delta S_I = n\, C_p \ln(T / T_0) \quad \textit{and} \quad \Delta S_{II} = n\, R \ln(P_0 / P)$$

In the first step, ΔS is *less* than zero. This step involves the cooling of the air mass and a reduction of its entropy. In the second step, ΔS is *greater* than zero. This step is the expansion of the air mass. The sum of the ΔS's must be zero because the overall process, the sum of the two steps, is *isentropic*:

$$C_p \ln(T / T_0) + R \ln(P_0 / P) = 0$$

c) According to the problem, $\ln (P / P_0)$ is approximately equal to $-Mgh / RT$. This means that:

$$-\ln (P / P_0) = \ln (P_0 / P) = Mgh / RT$$

Substitute this (approximate) equality into the final expression in part b) and rearrange to give:

$$T \ln (T / T_0) = -Mgh / C_p$$

All of the quantities on the right-hand side of this equation are given in the problem, so:

$$-Mgh / C_p = (0.029)(9.8)(5.9 \times 10^3) / 29 = -57.8 \text{ K}$$

This means that T, the temperature on top of the mountain, fulfills the equation:

$$T \ln (T / T_0) = -57.8 \text{ K}$$

where T_0 is the sea-level temperature (311 K). It is easiest to determine T by successive approximations. Guessing a few values for T and using a calculator quickly show that T = 246 K (−27° C) satisfies the equation.

---

35. In problem 8, a 1 mol piece of iron at 100° C is plunged into a large reservoir of water at 0° C. It loses 2510 J to the water, and its temperature falls from 373 K to 273 K. Its entropy decreases. The change is:

$$\Delta S_{Fe} = n\, C_p \ln (T_2 / T_1)$$

$$\Delta S_{Fe} = (1.00 \text{ mol})(25.1 \text{ J mol}^{-1} \text{K}^{-1}) \ln(373 / 273) = -7.83 \text{ J K}^{-1}$$

a) The piece of iron is first cooled from 100 to 50° C and then from 50 to 0° C using two water reservoirs. It loses 1255 J of heat to the first reservoir and 1255 J of heat to the second. The entropy change of the first reservoir, which *absorbs* 1255 J of heat and which is so big it stays at 348 K is:

$$\Delta S_I = q / T = (1255 / 323) \text{ J K}^{-1} = 3.88 \text{ J K}^{-1}$$

The entropy change of the second reservoir, which also absorbs 1255 J of heat but at 273 K, is larger:

$$\Delta S_{II} = q / T = (1255 / 273) \text{ J K}^{-1} = 4.60 \text{ J K}^{-1}$$

These two reservoirs are the surroundings of the iron:

$$\Delta S_{surr} = 3.88 + 4.60 = 8.48 \text{ J K}^{-1}$$

ΔS of the iron is still -7.83 J K^{-1} because only the path by which it cooled has changed. It still ends up in the same final state.

$$\Delta S_{univ} = 8.48 \text{ J K}^{-1} - 7.83 \text{ J K}^{-1} = 0.65 \text{ J K}^{-1}$$

b) Each of the four reservoirs absorbs 627.5 J, one-fourth of the total heat the iron gives up. The entropy changes of the four reservoirs are:

$$\Delta S_{I} = (627.5 / 348)\ J\ K^{-1} = 1.80\ J\ K^{-1}$$

$$\Delta S_{II} = (627.5 / 323)\ J\ K^{-1} = 1.94\ J\ K^{-1}$$

$$\Delta S_{III} = (627.5 / 298)\ J\ K^{-1} = 2.11\ J\ K^{-1}$$

$$\Delta S_{IV} = (627.5 / 273)\ J\ K^{-1} = 2.30\ J\ K^{-1}$$

ΔS_{surr} is the sum of the ΔS's of the four reservoirs. It is 8.15 J K^{-1}. ΔS of the iron is still −7.83 J K^{-1}.

$$\Delta S_{univ} = 8.15\ J\ K^{-1} - 7.83\ J\ K^{-1} = 0.32\ J\ K^{-1}$$

c) Using four reservoirs makes the process more nearly reversible as evidenced by the smaller ΔS_{univ}. Making the process exactly reversible would require an infinite series of reservoirs each one absorbing an infinitesimal quantity of heat from the iron at a temperature infinitesimally less than the previous reservoir. In such a case, the ΔS's of all the reservoirs would add up to +7.83 J K^{-1}, and ΔS_{univ} for the process would be zero.

---

37. ***a)*** Several different ideal gases, each occupying its own original volume V_i and all at the same temperature and pressure, are allowed to mix. The aim is to get an expression for ΔS for this process, using the microscopic interpretation of entropy. The best approach is to compute ΔS for each of the gases *separately* and then add up all of these contributions. Work with the *i*-th gas. This gas starts at V_1 and expands to V_2. In both of these states its entropy depends on the number of microstates, Ω, in the following way:

$$S_1 = k_B \ln \Omega_1 \qquad S_2 = k_B \ln \Omega_2$$

The *change* in entropy of the *i*-th gas is:

$$\Delta S_i = S_2 - S_1 = k_B \ln (\Omega_2 / \Omega_1)$$

Both before and after the expansion the number of microstates available to one molecule of the gas is proportional to the volume. The number of microstates available to *all* the molecules is proportional to the volume *raised to a power* $n_i N_0$, the total number of molecules. In these equations N_0 is Avogadro's number, and n_i is the number of moles of the *i*-th gas. The change in entropy for the *i*-th gas now is:

$$\Delta S_i = k_B \ln (cV_2)^{n_i N_0} - k_B \ln (cV_1)^{n_i N_0}$$

$$\Delta S_i = n_i N_0 k_B \ln (V_2 / V_1)$$

Now, concentrate on the term (V_2 / V_1). By the ideal gas law (in fact, by Boyle's law), it is equal to (P_1 / P_2), the ratio of the original pressure of the *i*-th gas to its final partial pressure in the mixture. This latter pressure is, by Dalton's law:

$$P_2 = X_i P_{tot}$$

where X_i is the mole fraction of the *i*-th gas. But P_1, the original pressure of the *i*-th gas, *equals* P_{tot}. All of the gases started at the same pressure. Therefore:

$$V_2 / V_1 = P_1 / P_2 = 1 / X_i = X_i^{-1}$$

Substituting this result into the expression for ΔS_i gives:

$$\Delta S_i = -n_i N_0 k_B \ln X_i$$

Now, because $N_0 k_B$ is equal to R and n_i is equal to X_i n, where n is the total number of moles of gas:

$$\Delta S_i = -nR X_i \ln X_i$$

Finally, add up the contributions of all of the gases to get the overall ΔS:

$$\Delta S = \Sigma \Delta S_i = -nR \Sigma X_i \ln X_i$$

b) This part requires first the calculation of the total chemical amount of mixed gas and the mole fractions of O_2, N_2 and Ar in the mixture. Beyond that, there is nothing more than careful substitution in the formula derived in part a). Dividing 50 g by the molar weights of the three gases gives the chemical amount of each. The mole fraction of each is its chemical amount divided by the total chemical amount in the mixture. The results of these calculations are:

Gas	*Chemical Amount*	*Mole Fraction (X)*	*X ln X*
O_2	1.563 mol	0.3399	−0.3668
N_2	1.784	0.3879	−0.3673
Ar	1.252	0.2722	−0.3542

The total chemical amount of gas is, of course, the sum of the numbers in the second column of the above table. It is 4.599 mol. The sum of the entries in the *last* column of the table is the summation term in the equation:

$$\Delta S = -n R \Sigma X_i \ln X_i$$

Using R = 8.314 J $mol^{-1} K^{-1}$ gives a ΔS of 41.6 J K^{-1}. This is the entropy change of mixing at any temperature and pressure as long as the assumption of ideal gas behavior holds. It is not necessary to know that the mixing occurs as STP.

c) Separating the components of air is the reverse of mixing them. The entropy change of separation is therefore the negative of the entropy change of mixing. To solve the problem, compute ΔS_{sys} of mixing and then change its sign.

Table 3-1 (text p. 66) gives the volume percentages of the various gases in the air. Assuming that the ideal gas equation holds, the mole fractions (*X*'s) of the gases are just these numbers divided by 100. In the following table the mole fraction is written and the quantity *X* ln *X* is calculated for each gas:

Gas	*Mole Fraction (X)*	*X ln X*
N_2	0.78110	−0.19297
O_2	0.20953	−0.32747
Ar	0.00934	−0.04365
Ne	0.00001818	−0.0001976

Continuing the table to include atmospheric gases of smaller concentration than Ne will not provide X ln X values significantly larger than zero. The sum of the values in the last column is −0.56429. The entropy change of mixing is −nR times this number. With a volume of 100 L at 298.15 K the chemical amount of air is 4.087 mol, using the ideal gas equation. The entropy change of mixing therefore is +19.2 J K^{-1}. The entropy change of separation of the components *of the system* is −19.2 J K^{-1}

The problem asks simply for the "entropy change." The total entropy change of the universe cannot be calculated because there is no information about the surroundings of the system. It is of course certain that ΔS of the universe exceeds zero.

---

39. The entropy change in heating a system from T_1 to T_2 at constant pressure is:

$$\Delta S_{sys} = n\, C_p \ln (T_2 / T_1)$$

assuming that C_p is *constant* over the range of temperature. As the table given in the problem demonstrates, the molar heat capacity of ice (C_p) is a *function* of temperature. The problem gives 10 different C_p's at 10 different temperatures. The simplest way to overcome this complication is to compute the overall ΔS between −250° C (23 K) and 0° C (273 K) as the sum of 10 smaller ΔS's each computed, using the above formula, over smaller ranges of temperature. In each range we take C_p as the mean of the C_p's at the lower and higher temperatures. This is acceptable because only an *estimate* of ΔS is required.

T-Range (K)	Mean C_p (J $mol^{-1}K^{-1}$)	$C_p \ln T_2/T_1$ (J $mol^{-1}K^{-1}$)
23−73	6.02	6.953
73−93	13.25	3.208
93−113	15.75	3.068
113−133	18.06	2.943
133−173	21.69	5.703
173−213	26.46	5.504
213−243	30.87	4.068
243−262	34.33	2.584
262−271	36.26	1.225
271−273	36.89	0.271

The sum of the entries in the last column of the table is ΔS_{tot}. It is 35.53 J $mol^{-1}K^{-1}$, The molar entropy change of ice between −250° C and 0° C is about 35.5 J $mol^{-1}K^{-1}$.

It is instructive to plot C_p versus ln T as another means of displaying the data from the problem. The desired ΔS is the area under this curve between a temperature of 23 K and 273 K.

---

41. ***a)*** The process under consideration is:

$$S_{(rhombic)} \rightarrow S_{(monoclinic)}$$

Assume that ΔH° and ΔS° for this process do not depend on the temperature. Then, at all temperatures:

$$\Delta G = \Delta H^\circ - T\Delta S^\circ$$

Below 368.5 K, the ΔG of the phase transition must *exceed* zero, because it is not spontaneous at constant T and P. At 368.5 K the ΔG equals zero. $T\Delta S^\circ$ is *subtracted* from ΔH° in the above equation. If ΔS° were negative then $-T\Delta S^\circ$ could get only more positive with increasing T. But $-T\Delta S^\circ$ is known to become more negative. Therefore ΔS° is positive.

b) At 368.5 K, $\Delta G = 0$ so that $|\Delta H^\circ / \Delta S^\circ| = 368.5$ K. The absolute value of ΔH° is given as 400 J mol^{-1}. The absolute value of ΔS° therefore is 1.08 J mol^{-1}K^{-1}. Since ΔS° is positive, from part a), ΔS° is +1.08 J mol^{-1} K^{-1}, and ΔH° is +400 J mol^{-1}.

---

43. ***a)*** The equilibrium is:

$$\tfrac{1}{2}\, Cl_2\,(g) + \tfrac{1}{2}\, F_2\,(g) \rightleftarrows ClF(g)$$

The ΔG°_{298} for this reaction is related to the equilibrium constant at 298 K:

$$\Delta G^\circ_{298} = -RT \ln K$$

Since $K = 9.3 \times 10^9$:

$$\Delta G^\circ_{298} = -(8.314 \text{ J mol}^{-1} \text{ K}^{-1})(298 \text{ K}) \ln(9.3 \times 10^9) = -5.69 \times 10^4 \text{ J}$$

b) Use the same method to compute ΔG_{398}. Because the K at 398 K substantially bigger (it is reported as 2.6×10^{12}), ΔG_{398} is *more negative.* It is -9.46×10^4 J.

The goal is to calculate ΔH° and ΔS°. Since K increases with increasing temperature, anticipate that ΔH° will be *positive.* Higher temperature favors endothermic reactions. Furthermore, expect ΔS° to be small since the equilibrium has the same number of moles of gas on both sides.

Use the ΔG values to write two equations involving the two unknowns:

$$-5.69 \times 10^4 \text{ J} = \Delta H^\circ - (298 \text{ K})\, \Delta S^\circ$$

$$-9.46 \times 10^4 \text{ J} = \Delta H^\circ - (398 \text{ K})\, \Delta S^\circ$$

The equations come from substitution in $\Delta G = \Delta H^\circ - T\Delta S^\circ$ at the two temperatures. The equations are easily solved by subtracting one from the other. $\Delta S^\circ = 377 \text{ J K}^{-1}$ and $\Delta H^\circ = 5.54 \times 10^4$ J.

The answer for ΔH° also is attainable by direct substitution of the given *K*'s into the van't Hoff equation. However, since ΔS° has to be calculated anyway, it is probably simplest to use the above method.

Finally, the experimental *K* at 398 K that is given in the problem is incorrect. This has no effect on the main point of the problem but may trouble those who double-check their answers using sources like *The Handbook of Chemistry and Physics.* The correct *K* at 398 K is 3.33×10^7. Since this figure is *less* than *K* at 298 K, we conclude that ΔH° is really negative. Using the corrected K_{398} in the above equations gives ΔG_{398} equal to -5.73×10^4 J, ΔH° equal to -5.55×10^4 J, and ΔS° equal to $+4.5 \text{ J K}^{-1}$. This small result is much more in line with what is expected considering that there is 1 mol of gas on both sides of the equilibrium and no solids or liquids.

---

45. From problem 44, the molar enthalpy change associated with the ionization of water is $+55.87 \text{ kJ mol}^{-1}$. The equation:

$$-RT \ln K = \Delta H^\circ - T\Delta S^\circ$$

relates this ΔH° to the equilibrium constant and the standard entropy change of the reaction. From problem 44, $K = 1.139 \times 10^{-15}$ at 273.15 K:

$$-R(273.15) \ln(1.139 \times 10^{-15}) = 55.87 \text{ kJ mol}^{-1} - 273.15\,\Delta S^\circ$$

Completing the arithmetic gives $\Delta S^\circ = -81.5 \text{ J mol}^{-1}\text{K}^{-1}$. If the *K* at 60° C and an absolute temperature of 333.15 K are inserted into the same formula, the same ΔS° results. ΔS is not sensitive to temperature, within narrow ranges.

Both the ΔH° and ΔS° values are readily checked by referring to Appendix D. The ΔH of the autoionization of water is the ΔH_f° of H_3O^+ *(aq)* plus that of OH^- *(aq)* minus that of $H_2O(l)$. The ΔS° of the autoionization is a similar combination of the tabulated S° values.

Chapter 9

ELECTROCHEMISTRY

9.1 *ELECTROCHEMICAL CELLS*

In an electrochemical cell an oxidation-reduction reaction occurs without the necessity for direct contact between the reactants. Instead, the oxidation and reduction processes occur at *electrodes* which may be widely separated from each other. In the cell the electrodes are in contact with an *electrolyte*, often an aqueous solution. Electrons flow through a wire from the electrode where the oxidation occurs to the one where reduction occurs. The separated processes are *half-reactions*. The reduction half-reaction occurs at the *cathode* and the oxidation half-reaction occurs at the *anode.*

The transfer of electrons quickly stops unless some means of maintaining electrical neutrality within the electrolyte is provided. The design of electrochemical cells therefore includes a *salt bridge* to allow the transfer of counter-ions from the vicinity of the one electrode to the vicinity of the other. The diagram (*right*) shows the essential parts of a specific electrochemical cell. Electrons flow from the anode to the cathode. Magnesium is oxidized, going into solution in the anode compartment, and NO_3^- ions pass through the salt bridge to maintain electrical neutrality in the anode compartment. Silver ions are reduced in the cathode compartment, and silver plates out on the cathode.

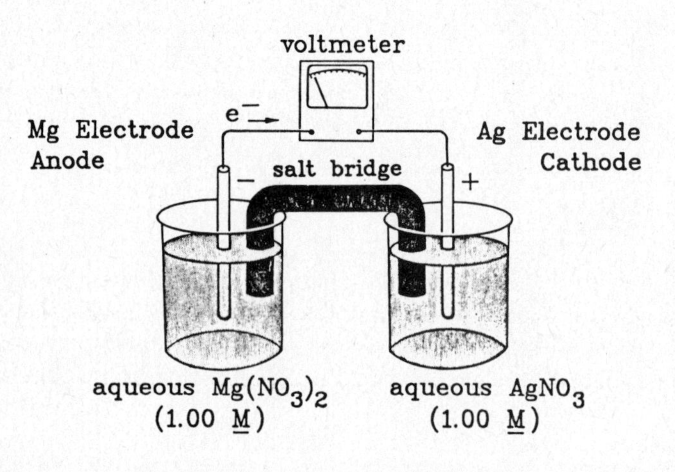

Magnesium-SilverCell

Originally, the solutions are both 1.00 M, but this quickly changes as electrons flow. The reading on the voltmeter is a measure of the intrinsic tendency of the reaction to proceed.

Electrical Current and its Properties

Usually, the only kind of work that chemical systems can do on their surroundings is pressure-volume work. Electrochemical cells are the major exception. They are able to produce or consume *electrical work*, in addition to pressure-volume work. A *galvanic* cell contains a chemical system that is not at equilibrium. It develops a *potential difference* between its electrodes. This potential difference ($\Delta\xi$) is measured in volts (V). It represents the intrinsic tendency of the cell to come to equilibrium. Completing an electrical circuit between the two electrodes allows electricity to flow from one to the other. If suitably harnessed, the electricity can perform work. The amount of electrical work absorbed is:

$$w_{elec} = -Q\Delta\xi$$

where Q is the quantity of electricity transferred and $\Delta\xi$ is the potential difference. This equation defines the joule in terms of electrical quantities because w is in joules when Q is in *coulombs* (C) and $\Delta\xi$ is in volts. *A joule is the amount of energy required to push one coulomb of charge through a potential difference of one volt.*

$$1\ \text{J} = 1\ \text{volt·coulomb}$$

The electron-volt, a much smaller unit of energy, is the amount of energy required to push 1 electron through a potential difference of 1 volt. A coulomb of electricity has the same amount of charge as 6.241×10^{18} electrons. A joule therefore is 6.241×10^{18} eV.

Electrical current is measured in amperes (A). A current of one ampere is one coulomb flowing through an electrical circuit every second:

$$I = Q / t$$

In practice, it is easy to measure electrical current (with an *ammeter*). Consequently, this definition crops up extensively in problems. **See problems 1b, 5a, 21.**

Galvanic and Electrolytic Cells

There are two types of electrochemical cells. The first, mentioned above, is the galvanic cell. It uses the drive of a spontaneous process toward equilibrium to generate a voltage that can perform electrical work. The second, the *electrolytic* cell, uses an outside source of electric current to force non-spontaneous chemical reactions to occur.

Both galvanic and electrolytic cells employ oxidation-reduction reactions. Any overall oxidation-reduction reaction is the sum of two *half-reactions.* For example, the overall reaction:

$$Cu^{2+}(aq) + Zn(s) \rightarrow Cu(s) + Zn^{2+}(aq)$$

is the sum of the half-reactions:

$$Cu^{2+}(aq) + 2\,e^- \rightarrow Cu(s) \qquad \text{reduction}$$
$$Zn(s) \rightarrow Zn^{2+}(aq) + 2\,e^- \qquad \text{oxidation}$$

Half-equations can never be confused with whole equations. They always show the electron (e^-) as a reactant or product.

Breaking a balanced equation down into half-equations is an important skill:

1. Determine the oxidation numbers of all elements in all compounds to learn which elements are oxidized and which reduced.

2. Write down the two processes separately.

3. Insert electrons so that the two sides of the resultant half-equations have the same net charge.

Also, writing half-equations helps in balancing oxidation-reduction reactions. See problem 37 in Chapter 2 (p. 36 of this Guide). The following are useful generalizations about half-equations:

- The oxidation half-reaction always involves the *loss* of electrons. It always takes place at the *anode* of an electrochemical cell.

- The reduction half-reaction always involves the *gain* of electrons. It always takes place at the *cathode* of an electrochemical cell.

- Electrons liberated at the anode pass through the outside circuit and are consumed by the reduction half-reaction at the cathode. The quantity of electrons liberated by the oxidation must equal the quantity absorbed by the reduction.

In solving problems it is vital to know the definitions of cathode and anode. Thus, in **problem 1a** the only way to know which half-reaction occurs at which electrode is to recognize which is an oxidation and to know that oxidation always occurs at the anode.

Do not bother to learn the sign of the electrical polarity of cathode and anode in electrochemical cells. It depends on whether the cell is galvanic or electrolytic. It is a mistake to memorize that "the cathode is negative and the anode is positive." The statement is true enough for electrolytic cells, but exactly wrong for galvanic cells.

Faraday's Laws

When channeled through an outside circuit, electrons are easily manipulated. Their rate of flow is readily measured with an ammeter and can be varied by changing the electrical resistance of the circuit. *The electrons are in effect just another chemical reactant.* Faraday's laws apply the rules of chemical stoichiometry to the electron as chemical reactant (or product). The laws are:

1. In any cell the mass of a given substance produced or consumed at an electrode is proportional to the quantity of electrical charge (Q) passed through the cell.

2. Equivalent weights of different substances are produced or consumed at an electrode by the passage of a given quantity of electrical charge through the cell.

Once the electron is accepted as a chemical reactant a mole of electrons assumes the same role in chemical calculations as a mole of sodium or $ZnCl_2$ or other substance. **See problem 5b.** Although electrons cannot be weighed out, they are easily tracked because they are charged. *A mole of electrons is 96485 coulomb.* This quantity of charge is the *faraday*:

$$1 \text{ faraday } (F) = 96485 \text{ C mol}^{-1}$$

Faraday's laws are essential in solving numerous problems. Consider the following key points in applying the laws:

- The quantity of charge passing through a cell is the average current times the time it flows. If *t* is in seconds and *I* in amperes then their product is the quantity of electricity (total electrical charge) passing through the cell in coulombs.

- To convert a quantity of electricity from electrical units (coulombs) to chemical units (moles) use the faraday. Memorize the fact that 1 mole is 96485 C.

- Balanced half-reactions show the chemical changes at a cell's electrodes. Do stoichiometry problems with half-reactions and electrons just like any other stoichiometry problems.

- An *equivalent weight* of a substance is the amount produced or consumed in an electrochemical cell by the passage of of 1 mole of electrons. The following table shows the relationship of molar weight and equivalent weight:

Half-Reaction	*Molar Weight of Product*	*Mass of Product per 96485 C* *
$Ag^{+}(aq) + e^{-} \rightarrow Ag(s)$	107.9 g mol^{-1}	107.9 g
$Cu^{2+}(aq) + 2\ e^{-} \rightarrow Cu(s)$	63.54 g mol^{-1}	31.8 g
$Al^{3+}(aq) + 3\ e^{-} \rightarrow Al(s)$	26.98 g mol^{-1}	9.0 g
$2\ Cl^{-}(aq) \rightarrow Cl_2(g) + 2\ e^{-}$	70.91 g mol^{-1}	35.5 g (11.2 L at STP)

* Equivalent weight of product

For examples of the application of Faraday's laws see **problems 5, 19b and 23.** A typical reversal problem occurs in **problem 37** in which the mass of a reacting substance is given and the quantity of charge must be calculated. **Problem 21** gives the chemical amount of a substance and an average current and asks for a time.

9.2 *FREE ENERGY AND CELL VOLTAGE*

Oxidation-reduction reactions are either spontaneous or not spontaneous. A spontaneous oxidation-reduction reaction in an electrochemical cell generates a measurable potential difference (voltage) between the two electrodes. This $\Delta\xi$ is related to the free energy change of the reaction in the cell:

$$\Delta G = -n\,F\Delta\xi$$

In this equation $\Delta\xi$ is the measured cell voltage, F is the faraday and n is the chemical amount of electrons transferred in the cell. ΔG comes out in J if $\Delta\xi$ is in V, F is in C mol^{-1}, and n is in mol. ΔG comes out in J mol^{-1} if n is regarded as the number of moles of electrons transferred per mole of chemical reaction. The difference echoes the distinction between an extensive and an intensive property. Running an oxidation-reduction reaction in a huge industrial-scale cell transfers many thousands of moles of electrons. Doing the same reaction in a small laboratory-scale cell transfers only a few tenths of a mole of electrons even though the balanced reaction is the same in both cases. The *extensive* property ΔG (measured in J) is large in the first case and small in the second. If n is viewed as the number of moles of electrons transferred *per mole of reaction* then its units are "mol mol^{-1}", making it unit-less and causing ΔG to come out in J mol^{-1}. A ΔG in J mol^{-1} refers to a chemical reaction rather than to events in an actual physical cell. ΔG in J mol^{-1} is an *intensive* property.

In an electrolytic cell the chemical reaction is non-spontaneous. This corresponds to a positive ΔG and therefore to a negative $\Delta\mathcal{E}$. A negative $\Delta\mathcal{E}$ means a positive *outside* voltage is required to force the reaction to run. The reverse reaction has a positive $\Delta\mathcal{E}$ and runs spontaneously.

It is not necessary to allow any current to flow in a galvanic cell to measure its voltage. If no current flows, then no chemical reaction occurs. The potential difference between the electrodes tells the *tendency* or drive for the chemical reaction to occur. It is thus well-named.

If current is allowed to flow, a chemical change take place. Only then does a galvanic cell have even a *chance* of doing any electrical work on its surroundings. If the current flows infinitely slowly, the chemical change goes reversibly and can produce the maximum electrical work. Continue to focus on the work produced. Recall that the work produced by a system is $-w$ because $+w$ is the work absorbed. Then:

$$(-w)_{elec,max} = -\Delta G = n F \Delta\mathcal{E}$$

This equation is used to solve **problem 17.** The operation of a practical cell is always irreversible and produces *less* than $(-w)_{max}$:

$$(-w)_{elec,irrev} < -\Delta G$$

If some galvanic cell has a ΔG of -25 kJ, then, by the above equations, it can produce a maximum of $+25$ kJ of electrical work. The catch is that extracting this maximum work would require reversible operation, which is an unattainable ideal. In actual operation, *less* than 25 kJ of electrical work would appear in the surroundings. Depending on how the cell was hooked up, as little as zero electrical work could appear. All of free energy change might be diverted to simple resistive heating, for example.

Standard States and Cell Voltages

The above equation applies to cell reactions going on at any combination of temperature and pressure as long as T and P are constant. It certainly also applies when the products and reactants in the equation are in their standard states at a temperature of 298.15 K and a pressure of 1 atm. Under these restricted circumstances the superscript zero is added to the symbols:

$$\Delta G^\circ = -n F \Delta\mathcal{E}^\circ$$

Solutes are in their standard states when their concentration is 1 M under the standard conditions. Gases are in their standard states when their partial pressure is 1.00 atm.

The above equation means that measurement of $\Delta\mathcal{E}^\circ$ values gives equilibrium constants:

$$\Delta G^\circ = -RT \ln K = -n F \Delta\mathcal{E}^\circ$$

This relationship is used in **problem 11.** In practice, $\Delta\mathcal{E}^\circ$ is measured in volts. Multiplication by F (which is 96485 C mol^{-1}) and n, the number of moles of electrons transferred per mole of reaction as written, gives ΔG° in J mol^{-1}. Division of this value by $-R$ (which is 8.314 J $mol^{-1}K^{-1}$) and T on the absolute scale gives $\ln K$.

Half-Cell Potentials

Appendix E (text) tabulates *half-cell* reduction potentials, $\mathcal{E}^\circ$. The data in this table occur in many problems, and the structure of the table must be understood. The following are important facts about this table:

1. The $\mathcal{E}^\circ$'s are standard *reduction* potentials. They refer to reduction half-reactions (half-reactions with electrons shown on the *left*). *The larger the reduction potential of a species, the greater is its relative tendency to be reduced.*

2. A species on the *left* of a given half-cell reaction and in its standard state will spontaneously oxidize a standard-state species on the *right* of any half-cell reaction located *below* it in the table.

3. A reduction potential becomes an oxidation potential when the direction of a half-cell reaction is reversed. Simply change the sign of the associated potential. All half-cell reactions may be written in the reverse direction. The larger the oxidation potential of a species then the greater is its relative tendency to be oxidized. This point is used in **problem 41.**

4. The standard reduction potentials are *intensive* properties. They are *not* affected when the coefficients in the half-reaction are changed. *Example.* The half-reaction $Cu^{2+}(aq) + 2\,e^- \rightarrow Cu(s)$ has the same $\mathcal{E}^\circ$ as $2\,Cu^{2+}(aq) + 4\,e^- \rightarrow 2\,Cu(s)$

5. A standard half-cell reduction potential, $\mathcal{E}^\circ$, cannot be directly measured. Such potentials are derived by measuring $\Delta\mathcal{E}^\circ$ values for different pairings of half-reactions and *assigning* a $\mathcal{E}^\circ$ of 0.000 V to the *reference* half-reaction: $H^+(aq) + e^- \rightarrow \frac{1}{2}\,H_2(g)$

The standard voltage of any galvanic cell is the *difference* between the reduction potential of the half-reaction taking place at the *cathode* and the half-reaction taking place at the anode:

$$\Delta\mathcal{E}^\circ = \mathcal{E}^\circ\text{ (cathode)} - \mathcal{E}^\circ\text{ (anode)}$$

The procedure for getting $\Delta\mathcal{E}^\circ$ in applications is:

1. Balance the overall oxidation-reduction equation.

2. Break down this chemical equation into a balanced reduction half-equation and a balanced oxidation half-equation.

3. Find these half-equations in Appendix E and note their standard reduction potentials. The oxidation half-reaction will be the *reverse* of one of the tabulated half-reactions. **Caution:** Find the correct half-equations, including the physical state of all reactants and products. Some half-equations resemble each other fairly closely, but have quite different standard reduction potentials.

4. Subtract the tabulated reduction potential, $\mathcal{E}^\circ$, of the reaction taking place at the anode from the tabulated reduction potential for the reaction taking place at the cathode.

This procedure is followed in **problems 3 and 15.**

Addition and Subtraction of Half-Cell Reactions

As just pointed out, when two half-reactions are to give a whole reaction, in which no electrons appear explicitly, the $\mathcal{E}^\circ$ values are combined by simple subtraction:

$$\Delta\mathcal{E}^\circ = \mathcal{E}^\circ(\text{cathode}) - \mathcal{E}^\circ(\text{anode})$$

If $\Delta\mathcal{E}^\circ$ comes out negative, then the reaction is *non-spontaneous* as written and the reverse reaction is spontaneous.

When two *half-reactions* are combined to make a new half-reaction the calculation is not so simple. Then:

$$\mathcal{E}^\circ_3 = \frac{n_1\mathcal{E}^\circ_1 - n_2\mathcal{E}^\circ_2}{n_3}$$

where the subscripts 1 and 2 refer to the reduction potentials and quantities of electrons transferred in the two half-reactions being combined, and where subscript 3 refers to the resultant half-reaction. See **problem 7a.** Note that this equation, if used when half-reactions combine to make a whole reaction, has $n_1 = n_2 = n_3$, and the simple subtractive relationship already cited is the result.

9.3 *CONCENTRATION EFFECTS AND THE NERNST EQUATION*

Standard cell potentials calculated from Appendix E are exactly correct only for a cell in which all of the reactants and products are present in their standard states (activity = 1). This is because cell potentials depend on the *activity* of the reactants and products in the cell reaction as well as their identity. To compute $\Delta\mathcal{E}$ (without a superscript) requires the concentrations (in mol L^{-1}) of all the solutes in the chemical equation and the partial pressures (in atm) of all the gases. Accept these numbers as equal to the activities of the solutes and gases. The activities of pure solids and liquids are 1. Then apply the *Nernst equation:*

$$\Delta\mathcal{E} = \Delta\mathcal{E}^\circ - (RT / nF) \ln Q$$

where Q is the reaction quotient of the chemical equation. The final term in the Nernst equation is a subtractive correction factor for the non-standard activities of the reactants and products in the reaction. Suppose a galvanic cell employs the reaction:

$$Mg(s) + 2\,Ag^{+}(aq) \rightarrow Mg^{2+}(aq) + 2\,Ag(s)$$

This is exactly the cell diagrammed on p. 192 of this Guide. $\Delta\mathcal{E}^\circ$ is $\mathcal{E}^\circ$(cathode) − $\mathcal{E}^\circ$(anode) which is 0.800 − (−2.375) = 3.175 V. The expression for Q for this reaction is:

$$Q = [Mg^{2+}] / [Ag^{+}]^2$$

and the Nernst equation is:

$$\Delta\mathcal{E} = 3.175\text{ V} - (RT/2F) \ln [Mg^{2+}] / [Ag^{+}]^2$$

Note that n is 2 because 2 moles of electrons are transferred per mole of the balanced oxidation-reduction reaction. If $[Mg^{2+}] = [Ag^{+}] = 1$ M, Q is equal to 1 and ln Q is zero. The observed potential difference, displayed on the voltmeter in the diagram on p. 192 of this Guide, then equals $\Delta\xi^\circ$, which is 3.175 V. If the anode and cathode are connected by a wire, a current spontaneously flows through the wire, and the reaction proceeds to the right. $[Mg^{2+}]$ increases and $[Ag^{+}]$ decreases. As they do, Q increases and $\Delta\xi$, the electrical potential of the cell, instantly becomes less than 3.175 V. As current continues to flow, the observed potential difference ultimately trickles all the way down to 0.00 V since in the Nernst equation an ever-larger correction factor is subtracted from 3.175 V.

At $\Delta\xi = 0$ the cell is *dead* as far as this reaction's further release of chemical free energy is concerned. ΔG is zero because $\Delta\xi$ is zero. A dead galvanic cell is at equilibrium. Meanwhile, neither $\Delta\xi^\circ$ nor ΔG° has changed.

At equilibrium the reaction quotient Q equals K, the equilibrium constant. Measurement of $\Delta\xi^\circ$'s of cells gives K's because:

$$0.00 = \Delta\xi^\circ - (RT / nF) \ln K$$

This is illustrated in **problem 11.**

In problem solving, a convenient form of the Nernst equation is:

$$\Delta\xi = \Delta\xi^\circ - (0.0592 / n) \log_{10} Q$$

This equation combines numerical values of R, T and F and converts from natural logarithms to base 10 logarithms. The units of the number 0.0592 are V. This specialized form should *not* be used unless the temperature is 298.15 K. Like ΔG, $\Delta\xi$ is strongly dependent on temperature. Taking the magic number approach (shoe-horning "0.0592" into every problem) often results in ignoring this dependence.

The Nernst equation occurs constantly in problems in electrochemistry. **Problem 9** is a simple example. As is generally the case, other problems involve reversal of the usual pattern of information given. Thus, in **problem 13** $\Delta\xi$ and $\Delta\xi^\circ$ are known and Q must be computed.

The Nernst equation applies to half-reactions and their standard half-cell potentials just as well as it does to whole reactions. For example, the half-reaction:

$$H^{+}(aq) \rightarrow \tfrac{1}{2} H_2(g)$$

has the Nernst equation:

$$\xi = \xi^\circ - (RT / nF) \ln (P_{H_2} / [H^{+}])$$

This exact expression is used in **problem 19a.**

9.4 *ACID-BASE POTENTIOMETRIC TITRATIONS*

The operation of the *pH meter* is an important application of the dependence of cell potential on concentration. A *reference* electrode and a H^+ *(aq)*-sensitive electrode are placed in the aqueous solution under investigation. The observed cell potential depends linearly on $[H^+]$ in the solution, according to the Nernst equation. At 25° C:

$$\Delta\mathcal{E} = \Delta\mathcal{E}^\circ(\text{ref}) - 0.0592 \log [H^+]$$

where $\Delta\mathcal{E}^\circ$ (ref) is a constant that depends on which half-reaction is selected for use in the reference electrode. In an acid-base potentiometric titration, the voltage observed with a pH meter is plotted as a function of the volume of titrant added. The plot has the same shape as a plot of pH versus volume of titrant.

9.5 *APPLICATIONS OF GALVANIC CELLS*

Galvanic cells (*batteries*) have many important applications. *Primary cells* generate electrical work from a chemical system but cannot be recharged. *Secondary cells*, or accumulators, are cells that can be recharged by using electrical energy from an external source to reverse the oxidation-reduction reaction that discharged them.

"Re-charging" does not involve putting electrons back into a cell. A cell does not in fact lose any electrons even if completely discharged. Instead it means forcing the electrons in the cell back to their former positions of higher chemical potential energy.

The text gives the overall cell reactions of two important non-rechargeable galvanic cells, the *Leclanché* cell and the *mercury* cell.

Rechargeable Batteries

The *lead storage* battery is a rechargeable battery. As it is discharged the following takes place:

$$Pb(s) + SO_4^{2-}(aq) \rightarrow PbSO_4(s) + 2\,e^- \quad \text{anode}$$
$$PbO_2(s) + SO_4^{2-}(aq) + 4\,H^+(aq) + 2\,e^- \rightarrow PbSO_4(s) + 2\,H_2O(\ell) \quad \text{cathode}$$

$$Pb(s) + PbO_2(s) + 2\,SO_4^{2-}(aq) + 4\,H^+(aq) \rightarrow 2\,PbSO_4(s) + 2\,H_2O(\ell)$$

Fuel Cells

Unlike a battery, a closed system to which additional quantities of reactants cannot be added, a *fuel cell* continuously "burns" fuel electrochemically. Fuel cells offer a theoretical advantage over traditional means of using fuels. When a fuel is burned in the usual way, the conversion of the heat released, $-\Delta H$, to work encounters thermodynamic limitations on its efficiency. In a fuel cell the free energy of the chemical reaction is converted directly to electrical energy, and this limitation is evaded.

9.6 *ELECTROLYSIS OF WATER AND AQUEOUS SOLUTIONS*

In pure water $H^+(aq)$ is not in its standard state (1 M) but instead has a concentration of 10^{-7} M. This *lowers* $\mathcal{E}$ for the half-reaction:

$$2\,H^+(aq) + 2\,e^- \rightarrow H_2(g)\ 1\text{ atm}$$

from 0.00 V, the standard half-cell potential, to −0.414 V.

Similarly, it *lowers* $\mathcal{E}$ for the reduction of oxygen to water:

$$\tfrac{1}{2}\,O_2(g)\,(1\text{ atm}) + 2\,H^+(aq) + 2\,e^- \rightarrow H_2O(l)$$

from 1.229 V to 0.815 V.

The two new half-cell potentials were calculated using the Nernst equation. Both answers make sense from the point of view of LeChatelier's principle. In the two reactions, reactant $H^+(aq)$ is present in less than 1 M (standard) concentration so there is a decreased drive to the right.

The overall cell voltage for the process composed by subtracting the second of these half-equations from the first is $\Delta\mathcal{E}^\circ = -1.229$ V. The process is:

$$H_2O(l) \rightarrow H_2(g)\,(1\text{ atm}) + \tfrac{1}{2}\,O_2(g)\,(1\text{ atm})$$

The negative $\Delta\mathcal{E}^\circ$ confirms that water does not spontaneously decompose to $H_2(g)$ and $O_2(g)$ at ordinary conditions. The superscript appears on $\Delta\mathcal{E}$ because all reactants and products are in their standard states. Instead, the decomposition of water requires an outside voltage of 1.229 V in an electrolytic cell. When such a cell is run, $H_2(g)$ forms at the cathode and $O_2(g)$ at the anode.

These facts have important applications in the electrolysis of aqueous solutions:

- A species in neutral aqueous solution can be reduced only if its reduction potential exceeds −0.414 V. (Note: The number −0.50, for example, is *less* than −0.414 because of the minus sign). If this condition is not met, then $H^+(aq)$ is reduced instead of the species in question.

- A species in neutral aqueous solution can be oxidized only if its reduction potential is *less* than 0.815 V. Otherwise $H_2O(l)$ is oxidized to $O_2(g)$ instead.

These criteria are applied in **problem 43.**

DETAILED SOLUTIONS TO ODD-NUMBERED PROBLEMS

1. ***a)*** In the electrolysis of molten KCl the half-cell reactions are:

$$K^+(l) + e^- \rightarrow K(l) \qquad \text{reduction; at the cathode}$$

$$Cl^-(l) \rightarrow \tfrac{1}{2} Cl_2(g) + e^- \qquad \text{oxidation; at the anode}$$

The sum of these two half-equations represents the overall cell reaction:

$$K^+ + Cl^- \rightarrow K(l) + \tfrac{1}{2} Cl_2(g)$$

b) A current of 2.00 A is the same as 2.00 C s^{-1}. 500 hours is 1.80×10^4 s. The amount of electricity passing through the cell is the product of the average current and the length of time it flows or 3.60×10^4 coulomb. Since 96,485 coulomb is a mole of electrons, 0.373 mol of electrons passes through the cell. Every mole of electrons causes 1 mole of K and ½ mole of Cl_2 to form. Therefore 0.0187 mol of Cl_2 and 0.373 mol of K form during the 5.00 hour electrolysis run. This is 13.2 g of Cl_2 (molar weight 70.91 g mol^{-1}) and 14.6 g of K (molar weight 39.102 g mol^{-1}). A tacit assumption is that the cell has at least 27.8 g of KCl in it.

---

3. ***a)*** From the description of the two half cells the reaction going on in the first is either:

$$Zn(s) \rightarrow Zn^{2+}(aq) + 2\,e^- \quad \text{or} \quad Zn^{2+}(aq) + 2\,e^- \rightarrow Zn(s)$$

and in the second either:

$$2\,e^- + Cl_2(g) \rightarrow 2\,Cl^-(aq) \quad \text{or} \quad 2\,Cl^-(aq) \rightarrow Cl_2(g) + 2\,e^-$$

The higher the reduction half-cell voltage the greater is the relative ease of reduction. Therefore, from the data in text Appendix E, $Cl_2(g)$ is more easily reduced to $Cl^-(aq)$ ($\mathcal{E}^\circ = +1.358$ volts) than $Zn^{2+}(aq)$ is to $Zn(s)$ ($\mathcal{E}^\circ = -0.763$ volts). It is therefore $Cl_2(g)$ that is *reduced.* The $Zn(s)$ is oxidized. Reduction *always* occurs at the cathode; the Cl_2 (Pt) is the cathode and the $Zn(s)$ is the anode.

b) The overall reaction is:

$$Zn(s) + Cl_2(g) \rightarrow Zn^{2+}(aq) + 2\,Cl^-(aq)$$

The tremendous ability of Cl_2 as an oxidizing agent is well known. That the reaction runs in this direction (and not the reverse) is thus general chemical knowledge.

c) The standard cell voltage is the reduction potential for the half-reaction at the anode subtracted from the reduction potential for the half-reaction at the cathode. Taking data from Appendix E, this is −0.763 V subtracted from 1.358 V or:

$$\mathcal{E} = 1.358 - (-0.763) = 2.121 \text{ V}$$

---

5. *a)* An ampere (A) is defined as a $C\ s^{-1}$. If a steady current of 0.800 A flows for 25.0 minutes it flows for 1.50×10^3 s. $0.800\ C\ s^{-1} \times (1.50 \times 10^3\ s)$ makes 1.20×10^3 C passing through the circuit. One faraday is 96,485 C, so 0.0124 faraday passes through the circuit.

b) A faraday is a mole of electrons. In the Zn half-cell the half-reaction is:

$$Zn(s) \rightarrow Zn^{2+}(aq) + 2\ e^-$$

which means that for every mole of electrons that passes through the cell ½ mol of solid Zn dissolves. For 0.0124 mol of e^- to pass the cell 0.00622 mol of Zn must dissolve. The 0.0124 is divided by 2 because 2 mol of electrons is transferred for every 1 mol of zinc. Zn has a molar weight of 65.37 g mol^{-1}. Hence 0.00622 g $\times$ 65.37 g mol^{-1} or 0.407 g of Zn dissolves from the anode. At the cathode 0.00622 mol of Cl_2 *(g)* is reduced to 0.0124 mol of Cl^- *(aq)*:

$$Cl_2(g) + 2\ e^- \rightarrow 2\ Cl^-(aq)$$

The product, 0.440 g of Cl^- *(aq)*, goes into solution leaving the weight of the inert Pt electrode unaffected.

---

7. *a)* The standard potential for the half-reaction:

$$Mn^{3+}(aq) + 3\ e^- \rightarrow Mn(s)$$

is *not* the simple sum of the half-cell potentials for the half-reactions:

$$Mn^{2+}(aq) + 2\ e^- \rightarrow Mn(s) \qquad \mathcal{E}^\circ = -1.029\ V$$

$$Mn^{3+}(aq) + e^- \rightarrow Mn^{2+}(aq) \qquad \mathcal{E}^\circ = 1.51\ V$$

even though the target half-reaction *is* the sum of these two half-reactions. Instead the potential is a *weighted average*:

$$n\mathcal{E}^\circ = n_1\mathcal{E}^\circ_1 + n_2\mathcal{E}^\circ_2$$

where the n's are the number of electrons transferred in the different half-reactions and the subscripts refer to the two half-reactions being combined. See page 198 of this Guide. Thus:

$$3\ \mathcal{E}^\circ = 2(-1.029) + 1(1.51)\ V \qquad \textit{ergo} \quad \mathcal{E}^\circ = -0.183\ V$$

b) The disproportionation reaction:

$$3\ Mn^{2+}(aq) \rightarrow Mn(s) + 2\ Mn^{3+}(aq)$$

is a combination of the reduction of Mn^{2+} *(aq)* to Mn*(s)* and the oxidation of Mn^{2+} *(aq)* to Mn^{3+} *(aq)*. As such it is the first of the following half-reactions subtracted from the second:

$$2\ Mn^{3+}(aq) + 2\ e^- \rightarrow 2\ Mn^{2+}(aq) \qquad \mathcal{E}^\circ = 1.51\ V$$

$$2\ e^- + Mn^{2+}(aq) \rightarrow Mn(s) \qquad \mathcal{E}^\circ = -1.029\ V$$

To attain balance in the net reaction all the coefficients in the first half-reaction had to be doubled. Doubling the quantities of reactants and products has no effect on the overall voltage of the disproportionation reaction. Cell voltage is an intensive property. It is the same whatever the size of the system. The cell potential is:

$$\Delta\xi^\circ = \xi^\circ(\text{reduction}) - \xi^\circ(\text{oxidation})$$
$$\Delta\xi^\circ = -1.029 - 1.51 = -2.539 \text{ V}$$

A formula like the one in part a) could be used:

$$n\xi^\circ = n_1\xi_2^\circ - n_2\xi_1^\circ$$

but all of the n's are equal to 2. Dividing through by 2 then gives:

$$\xi^\circ = \xi_2^\circ - \xi_1^\circ = -1.029 - 1.51 = -2.539 \text{ V}$$

The equilibrium constant for the disproportionation is related to its standard voltage by the equation:

$$RT \ln K = n F\Delta\xi^\circ$$

Substituting $n = 2$, $\Delta\xi^\circ = -2.539$ V, $R = 8.314$ J mol^{-1}K^{-1}, $T = 298.15$ K and $F = 96485$ C gives $K = 1.4 \times 10^{-86}$.

Solutions of Mn^{2+} *(aq)* do *not* spontaneously disproportionate to Mn*(s)* and Mn^{3+} *(aq)* because the standard voltage for the process is negative and large, making the equilibrium constant very small.

---

9. ***a)*** The silver ion is reduced, and nickel metal is oxidized:

$$Ag^+(aq) + e^- \rightarrow Ag(s) \quad \textit{and} \quad Ni(s) \rightarrow Ni^{2+}(aq) + 2\,e^-$$

The overall cell reaction is the sum of these two half-reactions with the first doubled so that the electrons cancel out:

$$Ni(s) + 2\,Ag^+(aq) \rightarrow Ni^{2+}(aq) + 2\,Ag(s)$$

b) The cell voltage depends on both the intrinsic tendency for electron transfer to occur and the activities of the reactants and products. The Nernst equation tells how:

$$\Delta\xi = \Delta\xi^\circ - (RT / nF) \ln Q$$

For this reaction $n = 2$ and $\Delta\xi^\circ = 0.080 \text{ V} - (-0.23 \text{ V})$ (the difference between the standard reduction potentials of the two half-reactions listed above). Consult Appendix E for such values. The reaction quotient is:

$$Q = [Ni^{2+}] / [Ag^+]^2 = 0.1 / (0.1)^2 = 10$$

Substituting the various constants:

$$\Delta\xi = 1.03 \text{ V} - (0.0592 \text{ V} / 2) \log 10$$

$$\Delta\xi = 1.00 \text{ V}$$

c) At equilibrium $\Delta\xi = 0$ and Q becomes equal to K:

$$0 \text{ V} = \Delta\xi^\circ - (0.0592 \text{ V}/2) \log K$$

$$\log K = 34.8 \qquad K = 6.3 \times 10^{34}$$

d) The equilibrium constant is so huge that the reaction by any reasonable standard goes entirely to completion. If all of the Ag^{+} *(aq)* is consumed, then the concentration of Ni^{2+} *(aq)* at equilibrium is 0.10 M (its original value) plus 0.05 M (from the reaction). Setting $[Ni^{2+}] = 0.15$ M and inserting it in the mass-action expression:

$$K = 6.3 \times 10^{34} = [Ni^{2+}] / [Ag^{+}]^2 \qquad \textit{hence} \qquad [Ag^{+}] = 1.5 \times 10^{-18} \text{ M}$$

---

11. The overall reaction consists of the oxidation of chromium(III) to chromium(VI) and the simultaneous reduction of chlorous acid to hypochlorous acid. The sum of two half-reactions can represent the reaction:

$$7\,H_2O(l) + 2\,Cr^{3+}(aq) \rightarrow Cr_2O_7^{2-}(aq) + 14\,H^{+}(aq) + 6\,e^{-}$$

$$3\,HClO_2(aq) + 6\,H^{+}(aq) + 6\,e^{-} \rightarrow 3\,HClO(aq) + 3\,H_2O(l)$$

Both of these half-reactions are tabulated in Appendix E along with their standard *reduction* potentials. The Cr(III) to Cr(VI) half-reaction has here been written as an oxidation. Its ξ° is for that reason the negative of the reduction potential found in Appendix E. The *oxidation potential* is -1.33 V. The second half-reaction appears in Appendix E as a reduction but *not* multiplied through by three. The $\Delta\xi^\circ$, the standard potential for the whole reaction, is the sum of -1.33 and 1.64 V or 0.31 V. The procedure of *reversing* one of the two half-reactions, *changing* the sign of that half-reaction's ξ° and then *adding* the half-cell voltages is equivalent to using the formula $\Delta\xi^\circ = \xi^\circ(\text{cathode}) - \xi^\circ(\text{anode})$.

The equilibrium constant is related to $\Delta\xi^\circ$ by the equation:

$$\Delta\xi^\circ = (RT / nF) \ln K$$

At 25° C this becomes:

$$\Delta\xi^\circ = (0.0592 / n) \log K$$

This reaction transfers 6 electrons, according to the half-equations. Because n is 6, log K is $6(0.31) / 0.0592 = 30.4$. K is 2.6×10^{31}.

If 2.00 L of 1.00 M $HClO_2$ *(aq)* and 2.00 L of 0.50 M $Cr(NO_3)_3$ *(aq)* are mixed then 2.00 mol of $HClO_2$ *(aq)* and 1.00 mol of Cr^{3+} *(aq)* are available for reaction. The 1.00 mol of Cr^{3+} *(aq)* only needs 1.50 mol of $HClO_2$ *(aq)* for complete reaction, as shown in the balanced equation. Therefore, $HClO_2$ is in excess. The enormous equilibrium constant means that the reaction converts essentially *all* of the Cr^{3+} *(aq)* to $Cr_2O_7^{2-}$ *(aq)* to reach equilibrium. The color of the solution therefore is *orange*, the color of the dichromate ion.

---

13. Reduction always take place at the cathode in electrochemical cells. Therefore the half-reaction at the cathode is:

$$I_2(s) + 2\,e^- \rightarrow 2\,I^-(aq)\,(1M) \qquad \xi^\circ = 0.535\text{ V}$$

At the anode H_2 gas (at 1 atm) is oxidized:

$$H_2(g) \rightarrow 2H^+(aq) + 2\,e^- \qquad \xi^\circ = 0.00\text{ V}$$

If all of the reactants and products were at unit activity, the voltage of this cell would be 0.535 − 0.00 = 0.535 V. The voltage that is observed is much larger than this, so at least one reactant or product is *not* present at unit activity. The Nernst equation allows the evaluation of such effects of concentration on voltage. The overall chemical reaction is:

$$I_2(s) + H_2(g) \rightarrow 2\,H^+(aq) + 2\,I^-(aq) \qquad \Delta\xi^\circ = 0.535\text{ V}$$

The corresponding Nernst equation at standard conditions is:

$$\Delta\xi = \Delta\xi^\circ - (0.0592\,/\,n)\log\{[I^-]^2\,[H^+]^2\,/\,P_{H_2}\}$$

In this case P_{H_2} = 1 atm, $[I^-]$ = 1 M, and n = 2:

$$0.841 = 0.535 - (0.0592\,/\,2)\ \log[H^+]^2$$

$$0.841 = 0.535 - 0.0592\log[H^+]$$

$$\log[H^+] = -5.17 \qquad \textit{and} \qquad pH = 5.17$$

---

15. ***a)*** The standard potential for the reaction is:

$$\Delta\xi^\circ = \xi^\circ(\text{cathode}) - \xi^\circ(\text{anode})$$

At the cathode, dichromate is reduced to chromium(III). At the anode, iodide is oxidized to iodine. The standard reduction potentials (Appendix E) for these two half-reactions are 1.33 V and 0.535 V, respectively. Therefore $\Delta\xi^\circ$ is 0.795 V.

b) For this process *Q*, the reaction quotient, is:

$$Q = \frac{[Cr^{3+}]^2}{[Cr_2O_7^{\,2-}][H^+]^{14}[I^-]^6}$$

At pH 0 the hydronium ion concentration, $[H_3O^+]$, is 1 M. Recall that hydronium ion is H^+ *(aq)* ion written so as to emphasize its association with H_2O. Hence, $[H^+]$ = 1 M. Also, from the statement of the problem, $[Cr_2O_7^{\,2-}]$ = 1.5 M and $[I^-]$ = 0.40 M. So:

$$Q = 1.628 \times 10^2\,[Cr^{3+}]^2$$

Using the Nernst equation at 25° C:

$$0.87 = 0.795 - (0.0592 / n) \log Q$$

where n = 6 because the balanced equation shows a transfer of 6 electrons. Solving for log Q:

$$\log Q = -7.601 \qquad \textit{and} \qquad Q = 2.5 \times 10^{-8}$$

Substitution of Q and the available concentration values into the reaction quotient expression gives:

$$1.6 \times 10^{2} [Cr^{3+}]^{2} = 2.5 \times 10^{-8} \qquad \textit{ergo} \qquad [Cr^{3+}] = 1.2 \times 10^{-5}\ M$$

---

17. If the fuel cell were to operate with 100 percent efficiency, then the electrical work absorbed per *mole* of reaction would be the $\Delta G°$ of formation of liquid water. This is true first because the chemical equation in the problem exactly represents the formation of liquid water from its constituent elements in their standard states at 298.15 K and second because:

$$\Delta G = w_{elec,max}$$

From Appendix D, the $\Delta G°_f$ of $H_2O(l)$ is -237.18 kJ mol^{-1}. The maximum electrical work absorbed per gram of water is 1 / 18.01 times this figure because the molar weight of water is 18.01 g mol^{-1}. The maximum work is -13.17 kJ g^{-1}. Therefore $+13.17$ kJ g^{-1} of electrical work is generated, assuming 100 percent efficiency. At 60.0 percent efficiency, 7.90 kJ g^{-1} is generated.

---

19. ***a)*** The product at the cathode could be either hydrogen gas from the reduction of 1.0×10^{-5} M H^{+} *(aq)* or nickel metal from the reduction of 1.00 M Ni^{2+} *(aq)*. (The candidates for reduction at the *cathode* are both *cations*. In electrolytic cells positively-charged cations always migrate toward the negatively-charged cathode). The reduction potential for 1 M Ni^{2+} *(aq)* going to nickel metal is -0.23 V, as tabulated in Appendix E. The reduction potential of the H^{+} *(aq)* must be adjusted from the tabulated value of 0.00 V because the H^{+} *(aq)* is not in its standard state (1 M) but instead is 1.0×10^{-5} M. The Nernst equation (at 25° C) for this half-reaction is:

$$\xi = \xi° - (0.0592 / 1) \log (P_{H_2} / [H^{+}])$$

$$\xi = 0 - 0.0592 \log [10^{-5}]^{-1} = -0.0592(5) = -0.296\ V$$

This reduction potential is algebraically less than -0.23 V (for the Ni^{2+} *(aq)* reduction). Therefore the Ni^{2+} *(aq)* is reduced first. The fact that the H^{+} *(aq)* is dilute rather than in its standard state lowers its reduction potential (makes it harder to reduce).

b) A current of 2.00 amperes for 10 hours is the same as 2.00 C s^{-1} for 3.6×10^{4} s. Therefore 7.20×10^{4} C passes through the cell. This is 0.746 faraday because 96485 C equals one faraday. *One* mole of nickel plates out at the cathode for every *two* faradays:

$$Ni^{2+}(aq) + 2\,e^- \rightarrow Ni(s)$$

so 0.373 mol of Ni plates out. The molar weight of Ni is 58.71 g mol^{-1}. Therefore, 21.9 g of *Ni(s)* plates out.

The volume of the electrolyte has to be large so that removal of this much nickel does not lower the concentration of $Ni^{2+}(aq)$ to the point that $H^+(aq)$ starts to be reduced.

c) If the pH is 1.0, then $[H^+] = 10^{-1}$ M, and the Nernst equation for the reduction of $H^+(aq)$ to $H_2(g)$ at 1 atm and 25° C becomes:

$$\mathcal{E} = 0.00 - 0.0592 \log [H^+]^{-1} = -0.0592 \text{ V}$$

Now, $H_2(g)$ forms at the cathode because $H^+(aq)$ is more concentrated than in part a). The reduction potential for the $H^+(aq)$ is algebraically greater than the potential for $Ni^{2+}(aq)$ reduction. Bubbles of H_2 gas will rise from the cathode.

---

21. The problem can be solved conveniently by a series of unit conversions. The maximum allowable concentration of Sn^{2+} is 10 ppm. Each 100 mL sample of rinse solution will weigh 100 g because it is a dilute solution and consists mostly of water (density = 1.00 g mL^{-1}). Then:

$$100 \text{ mL sample} \times \frac{1.00 \text{ g sample}}{1.00 \text{ mL sample}} \times \frac{10 \times 10^{-6} \text{ g } Sn^{2+}}{1 \text{ g sample}}$$

$$\times \frac{1 \text{ mol } Sn^{2+}}{118.69 \text{ g } Sn^{2+}} \times \frac{2 \text{ mol } e^-}{1 \text{ mol } Sn^{2+}} \times \frac{96485 \text{ C}}{\text{mol } e^-} \times \frac{1 \text{ s}}{2.50 \times 10^{-3} \text{ C}} = 65.0 \text{ s}$$

---

23. ***a)*** Silver plates out. The reduction half-reaction is therefore:

$$Ag^+(aq) + e^- \rightarrow Ag(s)$$

Reduction always occurs at the cathode. At the anode, *Co(s)* is oxidized to aqueous Co^{2+}:

$$Co(s) \rightarrow Co^{2+}(aq) + 2\,e^-$$

The balanced overall equation is the sum of these half-equations with the adjustment of multiplying the first by 2 so that the number of electrons gained equals the number of electrons lost:

$$2\,Ag^+(aq) + Co(s) \rightarrow 2\,Ag(s) + Co^{2+}(aq)$$

b) The molar weight of Co is 58.933 g mol^{-1} so the 0.36 g loss of cobalt corresponds to (0.36 g / 58.933 g mol^{-1}) = 6.1×10^{-3} mol of Co. For each mole of cobalt that dissolves, 2 mol of *Ag(s)* plates out (according to the balanced equation). The molar weight of Ag is 107.87 g mol^{-1} so 1.32 g of Ag plates out. This problem is just like those in Chapter 1. Once the balanced equation is available no knowledge of electrochemistry is needed to solve it.

c) As computed above, 6.1×10^{-3} mol of Co dissolves. The Co loses $2 \times 6.1 \times 10^{-3}$ mol of electrons in the process. One mol of electrons is 96485 C so the Co loses 1.18×10^{3} C. It does this over a period of 150 min or 150×60 s. This is an average current of 0.13 coloumb per second. The average current is 0.13 A.

---

25. The electrolytic reaction that produces Ti(*s*) is:

$$TiCl_4\,(l) \rightarrow Ti(s) + 2\,Cl_2\,(g)$$

The standard entropies of all the participants in the reaction are given. The fact that the value of S° for Ti(*s*) differs slightly in this problem from the value tabulated in Appendix D is not important. Indeed, experimental values like absolute entropies often vary slightly depending on the source. The standard entropy change of the reaction is:

$$\Delta S^\circ = 1\,S^\circ_{\,(Ti)} + 2\,S^\circ_{\,(Cl_2)} - 1\,S^\circ_{\,(TiCl_4)}$$

$$\Delta S^\circ = 30\ J\ K^{-1} + 2(223\ J\ K^{-1}) - 253\ J\ K^{-1} = 223\ J\ K^{-1}$$

ΔH° for the reaction is just the negative of ΔH°_f of one mole of liquid $TiCl_4$. Hence, ΔH° of the reaction as written is +750 kJ. ΔG is readily computed from the ΔH° and the ΔS° of the reaction. When T equals 100° C (373.15 K):

$$\Delta G = \Delta H^\circ - T\Delta S^\circ = 750\ kJ - 373.15\ K\ (0.223\ kJ\ K^{-1})$$

$$\Delta G = 667\ kJ$$

The free energy change of the reaction is related to its voltage:

$$\Delta G = -nF\Delta\mathcal{E}$$

This reaction transfers 4 mol of electrons every time it proceeds as written. F equals 96485 C mol^{-1} which is 96485 J V^{-1}mol^{-1}. Therefore:

$$\Delta\mathcal{E} = -667 \times 10^{3}\ J\ /\ (4\ mol \times 96485\ J\ V^{-1} mol^{-1}) = -1.73\ V$$

A voltage of +1.73 V will be required to force the reaction to run in the direction that is given. This assumes that Cl_2 gas is produced at 1 atm. If the outside pressure is less than 1 atm, a smaller minimum applied voltage will suffice.

27. ***a)*** The two half-equations given in the problem allow the overall cell equation to be written immediately:

$$Pb(s) + 2\,VO^{2+}(0.10\ M) + 4\,H^{+}(0.1\ M) \rightarrow 2\,V^{3+}(1.0 \times 10^{-5}\ M) + Pb^{2+}(0.01\ M) + 2\,H_2O(l)$$

This is simply the sum of the half-equations with the reduction multiplied through by 2 so that electrons do not appear in the net equation. The balanced equation also appears in the statement of part b) of the problem. The Nernst equation at 25° C for this reaction is:

$$\Delta\xi = \Delta\xi^\circ - (0.0592 / n) \log Q$$

where n is 2 and:

$$Q = \frac{[Pb^{2+}][V^{3+}]^2}{[VO^{2+}]^2[H^+]^4}$$

The concentrations of all of the ions in the reaction quotient, Q, are known from the statement of the problem. Substitution gives $Q = 10^{-6}$ so $\log Q$ is -6. Recalling that in electrochemical cells $\Delta\xi^\circ = \xi^\circ$ (cathode) $- \ \xi^\circ$ (anode) and that $\Delta\xi$ is stated to be 0.640 V:

$$0.640 \text{ V} = \xi^\circ(\text{cathode}) - \xi^\circ(\text{anode}) - (0.0592 / 2)(-6) \text{ V}$$

$$0.640 \text{ V} = \xi^\circ(\text{cathode}) - (-0.126 \text{ V}) + 0.1776 \text{ V}$$

$$\xi^\circ(\text{cathode}) = 0.336 \text{ V}$$

b) Equilibrium is reached when $\Delta\xi$, the voltage of the cell, dwindles to zero. Then, applying the Nernst equation at 25° C:

$$\Delta\xi = \Delta\xi^\circ - (0.0592 / n) \log K = 0$$

$$\log K = n\,\Delta\xi^\circ / 0.0592$$

In this problem $\Delta\xi^\circ = \xi^\circ$ (cathode) $- \ \xi^\circ$ (anode):

$$0.3364 \text{ V} - (-0.126 \text{ V}) = 0.4624 \text{ V}$$

$$\log K = 2\,(0.4624) / 0.0592 = 15.62 \qquad K = 4.2 \times 10^{15}$$

---

29. A gas confined at high pressure expands spontaneously to a low pressure if a path is available. In a pressure cell, the free energy change of this spontaneous expansion appears as electrical work.

In this case the gas is Cl_2. At the cathode, $Cl_2\,(g)$ is reduced to 1 M $Cl^-\,(aq)$. At the anode, 1 M $Cl^-\,(aq)$ is oxidized to Cl_2 gas:

$$\tfrac{1}{2}\,Cl_2\,(g) + e^- \rightarrow Cl^-\,(aq) \qquad \text{cathode}$$

$$Cl^-\,(aq) \rightarrow e^- + \tfrac{1}{2}\,Cl_2\,(g) \qquad \text{anode}$$

The cathode reaction *removes* $Cl_2\,(g)$ so the reduction must occur in the half-cell held at the *higher* $Cl_2\,(g)$ pressure, the 0.50 atm half-cell. The overall reaction is:

$$Cl_2\,(0.50 \text{ atm}) \rightarrow Cl_2\,(0.01 \text{ atm})$$

$\Delta\xi^\circ$, the *standard* potential for this reaction is 0.00 V. At 25° C the Nernst equation for this reaction is:

$$\Delta\xi = \Delta\xi^\circ - (0.0592 / 2) \times \log(0.01 / 0.50)$$

$$\Delta\xi = 0.00 - 0.050 = 0.050 \text{ V}$$

---

31. ***a)*** Writing the equation amounts to translating the problem's verbal description of the cell into a balanced chemical equation:

$$Fe^{3+}(aq) + \tfrac{1}{2} H_2(g) + CN^-(aq) \rightarrow Fe^{2+}(aq) + HCN(aq)$$

b) The overall process breaks down into *three* sub-processes:
the reduction:

$$Fe^{3+}(1\text{ M}) + e^- \rightarrow Fe^{2+}(1\text{ M}) \qquad \xi^\circ = 0.770 \text{ V}$$

the oxidation:

$$H_2\ (1\text{ atm}) \rightarrow H^+\ (1\text{ M}) \qquad \xi^\circ = 0.0 \text{ V}$$

and the acid-base interaction:

$$H^+(1\text{ M}) + CN^-(1\text{ M}) \rightarrow HCN(1\text{ M}) \qquad K = 2.02 \times 10^{10}$$

where the parenthetical notations emphasize that all species are in the standard states. The equilibrium constant of the third reaction is the reciprocal of the K_a value of HCN (text Table 6-2) since the reaction is the reverse of the acid ionization of HCN.

For the two half-reactions, knowing $\Delta\xi^\circ$ allows computation of ΔG° because $\Delta G^\circ = -nF\Delta\xi^\circ$. ΔG° for the combination of the oxidation and reduction is -74.3 kJ mol^{-1}. ΔG° for the *third* equation comes from the relationship:

$$\Delta G^\circ = -RT \ln K = -8.314 \text{ J mol}^{-1}\text{K}^{-1}\ (298.15 \text{ K}) \ln 2.02 \times 10^{10}$$

$$\Delta G^\circ \text{ (third step)} = -58.8 \text{ kJ mol}^{-1}$$

1 M $H^+(aq)$ appeared among the net products when the two half-reactions were added. The third equation shows $CN^-(aq)$ removing the $H^+(aq)$, tending to shift the combined reaction to the right. $\Delta\xi$ for the actual cell therefore, by LeChatelier's principle, *exceeds* $\Delta\xi^\circ$ for the simple combination of the two half-reactions. This is reflected in the fact that ΔG° for the three-part process is more negative than ΔG° for just the first two steps. It is the sum of -58.8 kJ mol^{-1} and -74.3 kJ mol^{-1} which is -132.3 kJ mol^{-1}.

c) The standard cell voltage of the overall reaction is related to its ΔG°:

$$\Delta G^\circ = -n\ F\Delta\xi^\circ$$

In this chemical equation, n is 1. F is 96485 coulomb mol^{-1}, so $\Delta\xi^\circ$ is 1.37 V.

d) According to the balanced overall equation, 1 mol of electrons (96485 C) passes through the cell for every 1 mol of $CN^-(aq)$ that is consumed. At a current of 0.50 ampere, it takes one second for 0.50 coulomb to pass through the cell. 96485 C takes 1.93×10^5 s or 53.6 hr, which, taken to two significant figures, is 54 hr.

---

33. ***a)*** The balanced equation for the titration reaction is:

$$2\ S_2O_3^{2-}(aq) + I_2(aq) \rightarrow S_4O_6^{2-}(aq) + 2\ I^-(aq)$$

The problem describes the two half-equations in words, and the above equation combines then.

b) The titration consumes 0.0564 L of $S_2O_3^{2-}$ solution which contains 0.100 mol $S_2O_3^{2-}$ per liter. This means that 0.00564 mol $S_2O_3^{2-}$ reacts. From the balanced equation ½ mol of I_2 reacts with each 1 mol of $S_2O_3^{2-}$ so 0.00282 mol of I_2 was originally present.

c) The standard half-cell potentials (Appendix E) for the reduction and oxidation half-reactions are:

$$2\ I^-(s) + 2\ e^- \rightarrow 2\ I^-(aq) \qquad \xi^\circ = 0.535\ V$$

$$2\ S_2O_3^{2-}(aq) \rightarrow S_4O_6^{2-} + 2\ e^- \qquad \xi^\circ = -0.09\ V$$

$\Delta\xi^\circ$ is the *sum* of these two numbers because the oxidation half-equation has been written as a *loss* of electrons and the sign of its ξ° reversed:

$$2\ S_2O_3^{2-}(aq) + I_2(s) \rightarrow S_4O_6^{2-}(aq) + 2\ I^-(aq) \qquad \Delta\xi^\circ = 0.445\ V$$

If $\Delta\xi^\circ$ is known then so is ΔG° because $\Delta G^\circ = -n F\xi^\circ$. In this reaction, n is 2 so ΔG° is -85.87 kJ mol^{-1}. Now there is a complication. The reaction consumes solid I_2, *not* $I_2(aq)$. The reaction in part a) consumes $I_2(aq)$. If the equation:

$$I_2(aq) \rightarrow I_2(s)$$

is added to the equation showing the consumption of $I_2(s)$, then the desired chemical equation results. When this is done, the ΔG° values are also added. The ΔG° for $I_2(aq) \rightarrow I_2(s)$ is -16.40 kJ mol^{-1}. The overall reaction:

$$2\ S_2O_3^{2-}(aq) + I_2(aq) \rightarrow S_4O_6^{2-}(aq) + 2\ I^-(aq)$$

therefore has $\Delta G^\circ = -85.87 - 16.40 = -102.27$ kJ mol^{-1}.

Using the fundamental relationship $\Delta G^\circ = -RT \ln K$ at 25° C (298.15 K):

$$\ln K = 41.26 \qquad \textit{and} \qquad K = 8.3 \times 10^{17}$$

---

35. ***a)*** The oxidation is:

$$OH^-(aq) + Mn(OH)_2(s) \rightarrow Mn(OH)_3(aq) + e^-$$

The reduction of $O_2(g)$ is:

$$O_2(g) + 2\ H_2O + 4\ e^- \rightarrow 4\ OH^-(aq)$$

The overall equation is the combination of these half-equations:

$$4\ Mn(OH)_2\,(s) + O_2\,(g) + 2\ H_2O \rightarrow 4\ Mn(OH)_3\,(s)$$

The reduction of $O_2\,(g)$ occurs at the cathode. From Appendix E, $\xi°$ for this reduction is 0.401 V. $Mn(OH)_2\,(s)$ reacts at the anode and, for this oxidation, the standard potential is the negative of the tabulated value, that is $\xi° = +0.40$ V. The overall standard voltage is 0.401 + 0.40 = 0.801 V. At 25° C this standard voltage is related to the equilibrium constant by the equation:

$$\Delta\xi° = (0.0592\ /\ n)\log K$$

In the overall equation n = 4 so K equals 1.32×10^{54}.

b) The reduction of $Mn(OH)_3\,(s)$ to $Mn^{2+}\,(aq)$ with $I^-\,(aq)$ produces $I_2\,(aq)$ as well. In acid solution the balanced equation is:

$$2\ Mn(OH)_3\,(s) + 2\ I^-\,(aq) + 6\ H^+\,(aq) \rightarrow 2\ Mn^{2+}\,(aq) + I_2\,(aq) + 6\ H_2O(l)$$

---

37. The discharge of the lead-acid battery involves the *oxidation* of spongy elemental lead at the anodes to Pb(II). Lead has a molar weight of 0.20719 kg mol^{-1} so 10 kg of Pb oxidized to Pb(II) furnishes $2 \times (10\ /\ 0.20719)$ mol of electrons. At 96485 C mol^{-1}, this is 9.314×10^6 C. An ampere-hour is a current of 1 ampere flowing for 1 hour. This is 1 C s^{-1} flowing for 3600 s or 3600 C. The discharge of the battery transfers 9.314×10^6 C which is thus equivalent to 2.6×10^3 ampere-hours.

---

39. The reaction is the oxidation of $CO(g)$ to $CO_2\,(g)$:

$$CO(g) + \tfrac{1}{2}\,O_2\,(g) \rightarrow CO_2\,(g)$$

Using the standard enthalpy changes of formation from Appendix D:

$$\Delta H° = -393.51\ kJ - (-110.52\ kJ) = -282.99\ kJ$$

$$\Delta S° = 213.63\ J\ K^{-1} - \tfrac{1}{2}\,(205.03\ J\ K^{-1}) - 197.56\ J\ K^{-1} = -0.086445\ kJ\ K^{-1}$$

If $\Delta H°$ and $\Delta S°$ are independent of temperature, then at 1200 K:

$$\Delta G = \Delta H° - 1200\,\Delta S° = -179.26\ kJ$$

The cell voltage depends on this ΔG:

$$\Delta G = -nF\Delta\xi$$

The number of electrons transferred is 2, so $\Delta\xi = 0.929$ V.

---

41. Once the protective layer of tin is broken, the iron body of a tin can rusts readily because Fe is more easily oxidized than tin. Its $\xi°$ for *oxidation* is algebraically larger than tin's. To see this, compare the two oxidation half-reactions:

$$Sn(s) \rightarrow Sn^{2+}(aq) + 2\,e^- \qquad \mathcal{E}° = 0.136\text{ V}$$
$$Fe(s) \rightarrow Fe^{2+}(aq) + 2\,e^- \qquad \mathcal{E}° = 0.409\text{ V}$$

Note that the voltages appearing to the right of the half-reactions are *oxidation* potentials. They are the negatives of the reduction potentials tabulated in text Appendix E. Sometimes it is convenient to reverse half-reactions and compare oxidation potentials instead of reduction potentials. The outcome is the same. A more negative reduction potential automatically means a more positive oxidation potential when any half-reaction is reversed.

When iron is coated with zinc (galvanized), it is in contact with a metal that is *more* active than it is. Zinc has a more positive standard oxidation potential than iron (+0.763 V) and, of course, a more negative standard reduction potential. Zn tends to be consumed first when the Zn and Fe are in contact and exposed to oxidizing conditions.

---

43. K^+ (0.05 M) cannot be reduced in neutral aqueous solution because its reduction potential is −2.92 V, algebraically far less than the critical value −0.414 V. −0.414 V is the reduction potential for the formation of H_2 *(g)* from H^+ *(aq)* in neutral water. See p. 201 of this Guide. H^+ (1×10^{-7} M) is reduced before K^+ can be reduced.

The reduction potential for the Br_2/Br^- (0.05 M) couple is 1.14 V. This reduction potential exceeds 0.815 V algebraically. In theory H_2O is oxidized, not 0.05 M Br^-. The critical value 0.815 V is the potential for the reduction of O_2 *(g)* (1 atm) to OH^- *(aq)* at a concentration of 10^{-7} M in water. In other terms, the potential for the *oxidation* of OH^- (1×10^{-7} M) to O_2 *(g)* is −0.815 V. It is thermodynamically easier to cause this process to occur than to oxidize 0.05 M Br^- *(aq)* to Br_2 *(l)* (oxidation potential −1.14 V).

In actual practice the rapid formation of O_2 *(g)* is prevented by the large *overvoltage* of O_2 on most electrode surfaces (a kinetic effect). Little current flows until the voltage is raised to the point that Br_2 starts to form.

Chapter 10

CHEMICAL KINETICS

10.1 *RATE LAWS*

Chemical kinetics is the study of the *rates* and *mechanisms* of chemical reactions. Most chemical reactions do not occur as represented in chemical equations but instead by a series of much simpler steps which add together to give the net equation. A *mechanism* is a sequence of such *elementary steps.*

The rate of the general chemical reaction:

$$a\,\mathrm{A} + b\,\mathrm{B} \rightarrow c\,\mathrm{C} + d\,\mathrm{D}$$

is determined by monitoring the *disappearance* of a reactant or the *appearance* of a product. It is defined as:

$$\text{Rate} = -\frac{1}{a}\frac{d[\mathrm{A}]}{dt} = -\frac{1}{b}\frac{d[\mathrm{B}]}{dt} = \frac{1}{c}\frac{d[\mathrm{C}]}{dt} = \frac{1}{d}\frac{d[\mathrm{D}]}{dt}$$

In this equation, the first d in the last term stands for the coefficient of the product D in the chemical equation, and the others are part of the derivative. Because the concentration of A falls as the reaction proceeds, the time rate of change of [A], $d[\mathrm{A}]/dt$, is naturally negative. The same holds for all reactants. The minus signs are put in to keep the rate positive.

The experimental determination of reaction rates requires a thermostat, because rates are quite sensitive to temperature, a clock to measure the time, and some means for monitoring concentration of at least one product or reactant. In some cases it is possible to *quench* (stop) a reaction by suddenly cooling it and then to analyze for the concentration of a reactant or product. An alternative is to monitor some physical property as a continuous function of time. The total pressure is often used. **See problem 21.** The color of the reaction mixture is another property frequently monitored.

Reaction Order

Rates of reactions almost always change with time. A typical observation is the slowing of the production of products as time wears on. One reason is that the observed rate of a reaction is really a *net rate,* the difference between the rate of the forward reaction and of the reverse reaction. If pure reactants are mixed, there is at first no reverse reaction because there are no products available to react.

But, even isolated from these effects, the forward rate of a reaction still may change with time. In general it depends on the concentrations of the reactants. The dependence is expressed in the *rate law* of the chemical reaction. For the general chemical equation written above, the rate law usually has the form:

$$\text{Rate} = -\ 1/a\ d[\text{A}]/dt = k[\text{A}]^m[\text{B}]^n$$

The exponents m and n may equal positive or negative whole numbers or fractions or zero. A species whose concentration has an exponent of zero has no effect on the rate of the reaction (although some of it does have to be present if it is a reactant). A species with a negative exponent in a rate law makes the reaction go more slowly as its concentration is raised. **See problem 17** for an illustration of both of these points.

The constant of proportionality, k, is the *rate constant.* Rate constants are strongly affected by the temperature. **See problem 11.** The exponents m and n determine the *order* of the reaction. m is the order with respect to reactant A, and n is the order with respect to reactant B.

The *overall order* of the reaction is the sum of all the exponents in the rate law.

The Units of k

The overall order of a reaction determines the units of its rate constant. A rate (not rate constant) is measured in moles per liter per second (mol $L^{-1}s^{-1}$). In the rate law for a first-order reaction, k is multiplied by a concentration (mol L^{-1}). The units of k must be s^{-1} to make the rate come out as required. In a second-order reaction k is multiplied by *two* concentrations (with overall units $mol^2 L^{-2}$). The units of k now must be L $mol^{-1}s^{-1}$ to make the rate come out right. **See problem 17a** for another example.

Determining a Rate Law

Rate laws are experimental results. In one kind of experiment used to get rate laws, the concentrations of all but one reactant are fixed and, in a series of runs, the *initial rate* is measured at different concentrations of the one reactant under study. The initial rate is the rate of the reaction observed before back-reaction among the products has a chance to confuse the issue. If doubling the concentration of the reactant under test doubles the initial rate, then the reaction is first-order in the test reactant. If the same doubling quadruples the rate, then the reaction is second order in that one reactant. **See problems 1 and 17a.**

Integrated Rate Laws

Rate laws are differential equations. The instantaneous rates of chemical reactions in general change as reactions proceed.

Consider the first-order reaction:

$$\text{A} \rightarrow \text{products}$$

Let c equal the concentration of A at any time, t, after the reaction starts. Let c_0 equal the original concentration of A (the concentration at time 0). Then:

$$c = c_0 e^{-kt}$$

Another form of this integrated rate law is:

$$\ln c - \ln c_0 = -kt$$

According to the equation, the change in ln c over equal intervals of time is always the same in a given first-order reaction. In other words, the equation predicts that in a first-order reaction a plot of ln c against time will be a straight line. **See problem 21b** for an application of this fact.

The *half-life*, τ, of a first-order reaction is the time required for the concentration of a reactant to decrease to half of its initial value. Big rate constants mean fast reactions which require but little time to consume half the initial concentration of a reactant. Thus the half-life is inversely proportional to the rate constant:

$$\text{half-life} = \tau = \ln 2 / k$$

Now consider the *second-order* reaction A → products. The rate law is:

$$\text{rate} = -d[A]/dt = k[A]^2$$

and the integrated rate law is:

$$1/c - 1/c_0 = kt$$

where c is the concentration of A at any time and c_0 is its original concentration. This integrated rate equation predicts that a plot of $1/c$ versus time will be a straight line. **See problem 21b.** Knowing the rate constant and the integrated rate law for a reaction it is possible to predict the concentrations of reactants at any time or the time required for some specified change in concentration. **See problems 3 and 19.**

Integrated rate laws for reactions of the same order differ to reflect the reactions' stoichiometry. Contrast the integrated rate law for the dimerization of tetrafluoroethylene (text Example 10-4, text p. 334) with the integrated rate law in **problem 3.** The extra factor of 2 in the former reflects the fact that the balanced equation has the form 2 A → products.

10.2 *REACTION MECHANISMS*

A reaction mechanism consists of a sequence of *elementary steps* which tell just which molecules must collide with each others to convert the reactants into products. A mechanism also includes an indication of the rates of all the steps. This indication consists of a numerical rate constant for each step or a word or two giving a *qualitative* idea of the relative rates of of the steps.

A mechanism may include *intermediates*, species produced in a early step and later consumed. It may include *catalysts*, species that come from outside, join in an early step in the mechanism and are regenerated in their original form by a later step. The sum of all of the steps of a reaction mechanism equals the balanced chemical reaction. **See problems 7 and 29a.**

Elementary Reactions

An elementary step is also called an elementary reaction. Elementary reactions involve the collision *in a single event* of the reactant particles to give the product particles. Such reactions have *molecularity.* Collision of two molecules to give products is a *bimolecular* elementary reaction. A *termolecular* elementary reaction involves the simultaneous collision of three particles. An *unimolecular* reaction involves only a single reactant molecule. Only elementary reactions have molecularity. The order of elementary reactions equals their molecularity. The vast majority of chemical reactions proceed *via* mechanisms of more than one elementary step. The concept of molecularity does not apply to non-elementary reactions.

In any mechanism, the product of the rate constants for the forward steps divided by the product of the rate constants for the reverse step is equal to the equilibrium constant of the overall reaction.

Rate Determining Steps

If one of the elementary steps in a mechanism is much slower than the rest, it acts as a bottle-neck and determines the overall rate of reaction. With this in mind the procedure for writing the rate law predicted by a mechanism is:

1. Write the rate law for the slow step. This requires nothing more than taking the coefficients of the species in the slow step as the exponents in the rate law.

2. Eliminate the concentrations of any intermediates that may occur in the slow-step rate law. Intermediates often appear in mechanisms but may never appear in experimental rate laws. Deducing a rate law from a mechanism is a waste of effort unless the result is in a form directly comparable with experiment. Many mechanisms have one or more fast equilibria preceding the slow step. **Problems 7 and 23 illustrate** how to get rid of concentrations of intermediates in writing rate laws from mechanisms.

Steady-State Approximation

If all the steps in a mechanism have about the same rate, then no single one of them is rate-determining. The *steady-state approximation* is useful in such cases. In this approximation, it is assumed that the concentrations of reactive intermediates remain constant as the reaction proceeds toward equilibrium. Suppose an intermediate is produced by the first step of a mechanism and consumed by the second. The net rate of change of its concentration is its rate of production minus its rate of consumption. *The steady-state approximation assumes that this net rate of change equals zero.* The assumption in this case is that the rate of production of the intermediate by the first step is equal to its rate of consumption by the second step. **See problem 9.** The net rate of change may involve three or more different terms. **See problem 25.**

In every case, the steady state approximation furnishes an equation that gives the concentrations of intermediates in terms of the concentrations of non-intermediates and rate constants. In this way it allows the elimination of intermediates' concentrations, as a rate law is deduced from a mechanism. **See problem 31** for an example of the approximation in action.

10.3 *COLLISION THEORY OF GAS PHASE REACTIONS*

Molecules must collide in order to react, but not all collisions lead to reactions. Most molecular collisions are not energetic enough. Only those collisions for which the collision energy exceeds some minimum can result in reaction. The minimum is the *activation energy*, E_a, of the reaction.

The *Arrhenius equation* gives the temperature dependence of rate constants (and thus of reaction rates themselves):

$$k = A\,e^{-E_a/RT}$$

The activation energy has the same units as RT, kJ mol^{-1}. The pre-exponential factor *A* is a constant having the same units as *k*. **See problem 11b.** If E_a were zero, then the exponent in the Arrhenius equation would equal zero and, because any number raised to the zero power is 1, *k* would equal *A*.

A useful form of the Arrhenius equation is:

$$\ln(k_1/k_2) = +(E_a/R)\,[1/T_2 - 1/T_1] = -(E_a/R)[1/T_1 - 1/T_2]$$

where k_1 is the rate constant at T_1 and k_2 is the rate constant at T_2. Note that if k_1 and k_2 are interchanged on the left in the above equation, then the signs in front of the (E_a/R) terms both reverse.

Clearly, measuring the rate constant at just two temperatures could give the activation energy. In practice, rate constants are measured at several different temperatures and the activation energy is determined from the slope of the line in a plot of ln *k* versus 1/T. **See problems 11, 13 and 27.**

The activation energy of an elementary step is always positive. For overall reactions consisting of two or more elementary steps, E_a can be negative as well as positive. A negative E_a means that the observed rate constant diminishes with increasing temperature. This occurs when a fast exothermic equilibrium (with equilibrium constant *K*) precedes the rate-determining step (with rate constant *k*) in the reaction's mechanism. The observed rate constant is such a case equals *Kk*. **See problems 7 and 29.** Although *k* rises with temperature, *K* for an exothermic reaction gets smaller and can outweigh the effect of *k* in determining the behavior of *Kk*.

10.4 *CATALYSIS*

A catalyst is a substance that enters into a chemical reaction and speeds it up, but is not itself consumed in the reaction. Catalysts do not appear in the balanced overall equation representing a reaction, but do appear in the reaction's rate law.

Suppose that the decomposition of ozone has the mechanism:

$$O_3 + Cl \rightarrow ClO + ClO \qquad \text{slow}$$
$$ClO + O \rightarrow Cl + O_2 \qquad \text{fast}$$

Atomic chlorine (Cl) is the catalyst. It does not appear in the overall reaction (which is $O_3 + O \rightarrow 2\,O_2$) but it does appear in the rate law (which is rate = $k[O_3][Cl]$). Catalysts accelerate chemical reactions by providing new reaction pathways which have lower activation energies. In the above example, if there were no Cl around, then the reaction would either go by a different mechanism or else just not go.

An *inhibitor* is the opposite of a catalyst. It *slows* the rates of chemical reactions.

In a sense a catalyst and intermediate are opposites, too. An intermediate is generated within a reaction's mechanism and consumed there; it may *not* appear in the rate law for the reaction. A catalyst joins a mechanism from without and is regenerated before the mechanism is over; it *may* appear in the rate law for the reaction.

In *homogeneous catalysis* the catalyst is present in the same phase as the reactants. In *heterogeneous catalysis* the reactants are in one phase and the catalyst is in another phase.

Enzyme Catalysis

Enzymes catalyze many chemical reactions that take place in living organisms. They are high molecular weight protein molecules that bind at their *active sites* to reactant molecules (*substrates* of the enzyme) and promote specific transformations of the reactant molecules. The kinetics of enzyme-catalyzed reactions involve no kinetic concepts that do not also appear in non-enzyme catalysis. **Problem 31** shows how the rate of an enzyme-catalyzed reaction is lowered by the presence of an inhibitor.

DETAILED SOLUTIONS TO ODD-NUMBERED PROBLEMS

1. ***a)*** From the table in the problem, doubling [A] quadruples the reaction rate as [B] is unchanged. Then, as [A] stays constant and [B] *triples*, the rate of the reaction triples. Assume that the sole differences among the three experiments are with the initial concentration of A and B. The rate law then is second-order in A and first-order in B:

$$\text{rate} = k\,[A]^2\,[B]$$

b) The initial rate of the reaction in $M\ s^{-1}$ is known in each of three experiments. The initial concentrations of A and B are also known in the three experiments. Solve the rate equation for k:

$$k = \text{rate} \,/\, [A]^2\,[B]$$

Substitution of any of the three sets of data gives $k = 1.9 \times 10^3\ M^{-2}\ s^{-1}$ or, equivalently, $1.9 \times 10^3\ L^2\ mol^{-2}\,s^{-1}$.

---

3. If the reaction is first-order in C_5H_5N and first-order in CH_3I, then:

$$\text{rate} = k\,[C_5H_5N][CH_3I] = -d\,[CH_3I]\,/\,dt = -d\,[C_5H_5N]\,/\,dt$$

Writing this equation requires a translation of the rate law from verbal to mathematical form. If equal volumes of 0.10 M C_5H_5N and 0.10 M CH_3I are mixed, then *after* the completion of the mixing but *before* the reaction starts, the concentration of each reactant is 0.050 M. As the reaction proceeds the 1:1 stoichiometry of the reaction guarantees that the concentrations of the two reactants decrease in equal measure. They are always equal; let their value at any time be c. The rate of change of c is:

$$dc\,/\,dt = -\,k\,c^2$$

Integrate this equation from the initial concentration c_o (=0.050 M) at time 0 to c at time t:

$$1\,/\,c_o \;-\; 1\,/\,c = -kt$$

The second-order rate constant k is $74.9\ L\ mol^{-1}s^{-1} = 74.9\ M^{-1}s^{-1}$. The problem asks the value of t when $c = 0.025$ M. Substituting and solving gives t = 0.27 s. It is *incorrect* blindly to employ text equation 10.3 (text p. 334), which has an additional factor of 2.

The concentrations of the reactants fall to one-half of their initial values in 0.27 s. This time is the *first* half-life of this reaction. The *second* half-life, the time required for the concentrations of the reactants to go from 0.025 M to 0.0125 M is 0.54 s, *twice* as long. The half-lives of second-order reactions vary with concentration.

---

5. The reaction is the neutralization of OH^- *(aq)* with NH_4^+ *(aq)*. Aqueous acid-base reactions are generally fast. This reaction is no exception because the room temperature rate constant is huge: $k = 3.4 \times 10^{10}\ M^{-1}s^{-1}$. The answer will be a very short time .

If 1.00 L of 0.0010 M NaOH and 1.00 L of 0.0010 M NH_4Cl are mixed, then *before* the reaction starts each reactant is 5.0×10^{-4} M. The kinetics are second-order overall:

$$\text{rate} = -d[OH^-]/dt = k[OH^-][NH_4^+]$$

Throughout the reaction $[OH^-] = [NH_4^+]$. Let this concentration be represented by c. Then:

$$-dc/dt = k\,c^2$$

Integrating this equation and inserting the initial condition gives:

$$-1/c + 1/c_o = -k\,t$$

where c_o is the original concentration of OH^- *(aq)*. For c equal to 1.0×10^{-5} M and k equal to $3.4 \times 10^{10}\ M^{-1}s^{-1}$, the equation becomes:

$$-1.0 \times 10^5\ M^{-1} + 2.0 \times 10^3\ M^{-1} = -(3.4 \times 10^{10}\ M^{-1}s^{-1})\,t$$

$$-9.8 \times 10^{-4}\ M^{-1} = -(3.4 \times 10^{10}\ M^{-1}\ s^{-1})\,t$$

$$t = 2.9 \times 10^{-6}\ s$$

---

7. **a)** The *rate-limiting* elementary step in a mechanism determines the overall reaction rate. In this case the slow step is:

$$C + E \rightarrow F$$

so a preliminary version of the rate law is:

$$\text{rate} = k_2[C][E]$$

Unfortunately, the expression involves the concentration of C, an *intermediate.* This is unacceptable. To eliminate [C] in the rate law, consider how C is formed. It arises in the first step of the mechanism, a fast equilibrium. For that first step:

$$k_1\ [A][B] = k_{-1}\ [C][D]$$

Solve this equation for the concentration of C and substitute into the preliminary rate law:

$$\text{rate} = k_2(k_1\ /\ k_{-1})\ [A][B][E][D]^{-1}$$

The overall reaction is the sum of the steps in the mechanism:

$$A + B + E \rightarrow D + F$$

The rate expression predicts that accumulation of product D in the system slows down the reaction.

b) The overall reaction in this case is:

$$A + D \rightarrow B + F$$

For the two fast equilibria:

$$k_1 [A] = k_{-1} [B][C] \quad \textit{and} \quad k_2 [C][D] = k_{-2} [E]$$

The last, slow step is the rate determining step:

$$\text{rate} = k_3 [E]$$

E is an intermediate and its concentration may not appear in the final rate expression. To eliminate [E] combine the second two equations to give:

$$\text{rate} = k_3 (k_2 / k_{-2}) [C][D]$$

This expression *still* contains the concentration of an intermediate (C now). Eliminate [C] using the first equation:

$$\text{rate} = k_3 (k_2 / k_{-2}) (k_1 / k_{-1}) [A][D] / [B]$$

The reaction is first-order in both A and D, is -1 order in B and first-order overall.

---

9. The overall reaction is $A + B + E \rightarrow D + F$. It proceeds first by equilibrium between A and B giving D and the intermediate C and then by the consumption of C in reaction with E to give F. The rate of appearance of C is $k_1 [A][B]$, and the rate of *disappearance* of C is $k_{-1} [C][D] + k_2 [C][E]$. This latter is a sum because C disappears both by *back* reaction (with D) and by *forward* reaction (with E). As the concentration of C becomes constant (the steady state approximation) its rate of disappearance equals its rate of appearance:

$$k_1 [A][B] = k_{-1} [C][D] + k_2 [C][E]$$

The rate of the reaction is the rate of appearance of a product, for example, F:

$$\text{rate} = d[F] / dt = k_2 [E][C]$$

Solve the previous equation for [C] and substitute the answer in this rate expression:

$$\text{rate} = \frac{k_2 k_1 [A][B][E]}{k_2 [E] + k_{-1} [D]}$$

If $k_2 [E]$ is much smaller than $k_{-1} [D]$ (which is the case if the first step of the reaction is a fast equilibrium) then this expression becomes:

$$\text{rate} = k_2 (k_1 / k_{-1}) [A][B][E][D]^{-1}$$

---

11. ***a)*** According to the Arrhenius equation, the rate constant of an elementary reaction depends on the absolute temperature and activation energy E_a:

$$k = A\,e^{-E_a/RT}$$

where *A* is a constant. Taking the natural logarithm of both sides:

$$\ln k = \ln A - E_a/RT$$

This means that a plot of ln *k* versus the reciprocal of T should be a straight line with a slope of $-E_a/R$ and an intercept (when $1/T = 0$) of ln *A*. Two points determine a line. A quick way to estimate E_a is to select any two of the four data points (such as the first two the problem gives), insert the values in the above equation:

$$\ln (5.49 \times 10^6) = \ln A - (E_a/R)(1/5000)$$

$$\ln (9.86 \times 10^6) = \ln A - (E_a/R)(1/10000)$$

and then solve for E_a (eliminating ln *A*). This procedure gives $E_a = 432$ kJ mol^{-1}. But selecting *another* pair of points (for example the *second* two) gives $E_a = 392$ kJ mol^{-1}. The discrepancy means that the experimental data do not fall exactly on a straight line. The best way to use all data is to perform a *least-squares fit*, mathematically determining the slope of the straight line that best fits the four data points. Many electronic calculators are equipped to complete the necessary calculations almost without effort. Based on the minimization of the sum of the squares of the deviations, E_a is 425 kJ mol^{-1}.

b) As $1/T$ goes to zero, ln *k* approaches ln *A*. The first two data points gave $E_a = 432$ kJ mol^{-1}, substituting gives ln $A = 25.9$ and $A = 1.8 \times 10^{11}$ M^{-1}s^{-1}. Using the last two data points (which gave $E_a = 392$ kJ mol^{-1}) makes ln $A = 25.5$ and A 1.18×10^{11} M^{-1}s^{-1}. The least squares fitting gives ln $A = 25.76$ and $A = 1.5 \times 10^{11}$ M^{-1}s^{-1}. This is the best answer. The units of *A* *must* be the same as the units of *k*.

---

13. The isomerization of cyclopropane is first-order. The Arrhenius equation tells how its rate constant depends on the absolute temperature. Let T_2 equal the temperature at which the half-life of the reaction is 2.0 minutes. Let T_1 = 773 K, the temperature at which the half-life is 19.0 minutes. Then, where k_1 and k_2 are the rate constants at T_1 and T_2:

$$\ln k_1 = \ln A - E_a/RT_1 \qquad \textit{and} \qquad \ln k_2 = \ln A - E_a/RT_2$$

If the half-life of the reaction is *shorter* at the higher temperature, then k_2 is proportionately larger:

$$k_2 = (19.0/2.00)\,k_1 \qquad \textit{or} \qquad \ln k_2 = \ln 9.50 + \ln k_1$$

This expression links the Arrhenius equation at the two temperatures:

$$\ln A - E_a/RT_2 = \ln 9.5 + \ln A - E_a/RT_1$$

$$-E_a/RT_2 = \ln 9.5 - E_a/RT_1$$

E_a is given (272×10^3 J mol^{-1}) and R is 8.314 J mol^{-1}K^{-1}. T_1 is 773 K. Inserting these values to solve for T_2 gives T_2 = 816 K, which is 543° C.

---

15. The reaction of OH^- with HCN is:

$$OH^-(aq) + HCN(aq) \rightarrow H_2O(l) + CN^-(aq)$$

This reaction is first-order in OH^- and HCN:

$$\text{rate (forward)} = k_f[OH^-][HCN]$$

At equilibrium the forward rate is exactly equaled by the *reverse* rate:

$$\text{rate (reverse)} = k_r[CN^-]$$

which means that:

$$k_f[OH^-][HCN] = k_r[CN^-]$$

$$k_r / k_f = [OH^-][HCN]/[CN^-]$$

Note that the units of the quantity on the right-hand side of the equation are mol L^{-1}. Numerically, the right-hand side of the equation equals the equilibrium constant K_b of the reaction:

$$CN^-(aq) + H_2O \rightleftarrows HCN(aq) + OH^-(aq)$$

K_b for CN^- is related to K_a, the acid dissociation constant of its conjugate acid HCN, by the equation $K_b = K_w / K_a$. K_w is 1.0×10^{-14}, and K_a is 4.93×10^{-10} (from text Table 6-2). Therefore:

$$k_r / k_f = 2.02 \times 10^{-5} \text{ mol L}^{-1}$$

$$k_r = (2.02 \times 10^{-5} \text{ mol L}^{-1})\, k_f$$

$$k_r = (2.02 \times 10^{-5} \text{ mol L}^{-1}) \times (3.7 \times 10^{9} \text{ L mol}^{-1}\text{s}^{-1})$$

$$k_r = 7.5 \times 10^{4} \text{ s}^{-1}$$

---

17. ***a)*** The verbal description of the rate law translates to the form:

$$\text{rate} = k[SO_2]^m[SO_3]^n[O_2]^0$$

where the rate's independence of O_2 concentration is displayed in the fact that any number to the zero power is 1.

Because tripling the SO_2 concentration triples the rate, m is 1 in the above expression. Increasing $[SO_3]$ diminishes the rate so n is negative. Diminishment by the factor 1.7 occurs when $[SO_3]$ is tripled. The square root of 3 is close to 1.7 so $n = -\frac{1}{2}$:

$$\text{rate} = k[SO_2][SO_3]^{-\frac{1}{2}}$$

The rate is expressed in mol $L^{-1}s^{-1}$ and the concentrations in mol L^{-1}. The units of k must, when multiplied by (mol $L^{-1})^{1/2}$, give mol $L^{-1}s^{-1}$. The units of k therefore are: $mol^{1/2}L^{-1/2}s^{-1}$.

b) Doubling [SO_2] and quadrupling [SO_3] should leave the rate of the reaction unchanged. The two effects cancel each other in the rate law.

---

19. Using the method which gives equation 10.3 in the text, the integrated form of the rate law for this reaction is:

$$1 / c = 1 / c_o + 2kt$$

The problem specifies values for all of the variables in this equation except c. By simple substitution c equals 2.34×10^{-5} M. At room temperature the concentration of I atoms falls by about 77 percent in the first two milliseconds after the start of the reaction.

---

21. *a)* The reaction is the gas-phase decomposition of di-*t*-butyl peroxide:

$$\text{DTBP}(g) \rightarrow 2\ \text{acetone}(g) + 1\ \text{ethane}(g)$$

In the reaction 3 mol of gas is produced for every 1 consumed. The total pressure of the gas mixture therefore rises as the reaction goes on. Assume that the reaction mixture at any time is a mixture of ideal gases which follows Dalton's law:

$$P_t = P_{DTBP} + P_{acetone} + P_{ethane}$$

Let x equal the fraction of the DTBP molecules which has decomposed at any time. At the start, $x = 0$ and $P_t = 0.2362$ atm. At later times:

$$P_{DTBF} = 0.2362(1 - x) \text{ atm}$$

$$P_{acetone} = 0.2362(2x) \text{ atm} \quad \textit{and} \quad P_{ethane} = 0.2362\ x \text{ atm}$$

Adding these three partial pressures gives the total pressure in terms of x:

$$P_t = 0.2362(1 + 2x) \text{ atm} \quad \textit{or} \quad x = (P_t / 0.4724) - 0.5$$

Substituting this expression for x into the equation for the partial pressure of DTBP:

$$P_{DTBP} = 0.2362(1.5) - P_t / 2$$

This result shows that the partial pressure of DTBP *diminishes* as the total pressure goes up. The following table shows the computed values for the pressure of DTBP as a function of time:

time(min)	P_{DTBP}*(atm)*	*time(min)*	P_{DTBP}*(atm)*
0	0.2362	26	0.1882
2	0.2310	30	0.1819
6	0.2236	34	0.1758
10	0.2158	38	0.1700
14	0.2080	40	0.1668
18	0.2017	42	0.1643
20	0.1982	46	0.1589
22	0.1949		

b) The partial pressure of DTBP and its concentration are in direct proportion. If the reaction is first-order in DTBP then, over any set time, the *natural logarithm* of the DTBP pressure will always change by the same amount. If the reaction is second-order in DTBP then, instead of ln P, the *reciprocal* of the DTBP pressure will exhibit a constant change over equal time intervals. The interval of 4 min is common in the table. From 2 to 6 min, $\Delta(1\ /\ P_{DTBP})$ is 0.1423; from 42 to 46 min, it is 0.2070. This is a substantial alteration. On the other hand, for the same pair of intervals, $\Delta(\ln P_{DTBP}) = -0.0323$ and -0.0334: very little change. The reaction is first-order in DTBP.

---

23. A correct mechanism for a reaction predicts the observed rate law although a mechanism that successfully predicts the observed rate law may *still* be wrong. Predicting the rate law is a *necessary* but not a *sufficient* condition for correctness in a mechanism.

The reaction of HCl with propane is first-order in propane and third order in HCl. This is an experimental fact. Mechanism (a) postulates the rapid formation of the intermediate H from 2 HCl's followed by slow combination of the H with CH_3CHCH_2. This mechanism thus predicts only second-order kinetics in HCl and must be wrong.

Mechanism (b) postulates *two* fast equilibria each involving HCl and the second also involving propane. The slow step is the combination of the two intermediates that these equilibria produce. The rate then is proportional to all the concentrations of all the reactants in the two fast equilibria. HCl occurs three times among these reactants and propane occurs once so this mechanism *is* consistent with the observed rate law.

Mechanism (c) involves HCl in the fast production of two different intermediates which then slowly combine to give the product (and to regenerate some HCl). It predicts second-order kinetics in HCl (2 HCl's consumed to make the slow-step reactants). It is therefore not consistent with observation.

---

25. The reaction is the photolytic decomposition of ozone:

$$2\ O_3 + \text{light} \rightarrow 3\ O_2$$

The mechanism involves the production of the intermediate O from O_3 (in the first step) and its consumption either to regenerate O_3 (the second step) or to make 2 O_2 by by pairing off with an O_3 (the third step). The change in [O] with time is:

$$d[O_3]/dt = k_1[O_3] - k_2[O][O_2][M] - k_3[O][O_3]$$

This equation states that the rate of change of the concentration of O is its rate of production minus its rate of consumption. The steady-state approximation assumes that [O] comes to a *steady* value. If [O] is steady it is unchanging and $d[O]/dt = 0$. Then:

$$k_1[O_3] - k_2[O][O_2][M] - k_3[O][O_3] = 0$$

All of this concerns the intermediate. The rate of the overall reaction is:

$$\text{rate} = {}^1/_3\, d[O_2]/dt = k_3[O][O_3]$$

Into this rate law, substitute the expression for [O] that can be obtained by solving the previous equation:

$$\text{rate} = \frac{k_3 k_1 [O_3]^2}{3\,k_2[O_2][M] + k_3[O_3]}$$

If the top and bottom of the fraction in this rate law are both divided by k_3 then:

$$\text{rate} = \frac{k_1[O_3]^2}{3\,(k_2/k_3)[O_2][M] + [O_3]}$$

and only the ratio k_2/k_3 affects the rate, not the individual values. This ratio tells how much O cycles back to O_3 relative to how much goes on to give the product.

---

27. ***a)*** From problem 18, the first-order rate constant for the isomerization of CH_3NC at 500 K is $6.57 \times 10^{-4}\ s^{-1}$. The activation energy for the isomerization is $1.61 \times 10^5\ J\ mol^{-1}$. Substitution of these values in the Arrhenius equation:

$$k = A\,e^{-E_a/RT}$$

allows computation of *A*. It is $4.3 \times 10^{13}\ s^{-1}$. The units of *A* must be the same as the units of *k*.

b) The activation energy and the Arrhenius factor do not change with temperature. Calculation of *k* at 300 K and 1000 K merely requires substitution of the two new temperatures into the Arrhenius equation. The *k* at 300 K should be smaller than $6.51 \times 10^{-4}\ s^{-1}$ and the *k* at 1000 K should be bigger:

$$k\,(\text{at } 300) = 4.0 \times 10^{-15}\ s^{-1} \qquad k\,(\text{at } 1000\ K) = 1.7 \times 10^5\ s^{-1}$$

The range of the dependence of *k* upon T is enormous. Reaction rates are quite sensitive to temperature.

---

29. ***a)*** The sum of all of the steps in a mechanism always equals the overall reaction. In this case the overall reaction is:

$$A_2 + 2\,B + 2\,CD \rightarrow 2\,AC + 2\,BD$$

The coefficients in the second step were doubled so that the intermediate A would cancel out in the addition. This meant the formation of 2 AB and necessitated the doubling of the third step to assure that no AB (also an intermediate) was among the final products.

b) The third step is the slowest and rate-determining step in the mechanism. Like all the steps in a mechanism it is an *elementary process*:

$$\text{rate} = k_3[AB][CD]$$

$$[AB] = K_2[A][B] \qquad \textit{and} \qquad [A]^2 = K_1[A_2]$$

By substituting first for [AB] and then for [A]:

$$\text{rate} = k_3 K_2 K_1^{1/2}[A_2]^{1/2}[B][CD]$$

where K_1 is (k_1 / k_{-1}) and K_2 is (k_2 / k_{-2}). As T increases the equilibrium constants of both of the fast equilibria *increase*. Also, the rate constant k_3 increases (according to the Arrhenius equation). Therefore the rate of the reaction increases.

---

31. The mechanism of enzyme catalysis, as modified to allow for the action of an inhibitor, is:

$E + S \rightleftarrows ES$	fast equilibrium	k_1 and k_{-1}
$ES \rightarrow E + P$	slow	k_2
$E + I \rightleftarrows EI$	fast equilibrium	k_3 and k_{-3}

Follow the general pattern of the derivation in Section 10-4, but allow for the complication of the inhibitor which should *decrease* the rate of formation of product. The total enzyme concentration, $[E]_0$ is:

$$[E]_0 = [E] + [EI] + [ES]$$

Let k_3 / k_{-3} equal K_3. Then:

$$[E]_0 = [E] + K_3[E][I] + [ES]$$

$$[E]_0 = [E]\{1 + K_3[I]\} + [ES]$$

$$[E] = \frac{[E]_0 - [ES]}{\{1 + K_3\}}$$

Now, make the steady-state assumption for [ES], the concentration of the intermediate:

$$0 = d[ES] / dt = k_1[E][S] - k_{-1}[ES] - k_2[ES]$$

Substitute the expression for [E} into the steady-state equation. Then solve for [ES]:

$$[ES] = \frac{k_1 [E]_0 [S]}{k_1 [S] + \{1 + K_3 [I]\} (k_{-1} + k_2)}$$

The rate of the reaction is k_2 [ES] so:

$$\text{rate} = \frac{k_2 [E]_0 [S]}{[S] + K_m \{ 1 + K_3 [I] \}}$$

where K_m is defined as $(k_{-1} + k_2) / k_1$). Any concentration of the inhibitor I increases the denominator of this expression and so lowers the rate.

Chapter 11

FUNDAMENTAL PARTICLES AND NUCLEAR CHEMISTRY

11.1 *BUILDING BLOCKS OF THE ATOM*

Several important experiments reveal the constituents of the atom and give numerical values for the parameters (like charge and mass) that characterize them. These experiments include:

The Thomson Experiment (Charge-to-Mass Ratio of the Electron)

J. J. Thomson and his predecessors knew that an electrical current flows across the gap between two electrodes inserted in an evacuated tube and held at a large potential difference. It does this although no air or other gas remains in the tube to conduct the current. The current is carried by *cathode rays*, so called because they emanate from the negatively-charged cathode. Cathode rays are *electrons*, the fundamental particles of electricity.

An electric field deflects a beam of cathode rays that passes through it. Similarly, a magnetic field established in the region of a beam of cathode rays deflects the trajectory of the beam. In the Thomson experiment a beam of cathode rays in an evacuated tube is passed through an electric and magnetic field simultaneously. By adjusting the strengths and directions of the two fields (E and H), Thomson attained a *balance* between their deflecting effects and caused the beam to pass undeviated through the tube. Knowing the strengths of the two fields allowed calculation of the velocity of the cathode rays ($v = E / H$). Turning off the magnetic field allowed the electric field alone to act on the cathode rays and deflect them. Measuring their displacement, s, as they left the field and the length, l, of the region of the field allowed calculation of the ratio of charge to mass of the electrons (cathode rays):

$$e/m = 2\,s\,E / l^2 H^2 = 1.758805 \times 10^{11}\ \mathrm{C\ kg^{-1}}$$

In the equation, the SI units of E are $\mathrm{N\ C^{-1}}$ and the SI units of H are $\mathrm{N\ s\ m^{-1} C^{-1}}$ (where N stands for the newton, the unit of force and C stands for the coulomb, the unit of charge). Refresh skills in dimensional analysis by verifying that the units of e/m are indeed $\mathrm{C\ kg^{-1}}$.

The Millikan Oil Drop Experiment (Charge of the Electron)

Judging from the large size of the ratio e/m, electrons are either very highly charged or very light. Determination of either e or m tells which, and the oil drop experiment determines e.

The experiment starts with a fine spray of oil droplets. Electrons are lost or gained by some droplets, either by friction or other means, giving them an electrostatic charge. The droplets fall under the influence of gravity but are caught and stopped by a properly arranged electrical field of variable strength.

If a droplet is held motionless, the electrical and gravitational forces acting on it are in balance. The gravitational force depends on the mass of the droplet, which can be determined from its terminal velocity as it falls through the air in the absence of the electric field, and g, the known acceleration of gravity. The electrical force depends on the known strength of the electric field and the charge on the oil droplet. The only unknown, the charge on the oil droplet, is thus determined.

Millikan observed that the magnitude of the charge on many different droplets was always an integral multiple of the same basic value. He suggested that different droplets carry different integral numbers of electrons, all with the same fundamental charge. The modern value is 1.602189×10^{-19} C. Using e/m in a quick computation gives m (also symbolized m_e) of the electron. It is $9.1095334 \times 10^{-31}$ kg.

***The Rutherford Experiment* (*Discovery of the Nucleus*)**

Rutherford and his co-workers investigated the scattering of α-particles (helium atoms with both 2 electrons removed) as beams of them impinged on thin foils of gold and other metallic elements. Most of the α-particles passed through the foils as if they were not there or suffered only slight deflection. Significantly, however, *the foils deflected some α particles through large angles.*

This unexpected result implied that the mass of the foils was concentrated in small, dense, positively-charged *nuclei* which were totally missed by most of the α-particles but closely approached by those few that ended up scattered through large angles. A detailed analysis based on this model predicted the number of scattered α-particles as a function of angle. The prediction matched with the experimental results.

The Rutherford model pictures the nucleus with charge $+Ze$ and radius on the order of 10^{-15} m surrounded by Z extra-nuclear electrons distributed at distances on the order of 10^{-10} m.

Protons and Neutrons

An atomic nucleus consists of Z protons which account for the nuclear charge by contributing one unit of positive charge ($+1.602 \times 10^{-19}$ C) each. The lightest atom ($Z = 1$) is hydrogen. Avogadro's number (1 mole) of hydrogen atoms weighs about 1.0 g so the mass of a proton is about 1.00×10^{-3} kg / 6.022×10^{23} or about 1.66×10^{-27} kg. The exact value of the proton mass is 1.67263×10^{-27} kg.

For elements other than hydrogen the atomic weight always *exceeds* the mass contributed by Z protons. Indeed, an early conclusion was: "the number of elementary charges composing the center of the atoms is equal to half the atomic weight." Rutherford accounted for the discrepancy in mass by assuming the existence of additional, electrically neutral particles, *neutrons*, in the nuclei of non-hydrogen atoms. Later experiments by James Chadwick detected the neutron. The neutron mass is 1.67493×10^{-27} kg, indeed close to the proton mass, and the neutron charge is zero.

Chemistry concerns itself with an atom's electrons and its interactions with other atoms' electrons. The positive charge on the atomic nucleus dictates the number of the negatively-charged electrons surrounding it and so dictates its chemistry. Thus Z, the atomic number, determines the chemical identity of a nucleus. The 100-plus different chemical symbols are each synonymous with a different Z. The *mass number*,

A of a nuclide is the sum of Z and the number of neutrons in the nucleus. It is the number of *nucleons* (protons and neutrons) comprising the nucleus. The mass number has but little influence on an atom's chemistry.

Isotopes

As just stated, a nuclear species is characterized by its atomic number Z and its mass number A. Nuclei having the same Z but different A's are *isotopes.* Such nuclei have the same number of protons but different numbers of neutrons. They differ only slightly in their chemical behavior. See p. 5 in this Guide.

A full symbolization of an atom or ion is comprised of the chemical symbol augmented by a left superscript, a left subscript and a right superscript. The left superscript is A, the particle's mass number, the left subscript is Z, the particle's atomic number and the right superscript is the particle's electrical charge, if any. For example, two isotopes of lithium are:

$${}^{6}_{3}\mathrm{Li} \quad \textit{and} \quad {}^{7}_{3}\mathrm{Li}$$

If the first lithium isotope lost an electron the result would be ${}^{6}_{3}\mathrm{Li}^{+}$.

Atomic Mass Units (Amu)

The masses of protons and neutrons are so small (about 10^{-27} kg) that it is wise to define a new unit of mass that is better scaled for use in talking about them.

- The *atomic mass unit* is defined as $^{1}/_{12}$ of the mass of a single neutral ^{12}C atom.

One mole of ^{12}C weighs exactly 12 g so 1 *amu* is $12 / N_0$ g or 1.660565×10^{-24} g.

The following table summarizes the symbols and masses of some important particles:

Particle	*Symbols*	*Mass (amu)*	*Charge (e)*
Proton	${}^{1}_{1}p$, ${}^{1}_{1}\mathrm{H}^{+}$	1.00727647	+1
Neutron	${}^{1}_{0}n$	1.0086649	0
Electron	${}^{0}e^{-}$, β^{-}	0.00054858	−1
Positron	${}^{0}e^{+}$, β^{+}	0.00054858	+1
Alpha	${}^{4}_{2}\mathrm{He}^{2+}$, ${}^{4}_{2}\alpha$	4.001506	+2

11.2 *NUCLEAR STABILITY*

The mass of an atom is not equal to the sum of the masses of the protons, neutrons and electrons that make it up. Instead, for all atoms the mass of the atom is *less* than the mass of the component electrons, protons and neutrons. A change in mass is related to a change in energy by the Einstein equation:

$$\Delta E = c^2 \, \Delta M$$

In problem-solving in nuclear chemistry, ΔM values come out in amu. The conversion of a ΔM (in amu) to a ΔE is so frequent that it is worthwhile to memorize the conversion factor. An amu is equivalent to 1.4924×10^{-10} J, but, more importantly:

1 amu is equivalent to 931.5×10^6 eV or 931.5 MeV.

An eV is an *electron-volt,* the energy required to accelerate an electron through a potential difference of one volt. (See page 193 of this Guide.) Energies of nuclear reactions are usually expressed in MeV, the megaelectron-volt or million electron-volts. **See problems 1, 3, 23.**

Since the formation of an atom from its component parts has a negative ΔM, a nuclear reaction **(see problem 1)** such as:

$$20\,{}^{1}_{1}H + 20\,{}^{1}_{0}n \rightarrow {}^{40}_{20}Ca$$

releases energy. The quantity of energy released tells how tightly the *nucleons* (protons and neutrons) in the product nucleus are bound.

- The *binding energy,* E_B, of a nucleus is the negative of the energy change of its formation from its component nucleons.

- The binding energy *per nucleon* is the binding energy of a nucleus divided by the sum of the number of neutrons and protons in the nucleus. **See problem 1.**

The above is a *nuclear equation.* Nuclear equations resemble ordinary chemical equations in requiring balancing. The criteria for balance however differ:

- In a balanced nuclear equation the sum of the subscripts (atomic numbers) must be same on the two sides of the equation.

- The sum of the superscripts (mass numbers) must be the same on the two sides of the equation.

- The net electrical charge represented on the two sides of the equation must be the same.

11.3 *NUCLEAR DECAY PROCESSES*

Spontaneous nuclear decay occurs only if the energy of the products is less than the energy of the reactants. Given the Einstein equation this implies that the total mass of the products must be less than the mass of the reactants:

$$\Delta E < 0 \qquad and \qquad \Delta M < 0$$

This criterion for spontaneity is equivalent to the thermodynamic criterion (at constant temperature and pressure) $\Delta G < 0$. In nuclear decay, a parent nucleus gives rise to a daughter nucleus, and energy leaves the system as kinetic energy of emitted particles and as electromagnetic radiation.

Although radioactive decay involves the conversion of one kind of *nucleus* into another, it is more convenient in problems to use the masses of whole atoms (nucleus plus electrons) in calculations. It is whole-atom masses that are measured (by mass spectrometry) and given in tables (text Table 11-1, text p. 374, for example). The following are important processes of nuclear decay:

- β^- *Decay.* This is the decay mode of *proton-deficient* nuclei because it increases the number of protons. A neutron in the unstable nucleus converts into a proton, emitting an electron (beta particle). In this decay mode a massless *antineutrino* ($\bar{\nu}$) is also emitted. The mass number *A* stays the same and the atomic number *Z* increases by one.

  The criterion for β^- decay is:

$$\text{M(daughter atom)} - \text{M(parent atom)} = \Delta M < 0$$

The masses in this equation are whole-atom masses (Table 11-1), not the masses of the atoms' bare nuclei. The mass of the β^- (electron) is *not* explicitly added in with the mass of the products because it is already counted in the mass of the neutral daughter atom. **See problem 21.** If ΔM is negative, then ΔE of the nuclear reaction is also negative, and the reaction is spontaneous. The antineutrino and the β^- share in carrying off a total of $-\Delta E$ in energy. If the antineutrino should get zero kinetic energy, then the β^- will have its maximum kinetic energy, $-\Delta E$.

- β^+ *Decay.* If a nucleus is neutron deficient it may emit a positron (β^+) and convert a proton into a neutron. The mass number *A* remains unchanged, but the atomic weight *Z decreases* by one. The process also emits a neutrino. **See problem 5.** Thus:

$$^{40}_{21}\text{Sc} \rightarrow {}^{40}_{20}\text{Ca} + \beta^+ + \nu$$

The criterion for β^+ decay is:

$$\text{M (daughter atom)} - \text{M (parent atom)} + 2\,\text{M}(e^{\pm}) = \Delta M < 0$$

where $M(e^{\pm})$ is 0.00054858 amu, the mass of the positron (and electron). The $2M(e^{\pm})$ term appears because using neutral atom masses leaves one electron and one positron unaccounted for. **See problem 3.** The energy equivalent of 2×0.00054848 amu is, using the conversion factor 931.5 MeV amu^{-1}, 1.02 MeV. This is a lot of energy and neglecting it gives seriously wrong answers. **See problem 23b.**

• *Electron Capture.* A neutron deficient nucleus may capture one of its extra-nuclear electrons, converting a proton into a neutron. *Example*:

$$^{7}_{4}Be + e^- \rightarrow \ ^{7}_{3}Li + \nu$$

This process achieves the same change as β^+ decay, that is, Z decreases by one while A stays the same. The energy (mass) criterion for electron capture is less stringent:

$$\text{M (daughter atom)} - \text{M (parent atom)} = \Delta M < 0$$

The additional 2 M($e^{\pm}$) term that appears in the case of β^+ decay is absent. The mass of the electron on the left-hand side of the nuclear equation does not appear explicitly because it is included in the mass of the neutral parent atom, and nothing but a neutrino is emitted.

• *α-Decay.* A nucleus may emit an alpha particle, $^{4}_{2}He^{2+}$ decreasing Z by 2 and A by 4. Thus:

$$^{226}_{88}Ra \rightarrow \ ^{222}_{86}Rn + \ ^{4}_{2}He^{2+}$$

The mass (energy) criterion for α decay is:

$$\text{M (daughter atom)} + \text{M } (^{4}_{2}\text{He atom}) - \text{M (parent atom)} = \Delta M < 0$$

Many problems use the above relationships. **Problem 21** varies the usual approach by giving ΔE for two processes and the mass of the parent atom and asking for the mass of the daughter atom. **Problem 19** uses none of the formal criteria but obliges the use of the fundamental criterion for spotaneous decay, a decrease in mass.

In nuclear decay problems use the conversion factor 931.5 MeV amu^{-1} to switch between mass units and energy units at will. As a practical matter, it may be easier to do arithmetic on the masses in amu, just because the numbers are smaller, and convert to MeV as a last step. In computing energy and mass differences, do not round off prematurely. Carry extra digits at the intermediate stages, and round off at the last step.

11.4 *KINETICS OF RADIOACTIVE DECAY*

All unstable nuclei decay by a first-order process. The kinetics of radioactive decay therefore follow the equation:

$$N = N_0 e^{-kt}$$

where N_0 is the number of nuclei originally present, N is the number present at time t and k is the rate constant. The half-life of a reaction is:

$$\tau = \ln 2 / k$$

Radioactivity is measured by detecting the high energy particles that the decay process produces. The average decay rate is the *activity A* of the sample, in terms of the number of disintegrations per second. The activity depends both on the number of atoms of the radioactive nucleus (i.e. the mass of the sample) and the rate constant of the decay:

$$A = kN$$

These definitions are used in **problems 7, 9 and 25.** *Specific activity* is the activity per specific quantity of sample (gram or mole). **See problem 27.** The unit of activity is the *curie.* One curie (Ci) is 3.7×10^{10} disintegrations per second. The activity and specific activity themselves decay with time:

$$A = A_0 e^{-kt}$$

Once A and k are known the number of nuclei at any time can be calculated. **See problem 9.**

In solving problems the following points are helpful:

- Having the half-life is the same as having the rate constant for the decay (and vice-versa) because the product of the two is ln 2 (0.69315). The units of k and τ are each other's reciprocals.

- Avoid confusing the units of time. Convert all half-lives (rate constants) in the same problem to the same unit of time (reciprocal time), preferably seconds (reciprocal seconds). **See problem 27.**

- All of the concepts of chemical kinetics (Chapter 10) apply to nuclear decay kinetics. In particular, the *steady state assumption* is prominent when a nuclide decays in a series of steps. **See problem 29b.**

- Many problems involve radioactive dating. ^{14}C dating is a major example. These problems offer only one additional facet that first-order chemical kinetics do not. *Activities* (the actual current rate of decay) rather than concentrations of reacting species are usually given. Uncertainties in counting the activities of radioactive samples create substantial uncertainties in estimated ages of different materials. **See problem 13.**

11.5 *RADIATION IN BIOLOGY AND MEDICINE*

The curie measures the radioactivity of a species in terms of the number of nuclear decay events per second. Different kinds of events emit different kinds of particles. These particles differ in energy and penetrating power. In biology and medicine the important factor is not the number of disintegrations per second in a radioactive source but the quantity of energy deposited in living tissue. The *rad* is therefore defined as the amount of radiation that deposits 10^{-2} J in a 1 kg mass.

The *rem* is the unit of radioactive *dosage.* It is the number of rads absorbed by tissue multiplied by a "fix-up factor", the *relative biological effectiveness,* that accounts for variables such as the type of tissue and type of radiation and the dose rate.

DETAILED SOLUTIONS TO ODD-NUMBERED PROBLEMS

1. ***a)*** The total binding energy of the calcium isotope equals the negative of the *energy equivalent* of the mass change in the formation reaction:

$$20\,{}^{1}_{1}H + 20\,{}^{1}_{0}n \rightarrow {}^{40}_{20}Ca$$

$$\Delta M = M[{}^{40}Ca] - 20\,M[{}^{1}H] - 20\,M[{}^{1}n]$$

$$\Delta M = 39.962589 - 20(1.00782504) - 20(1.00866497)\text{ amu}$$

$$\Delta M = -0.367209\text{ amu}$$

The energy equivalent of 1 amu is 931.502 MeV so this ΔM corresponds to ΔE = −342.056 MeV; E_B is +342.056 MeV. Since there are 40 nucleons, the E_B per nucleon is 342.056 / 40 = 8.55140 MeV

b) By similar reasoning, for ${}^{87}_{37}Rb$ the E_B is 757.860 MeV, and the binding energy per nucleon is E_B divided by 87 or 8.7110 MeV.

c) In ${}^{238}U$ there are 92 protons and 146 neutrons; ΔM in the assembly of the atom is −1.93420 amu and E_B = 1801.714 MeV which is 7.57023 MeV nucleon^{-1}.

---

3. Represent the positron emission as:

$${}^{8}_{5}B \rightarrow {}^{8}_{4}Be + \beta^{+} + \nu$$

The difference in mass between the two sides of the reaction *appears* to be:

$$\Delta M = 8.0053052 + 0.00054858 - 8.024612 = -0.0187582\text{ amu}$$

using the numbers from text page 374 for the boron atom, beryllium atom and positron. At this point a trap looms. The tabulated masses include the atoms' extranuclear electrons. Beryllium has only four such electrons but boron has five. The computation at this stage has "vanished" one electron without anything to show for it. The true change in mass is:

$$\Delta M = 8.0053052 + 2(0.00054858) - 8.024612 = -0.01820964\text{ amu}$$

This mass change has an energy equivalent of −16.962 MeV. Therefore 16.962 MeV is released by the reaction.

---

5. Positron emission by a nucleus always lowers the atomic number by one; electron emission raises the atomic number by one.

a) ${}^{39}_{17}Cl \rightarrow {}^{39}_{18}Ar + \beta^{-} + \tilde{\nu}$

b) ${}^{22}_{11}Na \rightarrow {}^{22}_{10}Ne + \beta^{+} + \nu$

c) ${}^{224}_{88}Ra \rightarrow {}^{220}_{86}Rn + {}^{4}_{2}He$

d) $^{82}_{38}Sr + e^- \rightarrow ^{82}_{37}Rb$

---

7. The original mass of the ^{209}Po is 0.0010 g. This is:

$$0.0010\ g \times \frac{1\ mole}{209\ g} \times \frac{6.022 \times 10^{23}\ atom}{mole} = 2.88 \times 10^{18}\ atom$$

The activity, *A*, the average rate of disintegration of the polonium, depends on the number of polonium atoms present:

$$A = -\,dN\,/\,dt = kN$$

The constant in this equation is a first order rate constant and equals ln 2 / τ where τ is the half-life. Substituting:

$$A = \ln 2\ (N\,/\,\tau)$$

$$A = \ln 2\ (2.88 \times 10^{18}\ /\ 103\ yr) = 1.94 \times 10^{6}\ yr^{-1}$$

Recalling that a year is 365.2 days and a day is 1440 minutes this result becomes $3.68 \times 10^{10}\ min^{-1}$, that is, 3.68×10^{10} atoms decaying per minute.

---

9. *a)* The initial observed *activity* or rate of decay of the fluorine is directly proportional to N_0 the number of atoms of fluorine originally present:

$$A_0 = kN_0$$

The rate constant *k* is (ln 2 / 29 s) = $2.39 \times 10^{-2}\ s^{-1}$. It follows that:

$$N_0 = A_0\,/\,k = A\tau\,/\ln 2$$

$$N_0 = 2.5 \times 10^{4}\ s^{-1} \times 29\ s\ /\ln 2 = 1.05 \times 10^{6}$$

b) The number of fluorine atoms that remains at any time *t* is:

$$N = N_0\, e^{-kt}$$

In this case *t* is 2 min or 120 s. Substitution of *k* and N_0 from part a) gives *N* equal to 5.96×10^{4}.

---

11. For both ^{235}U and ^{238}U first-order kinetics govern the radioactive decay:

$$N = N_0 e^{-kt}$$

If in the beginning the two isotopes were equally abundant, the current difference in abundance results entirely from ^{235}U's faster decay. (The half-life of ^{235}U is briefer by a factor of about 6.) The current abundance ratio is 137.7 to 1. In equation form:

$$N_0\text{ (U-238)} = N_0\text{ (U-235)} \quad and \quad N\text{ (U-238)} = 137.7\, N\text{ (U-235)}$$

This means that:

$$137.7 = e^{-k_2 t} / e^{-k_1 t}$$

where k_2 is the rate constant for the ^{238}U decay, and k_1 is the rate constant for the ^{235}U decay. Taking the natural logarithm of both sides:

$$\ln 137.7 = -k_2 t + k_1 t = t(k_1 - k_2)$$

For each isotope $k = \ln 2 / \tau$ so:

$$\ln 137.7 = t \ln 2\, (1/\tau_1 - 1/\tau_2)$$

The half-lives of the isotopes are both given. They are 7.13×10^8 years and 4.51×10^9 years respectively. Substitution in the equation gives t equal to 6.02×10^9 years. The supposed supernova occurred about 6 billion years ago. This is about 1.5 billion years before the estimated time of the formation of the solar system.

---

13. The *activity* of ^{14}C decays in the same way as the number of atoms of ^{14}C:

$$A = A_0\, e^{-kt}$$

For the papyrus , A is 9.2 disintegrations min^{-1}g^{-1} and A_0 is 15.3 disintegrations min^{-1}g^{-1} (if it is assumed that the activity of ^{14}C in the biosphere has not changed since Egyptian times). Also, k is $\ln 2 / \tau$ or ($\ln 2$ / 5730) yr^{-1}. Substituting and noting that the units of t and τ are the same:

$$9.2 \text{ min}^{-1}\text{g}^{-1} = 15.3 \text{ min}^{-1}\text{g}^{-1}\, e^{-(\ln 2 / 5730)t}$$

$$t = 4.20 \times 10^3 \text{ yr}$$

The uncertainties in the counting rates create uncertainty in the estimated age of the papyrus. Each rate is quoted as ± 0.1 disintegrations min^{-1}g^{-1}. Assuming the worst, A / A_0 varies between 9.3 / 15.2 and 9.1 / 15.4. In the first case t is 4.06×10^3 yr, and in the second t is 4.35×10^3 yr. Therefore, the age of the papyrus is $4.20 \pm 0.160 \times 10^3$ yr, a total range of four human life-times.

---

15. Convert the energy of the incoming 100 MeV neutron to SI units. It is 1.60218×10^{-11} J. The *velocity* of the neutron as it approaches the stationary proton is computed from the equation defining the kinetic energy:

$$K.E. = \tfrac{1}{2}\, m\, v^2$$

Using the mass of the neutron from text Table 11-1, the original velocity of the neutron is 1.383160×10^8 m s^{-1}. The original momentum of the neutron is just its mass times this velocity or 2.31670×10^{-19} kg m s^{-1}.

If the two particles had equal mass, one would expect the neutron to be stationary (with zero *K.E.*) after the collision, and the proton to move off with the same *K.E.* that the neutron came in with. Indeed, an excellent quick answer to the problem is: *K.E.* of the proton 100 MeV, and *K.E.* of the neutron 0.0 MeV. As it is, the neutron is slightly more massive than the proton and its velocity after the collision will be a small positive number.

Let v_p and v_n equal the velocities of the two particles *after* the collision. Let m_p equal the mass of the proton and m_n the mass of the neutron. Finally, let *K.E.* equal the total kinetic energy of the system after the collision, and let *P* equal the total momentum after the collision.

Kinetic energy and momentum are conserved during the collision so:

$$m_p v_p + m_n v_n = P = 2.31670 \times 10^{-19} \text{ kg m s}^{-1}$$

$$\tfrac{1}{2} m_p v_p^2 + \tfrac{1}{2} m_n v_n^2 = K.E. = 1.60218 \times 10^{-11} \text{ J}$$

Eliminating v_p between these two equations gives the quadratic:

$$[m_p m_n + m_n^2]\, v_n^2 - [2\, m_n P] v_n + [P^2 - 2 m_p\, K.E.] = 0$$

The quantities in brackets can be determined numerically:

$$[m_p m_n + m_n^2] = 5.60693 \times 10^{-54} \text{ kg}^2$$

$$[2\, m_n P] = 7.760621 \times 10^{-46} \text{ kg}^2 \text{ m s}^{-1}$$

$$[P^2 - 2 m_n\, K.E.] = 2.01952 \times 10^{-43} \text{ kg}^2 \text{m}^2 \text{s}^{-2}$$

The units confirm that v_n will come out in m s^{-1}. Substituting and dividing both sides of the equation by 5.60693×10^{-54} give the following:

$$v_n^2 - 1.38411 \times 10^{8}\, v_n + 1.31855 \times 10^{13} = 0$$

The post-collision velocity of the neutron will be small and positive. For a small v_n the first term in the quadratic is negligible relative to the second two terms and:

$$[1.38411 \times 10^{8}]\, v_n \simeq 1.31855 \times 10^{13}$$

The velocity of the neutron after the collision is therefore 9.52×10^4 m s^{-1}. Exact solution of the quadratic equation gives essentially the same answer. Using the quadratic formula, v_n is 9.550×10^4 m s^{-1} or, because there are two positive roots, 2.766×10^8 m s^{-1}. The large root is rejected.

The *energy* of the neutron after the collision is:

$$K.E.\text{(neutron)} = \tfrac{1}{2}\, m_n v_n^2 = 7.638 \times 10^{-18} \text{ J}$$

which converts to 47.7 eV or 4.77×10^{-5} MeV. The kinetic energy of the proton after the collision is 100 MeV minus this value or, effectively, 100 MeV.

---

17. ***a)*** The standard enthalpy change for the reaction:

$$N_2H_4(l) + O_2(g) \rightarrow N_2(g) + 2\,H_2O(g)$$

is the sum of the standard enthalpy changes of formation of the products less the sum of the standard enthalpy changes of formation of the reactants:

$$\Delta H^\circ = 2(-241.82) - 50.63 \text{ kJ} = -534.27 \text{ kJ}$$

where the numbers come from Appendix D (recall that elements in their standard states have ΔH_f° of zero).

b) The standard energy change of the system during the reaction is:

$$\Delta E^\circ = \Delta H^\circ - P\Delta V$$

The $P\Delta V$ term equals $\Delta n_g RT$, if it is assumed that the gases in the reaction are ideal and that the volume of the liquid hydrazine is negligible. As a mole of $N_2H_4(l)$ burns at T = 298.15 K, Δn_g of the system is +2 mol and Δn_g RT is 4.96 kJ. Hence, for the combustion of one mole of $N_2H_4(l)$:

$$\Delta E^\circ = \Delta H^\circ - P\Delta V = -534.27 \text{ kJ} - (4.96 \text{ kJ}) = -539.23 \text{ kJ}$$

c) ΔE for the reaction is −539.23 kJ per mole of hydrazine. From the Einstein relationship:

$$\Delta E = c^2\,\Delta M$$

The speed of light is 2.998×10^8 m s^{-1} so ΔM per mole of reaction is -6.00×10^{-12} kg = -6.00×10^{-9} g. The chemical reaction of 32 g of hydrazine and 32 g of oxygen occasions a mass loss of about 6 nanograms. This is too small to detect.

---

19. The equation representing the decay of the neutron is:

$${}^{1}_{0}n \rightarrow {}^{1}_{1}p + e^-$$

From Table 11-1 (text page 374), the mass of the products is 1.00782504 amu and the mass of the reactant is 1.00866497 amu. ΔM is −0.00083993 amu. The energy equivalent of 1 amu is 931.502 MeV so ΔE = −0.7824 MeV. As the neutron decays it emits an electron with a kinetic energy of 0.7824 MeV. (The proton is about 2000 times more massive than the electron and hardly recoils at all).

---

21. ***a)*** The nuclear reaction that produces ^{64}Ni is represented:

$${}^{64}_{29}Cu \rightarrow {}^{64}_{28}Ni + \beta^+ + \nu$$

The ΔE for this reaction is:

$$\Delta E = [M({}^{64}_{28}Ni) - M({}^{64}_{29}Cu)]\,c^2 + 1.02 \text{ MeV}$$

The term in square brackets is the difference in mass between the daughter and parent atoms. The 1.02 MeV is the energy equivalent of the mass of one positron plus one electron. It appears because a positron is emitted from the daughter-parent atom pair and a neutral daughter atom needs one *fewer* electron than the parent. ΔE is given in the problem as −0.65 MeV. Solving for the term in brackets shows it is equivalent to −1.67 MeV; that is, at 931.502 MeV amu^{-1}, equal to −0.00179 amu. The ^{64}Cu weighs 63.92976 amu so the ^{64}Ni weighs 63.92797 amu.

b) The equation representing the β^- decay of ^{64}Cu is:

$$^{64}_{29}Cu \rightarrow\ ^{64}_{30}Zn + \beta^- + \tilde{\nu}$$

For this process:

$$\Delta E = c^2\ [M(^{64}Zn) - M(^{64}Cu)]$$

The ΔE for the process is given as −0.58 MeV so the difference in mass between daughter and parent atoms is equivalent to −0.58 MeV. At 931.502 MeV amu^{-1} this is −0.00063 amu. It follows that the ^{64}Zn daughter weighs 0.00063 amu *less* than the parent or 63.92913 amu.

---

23. **a)** The formation of phosphorus-30 can be represented as:

$$15\ ^1_1H + 15\ ^1_0n \rightarrow\ ^{30}_{15}P$$

The mass of the product is 29.97832 amu, and the mass of the reactants is 15(1.00782504 + 1.00866497) or 30.24735 amu. The difference between these masses is −0.26903 amu, or, in the equivalent energy units, −250.60 MeV. The *binding energy* is the negative of this figure; the binding energy per nucleon is then 250.60 / 30 or 8.353 MeV per nucleon.

b) The equation for the decay of ^{30}P by β^+ emission is:

$$^{30}_{15}P \rightarrow\ ^{30}_{14}Si + \beta^+ + \nu$$

The criterion for spontaneity for this process is:

$$\Delta M = [M(^{30}_{14}Si) - M(^{30}_{15}P)] + 1.0972 \times 10^{-3}\ amu < 0$$

The term in brackets is 29.97376 −29.97832 amu or −0.00456 amu. ΔM for the process is therefore -3.46280×10^{-3} amu. The process is spontaneous. The change in energy of the system is:

$$\Delta E = (-3.46280 \times 10^{-3}\ amu) \times 931.5\ MeV\ amu^{-1} = -3.226\ MeV$$

The kinetic energy of the products is −ΔE and is distributed among them. The β^+ has its maximum kinetic energy when the other products get none; the value is +3.228 MeV.

The energy equivalent of the mass of the two electrons in the mass change equation is 1.02 MeV. Omitting this term introduces an unacceptable error (about 32 percent).

c) The most direct way to get the fraction of the P-30 atoms left after 450 s is to note that 450 s is exactly three 150 s half-lives. The fraction is then obviously $(\frac{1}{2})^3$ or $1/8$. The rate constant is $\ln 2/\tau$ for this first order decay process: $4.62 \times 10^{-3}\ s^{-1}$

---

25. Each ^{226}Ra atom disintegrates with the release of 4.79 MeV. The problem concerns 1.00 g of Ra-226. The number of atoms in 1.00 g of radium-226 is:

$$N = 1.00\ g \times (6.023 \times 10^{23}\ atom / 226\ g) = 2.665 \times 10^{21}\ atoms$$

The *activity* of the radium-226 is equal to $k\,N$:

$$A = k\,N = (\ln 2 / \tau)N$$

where k is the first-order rate constant and τ is the half-life of the decay. The half-life is 1622 yr (5.119×10^{10} s) so the activity of this sample is:

$$A = (\ln 2 / 5.119 \times 10^{10}\ s)(2.665 \times 10^{21}) = 3.61 \times 10^{10}\ s^{-1}$$

Every second 3.61×10^{10} radium atoms decay. Each decay event releases 4.79 MeV; energy appears at a rate of 1.73×10^{11} MeV s^{-1} or 6.22×10^{14} MeV hr^{-1}. The reaction releases:

$$6.22 \times 10^{14}\ MeV\ hr^{-1} \times 1.6022 \times 10^{-13}\ J/MeV = 99.7\ J\ hr^{-1}$$

In the calorimeter the 10.0 g of water absorbs 41.8 J K^{-1}. To absorb the radium's hourly production of heat, 99.7 J, the temperature rises by 99.7 J / 41.8 J K^{-1} = 2.39 K.

---

27. The activity of ^{14}C equals the rate constant for its decay multiplied by the number of atoms present. Consider 1.00 g of carbon from the biosphere. The total number of carbon atoms is:

$$N_{tot} = 1.00\ g \times (1\ mol / 12.01115\ g) \times (6.022 \times 10^{23}\ atoms\ mol^{-1})$$

$$N_{tot} = 5.014 \times 10^{22}\ atoms$$

The number of atoms of ^{14}C per gram of carbon is:

$$N = A / k = A\,(\tau / \ln 2)$$

where τ is the half-life of the ^{14}C and A is its activity. Expressing τ in minutes so that its units will cancel the units of A:

$$N = (15.3\ min^{-1}\ g^{-1})(3.014 \times 10^{9}\ min) / \ln 2 = 6.65 \times 10^{10}\ g^{-1}$$

1.00 g of carbon contains 6.65×10^{10} ^{14}C atoms out of a total of 5.014×10^{22} carbon atoms for a fraction of 1.32×10^{-12}.

---

29. ***a)*** In order to produce ^{228}Ra, ^{232}Th must emit an α particle in the first step of its decay:

$$^{232}_{90}\text{Th} \rightarrow \, ^{228}_{88}\text{Ra} + \, ^{4}_{2}\text{He}$$

In the next step, the ^{228}Ra emits a beta particle (an electron) to give ^{228}Ac:

$$^{228}_{88}\text{Ra} \rightarrow \, ^{228}_{89}\text{Ac} + \beta^- + \tilde{\nu}$$

The decay products over the two steps are an α particle, an actinium-228 atom, an electron and an anti-neutrino. These products have the same total mass as an Ac-228 (228.03117 amu), plus a He-4 atom (4.0026033 amu). The sum is 232.03377 amu. The mass of the beta particle (electron) is included in the neutral actinium atom's mass. The anti-neutrino is massless. The mass of the reactant is 232.038054 amu so ΔM for the process is -0.00428 amu. This mass has an energy equivalent of -3.98 MeV. 3.98 MeV is the energy *lost* in the process, taken away as kinetic energy by the product particles.

b) The thorium decays to radium in a first-order process and the radium goes on to decay to actinium in another first-order process. Let k_1 stand for the rate constant for the first step and k_2 the rate constant for the second. When the radium is present in a steady state amount then:

$$d[\text{Ra}] / dt = 0 = k_1 [\text{Th}] - k_2 [\text{Ra}]$$

$$[\text{Ra}] = (k_1 / k_2) [\text{Th}]$$

where the bracketed symbols signify the number of nuclei present. For any first-order process, the rate constant is ln 2 divided by the half-life. Applying this fact to the previous equation:

$$[\text{Ra}] = (\tau_2 / \tau_1) [\text{Th}] = (6.7 \text{ yr} / 1.39 \times 10^{10} \text{ yr})[\text{Th}]$$

The half-lives (the τ's) of the two steps are known. Therefore, [Ra] $= 4.8 \times 10^{-10}$ [Th]. The number of radium atoms is 4.8×10^{-10} times the number of thorium atoms.

---

31. The earth orbits the sun at a radius R of 1.50×10^{8} km. Imagine a sphere of this radius surrounding the sun. The surface area of this immense sphere is $4\pi R^2$. Radiation from the sun streams out in all directions, cutting through this sphere. The earth intercepts a small proportion of this radiation. The radius of the earth is r (6371 km). From the point of view of the sun the earth appears as a tiny disk of area πr^2. This disk, tiny in proportion to the area of the big sphere, intercepts a fraction f of the total radiation in proportion to the area of the big sphere that it covers:

$$f = \pi r^2 / 4\pi R^2 = \tfrac{1}{4}(r / R)^2 = 4.5 \times 10^{-10}$$

The surface area of the hemisphere of the earth that is exposed to the sun's rays at any time is $2\pi r^2$. Assume that the radiant flux of 0.135 J s^{-1}cm^{-2} is the *average* value over the earth's exposed hemisphere. Then the earth receives the radiant power:

$$P(\text{earth}) = 0.135 \text{ J s}^{-1}\text{cm}^{-2} \times 2\pi (6.371 \times 10^{8} \text{ cm})^2$$

$$P(\text{earth}) = 3.44 \times 10^{17}\ \text{J s}^{-1}$$

The *total* power output of the sun is P(earth) *divided* by the fraction of the sun's radiant power that hits the earth:

$$P(\text{sun}) = P(\text{earth}) / f = 7.6 \times 10^{26}\ \text{J s}^{-1}$$

The mass equivalent of energy is given by the equation $E = Mc^2$. To generate 7.6×10^{26} J each second, mass is converted. The equivalent mass is:

$$M = E / c^2 = 7.6 \times 10^{26}\ \text{J} / (3.0 \times 10^{8}\ \text{m s}^{-1})^2 = 8.5 \times 10^{9}\ \text{kg}$$

The sun burns (in the thermonuclear sense) 8.5 million metric tons of matter per second.

Chapter 12

QUANTUM THEORY AND ATOMIC STRUCTURE

A beam of light is electromagnetic radiation. It consists of electric and magnetic fields (symbolized by E and H respectively) oscillating perpendicular to the direction in which the beam propagates and perdendicular to each other. Several parameters characterize such radiation:

• *The wavelength.* The distance between successive peaks in the intensity of the oscillating electric field (or magnetic field) at any instant is λ, the wavelength of the electromagnetic radiation.

• *The frequency.* Electromagnetic radiation is a periodic or cyclic disturbance. As radiation passes a given point, a maximum in the electric field is followed by a minimum, then another maximum, etc. The number of complete cycles occurring each second is the frequency of the radiation, represented by the symbol ν, the Greek nu. The units of ν are reciprocal seconds, s^{-1}. A reciprocal second is also called a *Hertz*, Hz. Sometimes the unit "cps" for "cycles per second" is encountered. One cps is one reciprocal second.

The *reciprocal* of a wave's frequency is the *period* of the wave. The period is the time required for the wave to complete one cycle. Naturally, it has units of time.

• *The amplitude.* The amplitude of any wave disturbance is the maximum size of the excursion the wave makes during its oscillation. In classical wave theory the *intensity* of a wave is proportional to its amplitude. For electromagnetic waves:

$$\text{Intensity} \propto (E_{max})^2 + (H_{max})^2$$

• *The velocity of propagation.* A traveling wave completes ν oscillations per second and with each such cycle advances by a distance λ, the distance between its successive crests. It follows that:

$$\text{velocity of propagation} = \lambda\ \nu$$

The velocity of propagation of electromagnetic waves (more simply, *the speed of light*) in a vacuum is a universal constant, c. It is 2.9979×10^8 m s^{-1}. For light:

$$c = \lambda\ \nu$$

This equation is the basis for **problem 1** and occurs in many other problems. It is best to memorize it. There is clearly a simple inverse relationship between λ and ν: the higher the frequency of light then the shorter the wavelength. Visible light has wavelengths approximately ranging from 4×10^{-7} to 7×10^{-7} m (equivalent to 400 to 700 nm). See Figure 12-2, text p. 403. The frequency of visible light is, therefore, on the order of 10^{14} s^{-1}.

12.1 *PARADOXES IN CLASSICAL PHYSICS*

Classical physics (19th century physics) treated electricity, magnetism and electromagnetic radiation separately from *mechanics* which dealt with the motions of particles and their interactions. Key concepts of mechanics include the ***kinetic energy***, the energy associated with the motion of particles, and the *potential energy*, *V*, the energy associated with their positions. For a single particle:

$$K.E. = \tfrac{1}{2}\, p^2 \,/\, m \quad = \tfrac{1}{2}\, mv^2$$

where *p* is the *momentum* of the particle, *m* is its mass and *v* is its velocity. The potential energy depends on the location of the particle. The total energy of the particle is the sum of its potential and kinetic energies:

$$\text{total energy} = K.E. + P.E.$$

Classical physics is quite successful in describing many phenomena. Nevertheless it makes wrong predictions in some important experiments:

- *Blackbody radiation.* A heated object such as an iron bar or tungsten lightbulb filament emits radiation. A *blackbody* is an idealized version of such an emitter. Classical physics predicts that the intensity of the light emitted by a blackbody depends directly on the absolute temperature and *inversely on the fourth power of the light's wavelength.*

$$\text{intensity} \sim T / \lambda^4$$

At long wavelengths this formula agrees with observation. At short wavelengths it gives values that are far too large. As λ gets smaller and smaller the predicted intensity goes to infinity. This result is called the *ultraviolet catastrophe.*

- *The photoelectric effect.* When a beam of light falls onto a metal or other material, *photoelectrons* are, under many circumstances, ejected from the surface. Classical physics predicts the maximum kinetic energy of such light-ejected electrons to depend on the intensity of the beam of light. Instead the maximum kinetic energy depends on the frequency of the light.

$$K.E. = h(\nu - \nu_0)$$

where h is a constant and ν_0 is a threshold frequency. Light of frequency less than ν does not cause the emission of photoelectrons, no matter how great its intensity.

- *Stability of the atom.* In the Rutherford model of the atom a small massive nucleus is surrounded by electrons, which are charged particles. If charges move in an orbit about the nucleus, then classical physics predicts that they must emit electromagnetic radiation. If they actually did this they would lose energy and quickly spiral into the nucleus. But, if they are not in motion, there is nothing to keep them from being pulled into the nucleus. Thus, according to classical physics, the Rutherford model of the atom cannot be correct. Yet, Rutherford's experimental results are conclusive.

Quantum Explanation of Paradoxes

The various paradoxes of classical physics are all resolved by the introduction of the central idea of quantum mechanics:

• • **Energy is not continuous, but instead is quantized in discrete packets.**

Radiation transmits energy. The quantum idea, when applied to radiation (light) means that the energy of light is absorbed, emitted or converted in individual packets, or quanta. The quantum of radiation is the *photon.* It is the particle of light. Its energy is directly related to its frequency:

$$E = h\nu = hc/\lambda$$

where h is Planck's constant (6.626×10^{-34} J s). This is the Planck equation. It occurs again and again in problems and important applications. For example, see **problems 3, 5, 29.**

Planck's constant is extremely small. Consequently, the grain-size of particles of light is small. For many purposes light is as good as continuous.

Photons have zero rest mass but *do* have momentum, p. Their energy is:

$$E = pc$$

The preceding two equations give the energy of light first in terms of frequency (a wave property) then in terms of momentum (a particle property). *In quantum theory light has both wave and particle character.*

Putting the notion of the ultimate graininess of energy into the equations dealing with blackbody radiation gives predictions in accord with experiment. The quantum theory averts the ultraviolet catastrophe.

The quantum explanation of the photoelectric effect treats light as a stream of particles or packets or photons each carrying a quantity of energy proportional to the frequency (not the amplitude) of the light:

$$E = h\nu$$

It requires a minimum energy to liberate a photoelectron against the forces holding it to its atom. This accounts for the observation or a minimum or threshold frequency for the liberation of photoelectrons. The *work function* is the name for the minimum energy. Any extra energy from the photon can manifest itself as kinetic energy of the photoelectron. **See problem 7.**

The Bohr Model of Hydrogen

An important step in understanding the stability of the Rutherford nuclear atom was Niels Bohr's model of the hydrogen atom.

Both atoms and molecules can emit and absorb light. They do so at various points in the electromagnetic spectrum, that is, at various wavelengths (various frequencies). Atoms emit and absorb at *lines of sharply defined, specific frequency.* Early spectroscopists found empirical formulas relating the observed frequencies of members of groups of emission lines in the spectrum of atomic hydrogen. Bohr's model gave the same formulas entirely on a theoretical basis. Its postulates were:

1. The electron in the hydrogen atom (and one-electron ions like He^{+}, Li^{2+}, etc.) revolves around the nucleus in a circular orbit.

2. Coulomb (electrostatic) attraction between the negatively-charged electron and the positively-charged nucleus provides the force needed to sustain the circular orbital motion of the electron.

3. The only orbits allowed are those for which the angular momentum of the electron is an integer times the constant $h / 2\pi$: That is, *angular momentum is quantized in units of* $h/2\pi$:

$$\text{angular momentum} = m_e vr = n\,(h / 2\pi)$$

In this equation n is the integer (or quantum number). The units of angular momentum are of course J s, the units of h. Since h is so small, the quantum of angular momentum is small. **See problem 31.**

4. Electrons make transitions from one orbit to another only by absorbing or releasing a quantity of energy equal to the energy difference between the two orbits. When this energy is absorbed or emitted in the form of light:

$$|\Delta E| = h\nu$$

These postulates, when combined with the ordinary rules of mechanics, give two major results:

- *The radius of the hydrogen atom is quantized:*

$$r = \frac{\epsilon_0 n^2 h^2}{\pi Z e^2 m_e} = \frac{n^2}{Z}\, 0.529 \times 10^{-10}\ \text{m}$$

- *The energy of the hydrogen atom is quantized:*

$$E = -\frac{Z^2 e^4 m_e}{8\epsilon_0^{\,2} n^2 h^2} = -(2.178 \times 10^{-18}\ \text{J})\frac{Z^2}{n^2}$$

where m_e and e are the mass and charge of the electron, ϵ_0 is the permittivity of free space, n is the quantum number and Z is the atomic number of the nucleus.

To study these equations set Z equal to 1. Then r has a minimum value of 0.529×10^{-10} m (called a_0, the Bohr radius) when $n=1$. It goes to infinity as n steps up to 2, 3, etc. E likewise has its minimum when $n=1$, but this value is *less than zero.* E rises toward a maximum of zero as n goes to infinity.

An infinite n corresponds to removal of the electron from the atom. Setting n equal to zero corresponds to having never put the electron in the atom in the first place. Positive integral values of n generate the ordinary *allowed states* of the atom.

Z is 1 for hydrogen, 2 for the one-electron ion He^+, 3 for the Li^{2+} ion, and so forth. **See problems 9 and 11** for a comparison of the results of the Bohr model for hydrogen and an one-electron ion.

12.2 *WAVES AND PARTICLES*

Louis de Broglie suggested that if light has both wave-like and particle-like properties, then perhaps matter (such as electrons, protons or even baseballs) has a wave aspect. The DeBroglie equation relates the wavelength of any moving particle to its momentum:

$$\lambda = h / p = h / mv$$

Because Planck's constant (h) is very small the wavelength of most moving objects is too short to measure. The wavelength only becomes important for subatomic particles which have small masses. **See problem 13.**

Heisenberg Indeterminacy Principle

The presence of a wave aspect in matter means that the positions and momenta of particles cannot be determined precisely. Instead there is a built-in indeterminacy. In the measurement of positions, the indeterminancy is on the order of the wavelength, λ, of the particle. For momenta it is on the order of h / λ. More accurately:

$$(\Delta p)(\Delta x) > h / 4\pi$$

where Δp and Δx are the indeterminancies in momentum and position respectively. **Problem 15** requires nothing more than substitution in this inequality and some attention to the units. **Problem 33** is a more complex calculation applying the indeterminancy principle to an important case.

The Schrödinger Equation

Erwin Schrödinger generalized the DeBroglie wave relation to apply to bound particles such as electrons held in atoms. His contribution, a differential equation, is the fundamental equation of *quantum mechanics.* Its exact form is not as important as some key points about it and its solutions:

- Particles bound in different systems have different forms of the Schrödinger equation. The text considers only two of the many possible bound systems: the particle in a one-dimensional box and the hydrogen atom.

- Since the equation is a differential equation, its solutions are mathematical functions, not numbers. These solutions are *wave functions.* They are symbolized by ψ (the Greek letter psi).

- For any bound particle there is a *family* of wave functions, not just one, satisfying the Schrödinger equation. All describe the particle with equal success. For example, in **problem 35,** substituting the positive integers for n in the function:

$$\psi(x) = (2 / L)^{\frac{1}{2}} \sin (n\pi x / L)$$

gives the (infinite) family of different solutions. The quantity n is a quantum number.

Table 12-2 (text p. 426) gives just *some* of the wave functions that satisfy the Schrödinger equation for the electron in one-electron species. As the table shows, solutions to the Schrödinger equation are often complex and tiresome to write out in full. They are frequently designated by a quantum number (or numbers) alone.

- For every different ψ that satisfies the Schrödinger equation in a given system, there is a corresponding energy (E). These are the *allowed energies* of the system. Most energies are *not* allowed. For example, in one-electron atoms and ions the allowed energies are given by the formula:

$$E = -[2.178 \times 10^{-18} \text{ J}]\, Z^2 / n^2 = -[1312 \text{ kJ mol}^{-1}]\, Z^2 / n^2$$

When Z is 1, the species is the hydrogen atom. When n is 1, the energy of the hydrogen atom is -2.178×10^{-18} J. When n is 2, the energy of the H atom is -0.5445×10^{-18} J. All intermediate energies, such as -1.00×10^{-18} J, are impossible.

- The wave function that corresponds to the lowest E is the *ground state.* All others are *excited states* of a particle.

- The value of a wave function differs at different points in space, depending on the coordinates of the points. Regions of space in which ψ is positive have *positive phase.* Regions in which ψ is negative have *negative phase.* Regions where ψ passes through zero and changes sign are *nodes.* Finding a wave function's nodes helps in visualizing its shape. **See problem 19.**

- The square of the wave function of a bound particle, ψ^2, evaluated at some point, is the probability of finding the particle in a small volume about that point. **See problem 37,** which shows how to evaluate ψ^2 and get actual probabilities. A plot of ψ^2 against the space coordinates gives a picture of the *probability density distribution* of the bound particle.

The Particle in a Box

A simple case in which the Schrödinger equation can be solved is that of a particle confined in a one-dimensional box. The wave functions that describe this system are all sine functions. They are:

$$\psi(x) = (2 / L)^{\frac{1}{2}} \sin (n\pi x / L) \qquad n = 1, 2, \ldots$$

The energy of the particle is:

$$E_n = \frac{h^2 n^2}{8\, mL^2}$$

where n is an integer, m is the mass of the particle and L in the length of the box.

The particle in a box problem has slight practical application although in **problem 17** it does give a useful answer to a chemical question. Its main value is to illustrate how the wave property of a confined particle is accounted for. Key points are:

- In a one-dimensional box, the wave functions that satisfy the Schrödinger equation are characterized by *one* quantum number. In two-dimensional and three dimensional boxes the solutions are characterized by *two* and *three* quantum numbers, respectively.

- If the box is large (L large) or if the particle is massive (m large), the allowed energies of the particle are small. More importantly, the *differences* between neighboring values of allowed energy are small. When the allowed energies are closely spaced, the quantization of the energy is not apparent. This is the case with objects bigger than atoms.

- The quantum number alone is often used as a short-hand designation. Thus, the $n=2$ wave function for a particle in a box is referred to as ψ_2 or simply "the n=2 function".

12.3 *THE HYDROGEN ATOM*

When the Schrödinger equation is set up and solved for the case of a single electron (charge e and mass m_e) moving in the vicinity of a central nucleus of charge $+Ze$, then a set of wave functions characterized by *three* quantum numbers (n, ℓ, m) arises. There are three quantum numbers because the hydrogen atom is three-dimensional. These wave functions are *orbitals.* They are very important in chemical bonding theory.

The solution of the hydrogen atom is in terms of *spherical coordinates*, r, θ and ϕ, with the origin at the nucleus, instead of x, y and z, the standard Cartesian coordinates. Spherical coordinates simplify the mathematics of the solution. They are related to Cartesian coordinates by the equations:

$$x = r\sin\theta\cos\phi \qquad y = r\sin\theta\sin\phi \qquad z = r\cos\theta$$

See text Figure 12-11. The resultant wave functions can all be broken down into parts:

- A radial part, $R_{n\ell}(r)$, which depends only the distance from the nucleus (given by the first of the coordinates, r). The form of $R(r)$ is determined by the first two quantum numbers, n and ℓ, which is why they appear as subscripts on the symbol for the function.

- An angular part, $\chi(\theta, \phi)$, which depends on the two angles θ and ϕ and is specified by the second and third quantum numbers, ℓ and m. See **problem 19** and text table 12-2.

The three quantum numbers, which arise naturally in the mathematics of the solution of the Schrödinger equation, occur in *sets* such as [$n=3$, $\ell=2$, $m=0$]. Each set signifies a different wave function. For example, the set just listed specifies the function:

$$\psi_{(n=3,\ \ell=2,\ m=0)} = R_{(3d)}\ \chi_{(d_{z^2})}$$

where the radial and angular functions on the right are listed in text Table 12-2. More importantly, each set tells the characteristic *energy and shape* of its wave function. Sets are *not* constructed by taking just any values for n, l and m. Instead, possible values of the three are interlocked in a strict pattern that must be learned by memory:

- *The principal quantum number*, n, may have any integral value from +1 to infinity. The quantum number n gives the energy:

$$E_n = -\frac{Z^2 e^4 m_e}{8 \epsilon_0 n^2 h^2}$$

It also tells the number of *nodes* that the wave function has:

Number of nodes = $n - 1$

The nodes of orbitals are a valuable physical reference for study. In the hydrogen atom wave functions, the nodes are always either *radial* or *angular*. Radial nodes are spheres of different radius with the nucleus at their center. They are called radial because their locations are specified by values of the coordinate r. Angular nodes are planes or curved surfaces. They are specified by values of the angular coordinates θ and ϕ. **See problem 19.** Angular nodes always contain the nucleus. Radial nodes encircle the nucleus but never contain it. Keeping track of the nodes helps to solve problems. **See problem 21.** *The more nodes an orbital has the higher its energy.*

- *The angular momentum quantum number*, l, has allowed values ranging from 0 to $(n-1)$. The quantum number n imposes a strict ceiling on l, which may *not* take on just any value. For example, when n is 1, then l may equal 0 only.

The value for l always equals the number of angular nodes in the orbital. The presence of angular nodes lends a distinct shape to orbitals, so l really tells the shape of the orbital. When $l=0$, the orbital has no angular nodes. All the nodes are automatically radial, and all orbitals with $l=0$ are therefore *spherically symmetrical.* When $l=1$, the orbital has 1 angular node. When $l=2$, it has 2 angular nodes, etc.

In referring to orbitals, the numerical value of l is replaced by a letter, according to the code:

l	0	1	2	3
letter	*s*	*p*	*d*	*f*

An (n,l) combination is referred to by prefixing the letter for the l value with the integer equal to n. *Examples*: 2*s*, 3*d*.

All *s* orbitals are spherically symmetrical. All *p* orbitals have 1 angular node, a plane. All *d* orbitals have 2 angular nodes. The shapes of *s*, *p* and *d* orbitals are important in chemistry and should be memorized. Study Figures 12-12, 12-13 and 12-15, text p. 427-429, noting the location of nodes especially.

- *The magnetic quantum number*, m, governs the behavior of the atom in external magnetic fields. It completes the description of an orbital. For a given value of l,

m may range from $-\ell$ through 0 and up to $+\ell$. The number of possible values for m is $(2\ell+1)$. For example, an f orbital ($\ell=3$) has 7 different possible values of m: $-3, -2, -1, 0, 1, 2, 3$. There is only one s orbital (because $\ell=0$), but there are 3 p orbitals:

$$p_x \quad p_y \quad p_z$$

and 5 d orbitals:

$$d_{xy} \quad d_{xz} \quad d_{yz} \quad d_{x^2-y^2} \quad d_{z^2}$$

In these designations, the subscripts are Cartesian cooordinates that tell the orientation of the single angular node (in the p case) and the two angular nodes (in the d case). In the hydrogen atom, orbitals with the same n and ℓ are equivalent except in orientation. Thus, m, which distinguishes among the members of the set, can be interpreted as telling the orientation in space of the orbital's angular nodes. It is indeed sometimes called the space quantum number.

The rules stating the possible values of the three quantum numbers give a pattern. As the table shows, *the number of wave functions for a given n is n^2*:

Principal Quantum Number, n	*Number of Orbitals of Each Type*					*Total Orbitals*
	s	*p*	*d*	*f*	*g*	
1	1	0	0	0	0	1
2	1	3	0	0	0	4
3	1	3	5	0	0	9
4	1	3	5	7	0	16
5	1	3	5	7	9	25
n						n^2

Studying the One-Electron Atom Wave Functions

Although the wave functions listed in text Table 12-2 are quite imposing, they can be understood by taking them apart and studying them piece by piece. Bear in mind that the wave functions *already* are in two parts. A complete wave function is the product of an angular part and a radial part.

Concentrate on one of the radial functions and prepare a table showing the values at different r of the several terms it contains. For example, take the $R(3s)$ function. Simplify matters by setting $Z=1$, which means dealing with the hydrogen atom itself and not one of the one-electron ions. Work out the numerical value of the cluster of constants that comprise the leading term of $R(3s)$. The function then becomes:

$$R(3s) = (3.0753 \times 10^{13}\ \mathrm{m}^{-3/2})\,(27-18\sigma+2\sigma^2)\,e^{-\sigma/3}$$

The following is a table of values showing how this function behaves:

r (Å)	σ	$(27-18\sigma+2\sigma^2)$	$e^{-\sigma/3}$	$R(r)$
0.0000	0.0000	27.00	1.0000	1.00×10^{15} $m^{-3/2}$
0.2645	0.5000	18.50	0.8465	5.80×10^{14}
0.5290	1.0000	11.00	0.7165	2.92×10^{14}
0.7935	1.5000	4.50	0.6065	1.01×10^{14}
1.0060	1.9019	0.00	0.5305	0.0
2.0000	3.7807	−12.46	0.2836	-1.31×10^{14}
3.0000	5.6711	−10.76	0.1510	-0.602×10^{14}
3.7548	7.0980	0.00	0.0938	0.0
10.000	18.904	401.4	1.834×10^{-3}	0.23×10^{14}
100.00	189.04	68100	4.30×10^{-28}	1.08×10^{-9}

First, note that the variable σ has no units. It is the distance r divided by the Bohr radius, a_0, which is 0.529×10^{-10} m. The term in the third column causes the overall function (last column) to become negative between r = 1.006 and 3.754 Å. $R(3p)$ changes sign (goes from positive to negative phase and vice-versa) at these distances. They are the locations of the two radial nodes.

Despite the fact that the term in the third column gets large with increasing r, the exponential term in the fourth column gets so small that it forces the overall value of $R(r)$ toward zero at large r. Study the above numbers in combination with text Figure 12-14.

Finally, the unit of $R(r)$ is meters to the minus three-halves. It is hard to associate any physical meaning with this unit. When squared it becomes m^{-3} and if m^{-3} is multiplied by a volume, the result is a dimensionless number, a *probability.* (See next section.)

Note the similarity of the one electron wave functions and the Maxwell-Boltzmann distribution (p. 42 of this Guide). It is worthwhile to graph the $3s$ and other wave functions.

Size of Orbitals and Atoms

The Heisenberg indeterminancy principle and the wave nature of the electron make it impossible to know exactly where an electron in an atom is located. Instead it is necessary to speak of *probabilities.*

The probability, p, of finding an electron at a point (r,θ,ϕ) in a hydrogen atom is:

$$p(r,\theta,\phi) = [\psi(r,\theta,\phi]^2 = [R(r)]^2\ [\chi(\theta,\phi)]^2$$

The quantities in parentheses are the coordinates of the point. Different ψ's give different probability distributions.

Another probability function is quite useful in chemistry. The *radial probability density function* tells the probability of finding the electron in a thin shell of thickness Δr at a distance r from the nucleus irrespective of direction. It is defined as:

$$\text{Radial Probability Density} = 4\pi^2\ r^2\,[\psi(r,\theta,\phi)]^2\Delta r = r^2\,[R(r)]^2\Delta r$$

The radial probability density is proportional to the product $[\psi(r,\theta,\phi)]^2 \times r^2$. The presence of r^2, the square of the distance of the point from the nucleus, reflects the fact that a large-radius shell contains more volume than a small-radius shell with the same thickness. A large-radius shell is therefore more likely to contain the electron. **See problem 37.** In working with the radial probability density, remember:

- Probabilities are unit-less numbers between 0.0 and 1.0.

- The radial probability density equals zero when $r = 0$ for all orbitals.

- The radial probability density function for all orbitals has one or more relative maxima as r increases from zero.

- The function goes to zero between relative maxima at the locations of its radial nodes.

- At very large r the function approaches zero.

If the size of an orbital is defined by the r at which $r^2[R(r)]^2$ has its largest relative maximum then:

- The size of orbitals of the same ℓ (for example all s orbitals) increases with increasing n. By the same rule the $3p$ orbitals are bigger than the $2p$.

- For a given value of n the orbital size *decreases* with increasing ℓ. The $3d$ is smaller than the $3p$, which is smaller than the $3s$.

Electron Spin

The electron behaves like a tiny magnet. In classical theory this magnetism is explained by imagining the electron to be a ball of charge spinning about its own axis like a top. Two directions of spin are possible. They are identified with two possible values of the *spin quantum number*:

$$m_s = +\tfrac{1}{2} \quad or \quad -\tfrac{1}{2}$$

The electron spin quantum number adds a *fourth* quantum number to n, ℓ and m, the three quantum numbers that characterize a bound electron's wave function.

12.4 *MANY-ELECTRON ATOMS*

When an atom contains more than one electron, the Schrödinger equation becomes too complicated for easy use. The usual description of the *electronic structure* of such atoms avoids the complexity of an exact solution by using the atomic orbitals of hydrogen as approximations of the actual orbitals. This is the reason for the heavy emphasis in chemistry on understanding hydrogen atom wave functions. Hydrogen atom orbitals are a guide to all other atoms' electron configurations.

In this approximation each electron in the many-electron atom is described by a set of four quantum numbers: n, ℓ, m, and m_s. Certain rules govern how the orbitals accommodate electrons:

• The *Pauli Exclusion principle* states that no two electrons in an atom may have the same set of of four quantum numbers. Every orbital, fully specified by the first three quantum numbers, holds at most two electrons, one with $m_s = +\frac{1}{2}$ and the other with $m_s = -\frac{1}{2}$

• The ground-state electron configuration of many-electron atoms comes from filling the atomic orbitals with a maximum of two electrons each in order of increasing energy. This order is:

1s 2s 2p 3s 3p 4s 3d 4p 5s 4d 5p 6s 4f 5d 6p 7s ...

In this list the numbers are *n*, the principal quantum number. The *p* orbitals come in sets of 3 (of equal energy), *d* orbitals come in sets of 5 and *f* orbitals come in sets of 7. Only the *s* designation refers to just 1 orbital.

• A *subshell* is a group of orbitals with the same *n* and *ℓ*. For example, the 3*d* subshell has 5 orbitals.

• A *shell* of orbitals in a many-electron atom is a set of orbitals with similar energies. For example, in carbon the 1*s* orbital is much lower in energy than all other orbitals and is a distinct shell. The 2*s* and 2*p* subshells are relatively close together in energy and comprise a second shell. Orbitals of similar energy often are just orbitals with the same *n*. Sometimes however orbitals of different *n* are in the same shell. *Example*: In cobalt the 4*s*, 3*d* and 4*p* orbitals are in the same shell.

In chemistry, the most important shell is the highest energy or *outermost* shell. Electrons in this shell determine the chemical behavior of an atom. The orbitals of an atom's outermost shell are its *valence orbitals*. Electrons in these orbitals are *valence electrons*.

• *Hund's rule* states that, when electrons are added to a set of orbitals of equal energy, a single electron enters each one of the set before the second electron enters any of them. The spin quantum numbers (m_s) of the electrons stay the *same* as long as possible.

• When two electrons in an atom have the same spin quantum number their spins are *parallel*. The two must be in different orbitals, according to the Pauli principle. Two electrons in the same orbital have spin quantum numbers $+\frac{1}{2}$ and $-\frac{1}{2}$ respectively. The two electrons are then *paired*.

Paired electrons compensate for each other's magnetism. If all the electrons in an atom occur in pairs, then the atom is *diamagnetic*. If one or more electrons are not in pairs then their spins add together instead of cancelling each other out. An atom with one or more unpaired electrons is *paramagnetic*.

Many problems are based on proposed violations of the rules that relate the quantum numbers of the electrons in many-electron atoms. **See problem 21.**

An important exercise is to represent the ground-state electron configurations of different atoms and ions. **See problem 23.** In writing electron configurations the number of electrons in each subshell appears as a right superscript (and is often mistaken for an exponent). The maximum superscript is 2 for an *s* subshell (1 *s* orbital), 6 for a *p* subshell (3 *p* orbitals), 10 for a *d* subshell (5 *d* orbitals, and 14 for an *f* subshell (7 *f* orbitals). Noble gas electron configurations are *abbreviated by writing the symbol of the noble gas in brackets.*

Another important exercise is writing an element's valence electron configuration given its position in the periodic table. The number of the period that the element occupies gives the maximum value of n in such a configuration. For non-transition elements, the number of valence electrons equals the group number (number at head of the column in the periodic table).

12.5 EXPERIMENTAL MEASURES OF ORBITAL ENERGIES

Ionization Energy

The *ionization energy* (*IE*) of an atom is the minimum energy necessary to remove an electron from it:

$$X(g) \rightarrow X^{+}(g) + e^{-} \qquad \text{Ionization Energy} = \Delta E$$

Ionization energies are measured by photoelectron spectroscopy. The positively-charged ion produced by ionization may itself be able to lose an electron. Thus an atom with n electrons has a series of n ionization energies, each larger than the previous: IE_1, IE_2, IE_3....

In general the ionization energy *increases* from left to right across a period (row) in the periodic table and *decreases* from top to bottom within a group (column). **See problem 27. Problem 25** concerns a hypothetical universe and tests not only an understanding of these trends but also the *exceptions* which are caused by the extra stability of half-filled subshells. For example, nitrogen, with a half-filled $2p$ subshell, has a higher IE_1 than the neighbor to the right, oxygen. The orbital description of many-electron atoms explains these trends and exceptions.

Electron Affinity

Energy is usually *released* when an electron is added to an isolated atom to form a negative ion. When a process releases energy it has a negative ΔE. The *electron affinity, EA*, of an atom is the energy *gained* by the system as the electron is added:

$$X(g) + e^{-} \rightarrow X^{-}(g) \qquad \text{Electron Affinity} = -\Delta E$$

The use of the word "affinity", the tendency to gain, obliges the negative sign. A negative *EA* means that an atom repels the extra electron and the two must be forced together.

The periodic trends in *EA* parallel the changes in *IE*. Both atomic properties increase from left to right in a period and from bottom to top in a group.

If a second electron joins an originally neutral atom, there is a second electron affinity. EA_2 is *always* negative. The negative ion formed by the atom's gain of the first electron repels additional negative charge.

Electronegativity

Atoms with low *IE*'s and low *EA*'s readily lose their own electrons and are poor at accepting new electrons. Such atoms lie to the lower left of the periodic table. Atoms with high *IE*'s and high *EA*'s strongly resist giving up their own electrons and are good at accepting new electrons. These atoms lie to the upper right of the periodic table. The first kind of atom is an electron donor and the second kind an electron acceptor.

A new atomic property quantifies the progression in the periodic table from good electron donor to good acceptor. The *electonegativity* of an atom is:

$$\text{Electronegativity} \propto \tfrac{1}{2}(IE_1 + EA_1)$$

This definition is due to Robert Mulliken, and values attained with this formula are Mulliken electronegativities. Pauling electronegativities (Chapter 13) are similar in general meaning and more widely used. Atoms to the left and bottom of the periodic table, having low electronegativities, are *electropositive.*

DETAILED SOLUTIONS TO ODD-NUMBERED PROBLEMS

1. The FM radio station is broadcasting at a frequency of 98.6 MHz. This frequency (symbolized ν) times the wavelength (λ) of the waves is the speed of light (speed of propagation of the electromagnetic radiation):

$$c = \lambda \nu$$

In a vacuum c is 2.9979×10^{8} m s^{-1}. In air it is only *very* slightly less. Therefore:

$$\lambda = c / \nu = 3.00 \times 10^{8} \text{ m s}^{-1} / 9.86 \times 10^{7} \text{ s}^{-1} \quad \textit{and} \quad \lambda = 3.04 \text{ m}$$

---

3. The barium atoms gain energy in the flame. As they leave the flame they *decrease* their energy by 3.6×10^{-19} J each by emitting photons. The frequency of the photons is related to E:

$$E = h\nu$$

where h is Planck's constant. But $\nu = c / \lambda$ so:

$$E = hc/\lambda \quad \textit{and} \quad \lambda = hc/E$$

Using SI units the wavelength comes out in meters:

$$\lambda = [6.626 \times 10^{-34} \text{ J s} \times 2.9979 \times 10^{8} \text{ m s}^{-1}] / 3.6 \times 10^{-19} \text{ J}$$

$$\lambda = 5.5 \times 10^{-7} \text{ m} = 550 \text{ nm}$$

This wavelength is in the green region of the visible spectrum.

---

5. ***a)*** The energy change of an atom of sodium and the wavelength (λ) of the radiation its emits are inversely related:

$$\Delta E = hc/\lambda$$

If λ is 589.3 nm (or 5.893×10^{-7} m) then substitution of $h = 6.626 \times 10^{-34}$ J s and $c = 2.998 \times 10^{8}$ m s^{-1} gives $\Delta E = 3.371 \times 10^{-19}$ J.

b) A mole of sodium atoms consists of Avogadro's number of sodium atoms. The energy change per mole is:

$$3.371 \times 10^{-19} \text{ J atom}^{-1} \times 6.022 \times 10^{23} \text{ atom mol}^{-1} = 2.030 \times 10^{5} \text{ J mol}^{-1}$$

c) The sodium arc light puts out 1000 W (watt) in the form of radiant energy. 1000 W is 1000 J s^{-1}. Assume that all of this energy is emitted at the sodium D-line. Then:

$$1000 \text{ J s}^{-1} \times \frac{1 \text{ mol}}{2.030 \times 10^{5} \text{ J}} = 4.926 \times 10^{-3} \text{ mol s}^{-1}$$

---

7. *a)* The energy of the photons of the light is directly proportional to the frequency and inversely proportional to the wavelength of the light:

$$E = h\nu = hc/\lambda$$

where h is Planck's constant, ν is the frequency and λ the wavelength. The energy of 250 nm light is therefore:

$$E = 6.62 \times 10^{-34}\ \text{J s} \times 3.00 \times 10^{8}\ \text{m s}^{-1} / 250 \times 10^{-9}\ \text{m}$$

$$E = 7.94 \times 10^{-19}\ \text{J}$$

The work function of a metal is the minimum amount of energy required to free a photo-electron from its surface. For this metal it is 7.21×10^{-19} J. As photons hit the surface they liberate electrons with a maximum kinetic energy of 7.94×10^{-19} J $- 7.21 \times 10^{-19} = 0.73 \times 10^{-19}$ J.

b) To get the maximum speed of the photoelectrons use the definition of kinetic energy:

$$K.E. = \tfrac{1}{2} mu^2$$

The electron mass is m_e, 9.11×10^{-31} kg. Taking the kinetic energy from part a) and substituting gives $u = 4.0 \times 10^{5}\ \text{m s}^{-1}$.

---

9. The B^{4+} ion is a hydrogen-like ion. Like H, it has only one electron. Unlike H, its nuclear charge (Z) is 5 instead of 1. The questions about it can be answered by substitution in the Bohr equations (p. 250 in this Guide). It is also instructive to answer them by comparison to the answers for the H atom.

The closer an electron is to the nucleus the faster it must move. The $n=3$ radius (r_3) of the electron's orbit in the B^{4+} ion is $^1/_5$ that in an H atom in the same state. The velocity (v) of the electron must be 5 times larger because $m_e vr$, the angular momentum, equals the same constant for the two species.

The energy of the B^{4+} ion is 25 times more negative than that of the H atom because, in the Bohr model, E depends on $-Z^2$. The energy is more negative because the negatively-charged electron is orbiting 5 times *closer* to a nucleus with a positive charge that is 5 times *larger.*

For an H atom, r_3 is $3^2 a_0$ or $9a_0$ which is 4.761×10^{-10} m. For the B^{4+} ion, r_3 is $^1/_5$ this or 0.9522×10^{-10} m.

For an $n=3$ H atom:

$$v = nh / 2\pi m_e r = 7.294 \times 10^{5}\ \text{m s}^{-1}$$

so for the B^{4+} ion v is 5 times larger which is $3.647 \times 10^{6}\ \text{m s}^{-1}$.

Notice once again that the *product* vr is the same in the two species. The angular momentum of the electron, which equals this product times the electron mass, is therefore also the same in the two species. It is 3.163×10^{-34} J s.

For an H atom in the n=3 state:

$$E = -2.178\times 10^{-18}\ \text{J} / 3^2 = -2.420\times 10^{-19}\ \text{J}$$

For B^{4+} in the n=3 state E is 25 times more negative:

$$E = 25\times(-2.420\times 10^{-19}\ \text{J}) = -60.50\times 10^{-19}\ \text{J}$$

The negative of this answer is the energy needed to strip the electron away from a single B^{4+} ion in the n=3 state. It is $+60.50\times 10^{-19}$ J. For a mole of B^{4+} ions the energy is Avogadro's number times larger:

$$E = (6.022\times 10^{23}\ \text{mol}^{-1})\times 60.50\times 10^{-19}\ \text{J} = 3.643\times 10^{6}\ \text{J mol}^{-1}$$

The energy emitted in an n=3 to n=2 transition by a B^{4+} ion has the magnitude of the *difference* between energy levels:

$$E_{n=3} = -(2.178\times 10^{-18}\ \text{J})\ Z^2 / 3^2 = -Z^2 / 9\ (2.178\times 10^{-18}\ \text{J})$$

$$E_{n=2} = -(2.178\times 10^{-18}\ \text{J})\ Z^2 / 2^2 = -Z^2 / 4\ (2.178\times 10^{-18}\ \text{J})$$

The difference is the final energy minus the initial:

$$\Delta E = E_2 - E_3 = (-{}^1/_4 + {}^1/_9)\ Z^2\ (2.178\times 10^{-18}\ \text{J})$$

In this case Z=5 so $\Delta E = -0.1389\times 25\times 2.178\times 10^{-18}$ J which is -7.563×10^{-18} J. This is the energy change of the *ion*, which has fallen to a lower energy state. The energy *emitted* is $+7.563\times 10^{-18}$ J. Dividing this energy by h gives ν, the corresponding frequency, equal to $1.141\times 10^{16}\ \text{s}^{-1}$. The product of the frequency (ν) and wavelength (λ) equals c. Hence:

$$\lambda = 2.9979\times 10^{8}\ \text{m s}^{-1} / 1.141\times 10^{16}\ \text{s}^{-1} = 2.626\times 10^{-8}\ \text{m}$$

---

11. The problem can be solved by substitution into the equations of the Bohr model of the single-electron atom. In such an approach the fact that hydrogen's n=3 to n=2 emission occurs at 656.2 nm is not needed. A less laborious solution uses that fact.

Energy levels in hydrogen-like atoms are characterized by $-Z^2/n^2$ times a constant. The 3 → 2 transition for the neutral H atom (Z=1) corresponds to an energy jump proportional to $(1^2/2^2 - 1^2/3^2) = 0.13889$. In the Li^{2+} ion, Z=3, and the 4 → 2 energy jump is proportional to $3^2/2^2 - 3^2/4^2 = 1.6875$. The energy jump is bigger in the Li^{2+} transition by the factor (1.6875 / 0.1389) = 12.150. Therefore the wavelength of the emitted light in the Li^{2+} transition is 12.150 times *shorter* than 656.2 nm. This is 656.2 / 12.150 = 54.01 nm.

---

13. ***a)*** The velocity of the electron is 1000 m s^{-1}. Its mass is 9.109×10^{-31} kg. Substitution in the DeBroglie equation gives:

$$\lambda = \frac{h}{m_e v} = \frac{6.626\times 10^{-34}\ \text{J s}}{(9.109\times 10^{-31}\ \text{kg})(1000\ \text{m s}^{-1})} = 7.274\times 10^{-7}\ \text{m}$$

b) A proton weighs 1.672649×10^{-27} kg which is 1.836×10^{3} times more than the electron. At the same velocity its wavelength is therefore 1.836×10^{3} times *shorter* or 3.962×10^{-10} m.

c) An electron with kinetic energy of 1 electron volt (1.602×10^{-19} J) has a speed given by:

$$v = [2(K.E.) / m]^{\frac{1}{2}}$$

Substitution gives v equal to 5.931×10^{5} m s^{-1}, and $m_e v$, the momentum of this electron is 5.402×10^{-25} kg m s^{-1}. The wavelength of the electron is Planck's constant divided by this momentum or:

$$6.626 \times 10^{-34} \text{ J s} / 5.402 \times 10^{-25} \text{ kg m s}^{-1} = 1.226 \times 10^{-9} \text{ m}$$

d) The 200 g baseball will have a much *shorter* wavelength because its momentum is much larger even at low velocities. 75 km hr^{-1} is 20.8 m s^{-1}. Substitution gives a wavelength of 1.6×10^{-34} m.

---

15. *a)* According to the Heisenberg indeterminacy principle, the indeterminacy in the *momentum* of a particle multiplied by the indeterminacy in its *position* always exceeds a small constant:

$$(\Delta p)(\Delta x) > h / 4\pi$$

In this problem Δx is 1×10^{-9} m. Therefore, Δp exceeds 5.3×10^{-26} kg m s^{-1}. From the classical definition of momentum, Δp equals $\Delta(mv)$. Therefore, because m_e is 9.109×10^{-31} kg, Δv is 5.8×10^{4} m s^{-1}, at a minimum.

b) A helium atom has a mass of 4.003 amu which is 7.297×10^{3} times *larger* than the mass of an electron. See text Table 11-1. The indeterminacy of the momentum of the He atom equals the Δp just calculated in part a). The minimum Δv for the He atom therefore is 7.297×10^{3} times smaller than the minimum Δv for the electron or 7.9 m s^{-1}. Although much smaller than the answer for the electron, this is still faster than many people can run.

---

17. The energy of a particle in a one-dimensional box is given by the formula:

$$E_n = n^2 [\, h^2 / 8\, m\, L^2 \,]$$

where the quantities within the brackets include a fundamental constant, *h*, a quantity characterizing the particle (its mass, *m*), and a quantity characterizing the box (its length, *L*). The particle is an electron and the length of the box is 1.34 Å. If all of the bracketted quantitites are expressed in SI units ($h = 6.626 \times 10^{-34}$ J s, $m = 9.109 \times 10^{-31}$ kg, $L = 1.34 \times 10^{-10}$ m), then the term in brackets comes out to be 3.35×10^{-18} J.

The energy of any quantum state of the electron in this box corresponds to n^2 times this number. Thus, the ground state, E_1 is 3.35×10^{-18} J. When $n=2$ the energy is 2^2 times this value or 13.4×10^{-18} J. Similarly, E_3 is 30.2×10^{-18} J.

To excite an electron in this box from $n=1$ to $n=2$ requires energy equal to the energy difference between the $n=1$ and $n=2$ quantum levels. This is 10.1×10^{-18} J. If this energy is supplied by one photon then:

$$10.1 \times 10^{-18} \text{ J} = hc/\lambda$$

where λ is the the photon's wavelength and c its speed (2.9979×10^{8} m s^{-1}). Substitution gives λ equal to 1.97×10^{-8} m. This wavelength, 199 Å, occurs in the ultraviolet region of the spectrum.

---

19. The wave function $\psi(2p_z)$ is the product of a *radial* part, $R(2p_z)$ and an angular part $\chi(2p_z)$. When $\psi(2p_z)$ is written out it is imposing. See Table 12-2 of the text. Its radial part contains an exponential dependence on $-r$. Its angular part contains a $\cos\theta$ dependence. These two coordinates plus a third, ϕ, are the three spherical coordinates.

Only when the *square* of this function equals zero does the probability of finding the $2p_z$ electron go to zero. This happens when $r=0$ (at the nucleus) and also when $\theta = \pi/2 = 90°$ (because $\cos 90° = 0$). The coordinate θ equals 0° along the z axis and equals $\pi/2$ at right angles to z, in the xy plane. Therefore, the $2p_z$ wavefunction equals zero everywhere in the xy plane as does its square. This plane is a nodal plane.

The d_{xz} orbital has a $\sin\theta\cos\theta\cos\phi$ dependence in its angular part. This function (and its square) are zero whenever $\theta = \pi/2$ (in the xy plane) *or* whenever $\phi = \pi/2$ (in the yz plane). These two planes are the orbital's nodal planes.

The $d_{x^2-y^2}$ orbital has a $\sin^2\theta\cos 2\phi$ angular dependence. This term is zero when $\phi = \pi/4$ (45°), $3\pi/4$ (135°) $5\pi/4$ (225°) and $7\pi/4$ (315°). The values of $\pi/4$, $5\pi/4$ for ϕ define a plane containing the z axis and half-way between the x and y axes. The other pair of ϕ's ($3\pi/4$ and $7\pi/4$) define a plane at right angles to the first plane. These are the nodal planes. The $d_{x^2-y^2}$ orbital also goes to zero at $\theta = 0$. This happens only along the z axis, the line at the intersection of the two nodal planes just identified.

---

21. Combination *a* is not allowed because l must be less than n. It may not equal n. Combination *c* has $m > l$ which is not allowed. Combination *d* has $l < 0$ which is not allowed. Only combination *b* is allowed.

The rules are easy to apply when physical significance is attached. $(n-1)$ equals the total number of all nodes of the wave function. The quantum number l tells the number of angular nodes. Since the number of angular nodes cannot exceed the total number of nodes, l may equal $n-1$ at the most. Its *minimum* is zero because -1 planar nodes has no physical meaning. The third quantum number, m, is related to the orientation of the angular nodes and *may* be negative.

---

23. Be^{+} $1s^2 2s^1$; C^{-} $1s^2 2s^2 2p^3$; Ne^{2+} $1s^2 2s^2 2p^4$; Mg^{+} [Ne] $3s^1$; P^{2+} [Ne] $3s^2 3p^1$; Cl^{-} [Ne] $3s^2 3p^6$; As^{+} [Ar] $3d^{10} 4s^2 4p^2$; I^{-} [Kr] $4d^{10} 5s^2 5p^6$

All of these electron configurations are *ground-state* (lowest energy) configurations. Be^+, C^-, Ne^{2+}, Mg^+, P^{2+} and As^+ all have at least one unpaired electron (they have incompletely filled subshells) and should be paramagnetic.

---

25. If only one electron could occupy each orbital in many-electron atoms then the configurations:

$1s^1$ *and* $1s^1 2s^1 2p^3$ *and* $1s^1 2s^1 2p^3 3s^1 3p^3$

would be closed shell electron configurations. Neutral atoms with Z = 1, 5, 9 respectively would have these ground state electron configurations.

In a periodic table based on this system, the $Z=1$ element would have the highest ionization energy, just as He, the element with the first closed shell configuration in our periodic table, has the highest *IE*. The *IE* of the $Z=2$ element would be less and the *IE* would drop again for $Z=3$, just as the *IE* for borron is less than the *IE* for beryllium. The *IE* would rise for $Z=4$ and attain a relative high (not as high as $Z=1$) at $Z=5$. For $Z=6$, the *IE* would again diminish and diminish again for $Z=7$. At $Z=8$ it would rise once more and go on to another relative maximum at $Z=9$.

---

27. Sr has a higher first ionization energy than Rb. Radon has a higher first *IE* than polonium. Xe has a higher first *IE* than Cs. Sr has a higher first *IE* than Ba. Xe has a higher first *IE* than Bi. All of the answers come from application of the periodic trends.

---

29. The Planck equation:

$$E = h\nu = hc/\lambda$$

relates the energy and wavelength of all radiation. Planck's constant is 6.626 $\times 10^{-34}$ J s, and the speed of light is 3.00×10^8 m s^{-1}. Substitution gives E equal to 9.94×10^{-28} J for the 200 m radio photon.

This wavelength is 12 orders of magnitude longer than the wavelength of the X-ray photon. Thus, E is 12 orders of magnitude larger (9.94×10^{-16} J) for the X-ray photon.

Multiplying these results by N_0 gives the energy per mol of photons of each type: 5.98×10^8 J mol^{-1} for X-rays and 5.98×10^{-4} J mol^{-1} for radio waves. The latter is *less* than the range of energies of chemical bonds (~ 10^6 J mol^{-1}) and the former is more. X-rays break and make chemical bonds. Radio waves do not.

---

31. The angular momentum of the earth in its orbit around the sun is quantized in units of $h/2\pi$:

$$mvr = nh/2\pi$$

For the earth, m and r are both given. The velocity, v must be calculated. The earth is *gravitationally* attracted by the sun. The force is:

$$F_{grav} = \frac{G\,Mm}{r^2}$$

where G is the universal gravitational constant, and M is the mass of the sun. Both these values are in standard references. G is 6.670×10^{-11} N m^{2}kg^{-2} and M is 2.00×10^{30} kg. If the earth's orbit is circular, then the gravitational force acting on the planet is:

$$F_{grav} = m\,v^2 \,/\, r$$

Combining the preceding two equations gives:

$$v^2 = GM \,/\, r$$

Note that the mass of the earth, m, cancels out. According to the problem, the radius of the earth's orbit, r, is 1.5×10^{11} m. Substitution gives the velocity of the earth in its orbit as 3.0×10^4 m s^{-1}.

The angular momentum of the earth can now be calculated. It is mvr which is 2.7×10^{40} J s. Presumably, this large angular momemtum is quantized in units of $h \,/\, 2\pi$. If so, the quantum number, n,is 2.5×10^{74}. Since n is huge, ± 1 unit of angular momentum makes no meaningful change in the angular momentum.

---

33. *a)* If an electron is confined within a nucleus with a radius of 1×10^{-15} m, then the indeterminacy in its position is at most 2×10^{-15} m. If Δx exceeded this, the diameter of the nucleus, the electron could hardly be said to be confined. The Heisenberg indeterminancy principle states:

$$\Delta p\, \Delta x > h \,/\, 4\pi$$

The right-hand side of this inequality is 5.27×10^{-35} J s. It follows that Δp_{min}, the *minimum* indeterminancy in the momentum of a confined electron, is 2.63×10^{-20} kg m s^{-1}. If it is supposed that the mass of the electron is its rest mass, 9.11 x 10^{-31} kg, then, using the definition of momentum, $p = mv$:

$$\Delta v_{min} = \Delta p_{min} \,/\, m = 2.89 \times 10^{10} \text{ m s}^{-1}$$

The minimum indeterminancy in the velocity of the confined electron exceeds the speed of light. Since this is impossible, the electron cannot be confined in a nucleus.

There is another approach. The energy of this electron is:

$$E \simeq p\,c$$

This equation is analogous to the equation relating the energy and momentum of the photon. It follows that:

$$\Delta E = c\Delta p$$

Since Δp for the electron confined in the nucleus is 2.63×10^{-20} kg m s^{-1}, the minimum indeterminancy in the energy of the bound electron is 7.89×10^{-12} J. This is about 49 MeV. (A joule is 1.602×10^{-13} MeV.) But the binding energy of the electron in a nucleus is only about 7 MeV. Because the indeterminancy in the energy of the electron is larger than the binding energy, it can easily leave the confines of the nucleus at any time. It is effectively not bound.

b) The minimum indeterminancy in the momentum of a proton or neutron confined in the nucleus is again 2.63×10^{-20} kg m s^{-1}. The mass of a proton or neutron (about 1.67×10^{-27} kg) is greater than the mass of an electron. Hence, the minimum indeterminancy in the velocity of the confined proton or neutron is 1.57×10^{7} m s^{-1}, which is only 5 percent of the speed of light. The proton and neutron can be confined in the nucleus.

---

35. The wave functions for the particle in a box are sketched in text Figure 12-10. The $n=2$ wave function has just one node, half-way between $x=0$ and $x=L$. This node results results because the function $\sin 2\pi x / L$ equals zero when x equals L / 2. The function, on the other hand, has maxima at $x = L/4$ and 3L / 4. Figure 12-10 (text p. 421) reveals the symmetry of the function.

The probability of finding the particle at any point x between 0 and L is equal to ψ^2 evaluated at x. A plot of ψ^2 versus x has a zero at $x = L/2$ and symmetrical maxima at $x = L/4$ and 3L / 4. Passing from $x=0$ to $x = L/4$ covers one-fourth of the area under the curve defined by ψ^2, the probability distribution function. This is evident from the symmetry of the function. The answer is therefore ¼.

The same answer comes from writing out the particle in a box function for $n=2$, squaring it, and integrating from $x = 0$ to $x = L/4$. In doing this it is helpful to know that:

$$\int \sin^2 x \, dx = \tfrac{1}{2}x - \tfrac{1}{4}\sin 2x$$

---

37. ***a)*** The probability of finding an electron in the vicinity of any point is equal to the square of the wave function of the electron evaluated at that point. Thus, the probability of finding the $1s$ electron of the H atom at a point at the distance r from the nucleus is:

$$\psi^2 = (1/\pi a_o^3)\, e^{-2r/a_o}$$

The units of this function are m^{-3}; the portion in parentheses equals 2.15×10^{30} m^{-3} and the exponential term is unitless. At r equals 0 the exponential term equals 1 and ψ^2 is 2.15×10^{30} m^{-3}. At r equals 1.0 pm e^{-2r/a_o} is 0.9629 and ψ^2 is 2.07×10^{30} m^{-3}.

Actually, the probability of finding the electron *exactly at* a mathematical point is zero because a point has no volume to accommodate the electron. The small sphere centered at the nucleus has however a volume of 1 pm^3 (1.0×10^{-36} m^3). Over the very short distance between the center and surface of this sphere, ψ^2 stays nearly constant at 2.15×10^{30} m^{-3}. It follows that the probability of finding the electron within the small sphere is about:

$$p = 2.15 \times 10^{30}\ \text{m}^{-3} \times 1.00 \times 10^{-36}\ \text{m}^3 = 2.15 \times 10^{-6}$$

A slightly improved value comes by computing the radius of the 1 cubic picometer sphere centered at the nucleus (using the formula for the volume of a sphere the answer is 0.620 pm), evaluating ψ^2 at this r (it is 2.10×10^{30} m^{-3}) and taking ψ^2 throughout the sphere to equal the average of 2.10×10^{30} m^{-3} and 2.15×10^{30} m^{-3}. This improved answer is $p = 2.13 \times 10^{-6}$.

b) At a distance a_o or 52.9 pm (0.529×10^{-10} m) from the nucleus, the value of the function ψ^2 is *less* than it is at the nucleus. The exponential part of the function drops rapidly as r increases:

$$\psi^2(\text{at } a_o) = (2.15 \times 10^{30}\ \text{m}^{-3})\, e^{-2r/a_o} = 2.91 \times 10^{29}\ \text{m}^{-3}$$

Assume that ψ^2 is constant throughout the 1 pm^3 volume which the problem specifies. The chance of finding the electron at 52.9 pm in a fixed direction is:

$$p = 2.91 \times 10^{29}\ \text{m}^{-3} \times 1.0 \times 10^{-36}\ \text{m}^3 = 2.91 \times 10^{-7}$$

c) A spherical shell of thickness 1 pm and radius 52.9 pm has a volume:

$$V_{shell} = 4\pi r^2\, \Delta r = 4\pi\, (52.9\ \text{pm})^2 \times 1\ \text{pm}$$

$$V_{shell} = 3.52 \times 10^4\ \text{pm}^3 = 3.52 \times 10^{-32}\ \text{m}^3$$

This substantially larger volume naturally has a greater probability of holding the electron than does the tiny 1 pm^3 volume element considered in part b):

$$p = 2.91 \times 10^{29}\ \text{m}^{-3} \cdot 3.52 \times 10^{-32}\ \text{m}^3 = 0.0102$$

---

39. In chromium(IV) oxide the Cr^{4+} ion has the ground-state electron configuration: [Ar]$3d^2$. The neutral Cr atom has lost its 4s electron and three of its five 3d electrons. The two remaining 3d electrons are, at lowest energy, unpaired so the CrO_2 has two unpaired spins per Cr.

---

41. K should have a larger radius than Na because its outermost electron occupies a 4s orbital whereas Na's outermost electron occupies a closer 3s orbital. The Cs atom is larger than the Cs^+ because as Cs^+ gains an electron it goes into the more distant n=6 shell.

Rb^+ and Kr are isoelectronic. The larger species is the one with smaller nuclear charge: Kr. Ca has two 4s electrons which are at about the same distance from the nucleus as K's one 4s electron. Since K has Z=19 (less than Ca's Z of 20) it is larger. Cl^- is larger than Ar. The two are isoelectronic and Cl^- has the smaller Z.

Chapter 13

CHEMICAL BONDING AND MOLECULAR STRUCTURE

13.1 *MOLECULAR STRUCTURE*

Different parameters characterize different aspects of molecular structure. It is important to understand their physical meaning and to get an idea of the units used to measure them. They include:

- *Bond Lengths.* The distance from any nucleus in a molecule to any other is an *interatomic distance.* Interatomic distances have no upper limit because molecules can be arbitrarily large. Interatomic distances are labelled *bond lengths* when the two atoms under consideration are nearest-neighbors or near-neighbors in a molecule and therefore have a chemical bond between them. Bond lengths are on the order of 1×10^{-10} m. This distance has a special name, the Ångstrom (Å).

$$1 \text{ Ångstrom} = 0.1 \text{ nanometer (nm)} = 100 \text{ picometer (pm)}$$

It helps to get a few important bond lengths in mind: carbon to carbon bonds range from 1.20 to 1.54 Å; O to H bonds in various compounds range from 0.94 to 1.09 Å. Table 13-1 (text p. 448) gives the bond length in many diatomic molecules.

Bond lengths for a given type of bond do not change much from molecule to molecule. See Table 13-2, text p. 449. Bond lengths range up toward 3 Å. For example, the antimony to iodine distance in SbI_3 is 2.75 Å.

Bond lengths are symbolized R or R_e. The subscript (e for equilibrium) emphasizes that molecular vibrations can temporarily lengthen and shorten bonds.

- *Force Constants.* Molecules are not rigid. Rather, the atoms composing them vibrate about average positions. When a chemical bond is stretched or compressed from its equilibrium length, there is a restoring force. The *force constant, k,* tells this force's strength. If the bond is like a spring, then the force constant is its stiffness. A typical molecular force constant is 500 newton per meter. See text Table 13-1. If a bond with a k of 500 N m^{-1} is stretched 0.1 Å the restoring force is 5×10^{-9} N. Experiments in molecular spectroscopy (Section 13-6) allow the computation of force constants. **See problem 37.**

- *Bond Angles.* Any three atoms in a molecule define an angle, measured in degrees. Indeed, three points in space define *two* angles, the second being equal to 360° minus the first. *Bond angles* are interatomic angles defined by bonded atoms. They are always taken in the sense that makes them less than or equal to 180°.

- *Moments of Inertia.* Molecules rotate and tumble through space. These motions depend on their *moments of inertia.* A moment of inertia measures a body's resistance to being set in rotational motion about an axis. In general, molecules have three moments of inertia. Because theirs is a simple linear structure, diatomic molecules are a special case; they have a single non-zero moment of inertia, *I*:

$$I = \mu R_e^2$$

where μ is the *reduced mass* (see Section 13-6) of the diatomic molecule.

The SI unit for moments of inertia is the kg m^2. Typical values for molecular moments of inertia are on the order of 10^{-45} kg m^2. **See problems 33 and 35.**

Bond Enthalpies. Chemical bonds form because the atoms involved have a lower energy when close together than when far apart. Breaking a bond means moving the bonded atoms apart and requires energy. This energy is ΔE_d, the *dissociation energy*, of the bond. It is a quantitative measure of the strength of a bond.

Bond *enthalpies* are ΔH values for breaking bonds. They are related to bond energies in the same way any enthalpy change is related to an energy change: $\Delta H = \Delta E + \Delta(PV)$. **See problems 1d and 11.**

Bond energies and enthalpies are always positive. They are measured in J mol^{-1} or kJ mol^{-1} and sometimes in electron volts (eV). If a chemical bond between two atoms has a bond enthalpy of 1 eV, then it requires 96.48 kJ to break one mole (6.022×10^{23}) of these bonds at constant pressure.

The molar dissociation enthalpy of a diatomic molecule is *twice* the enthalpy required to produce 1 mol of atoms. Thus, it takes 243 kJ to dissociate 1 mol of Cl_2 *(g)*. It takes only 141.5 kJ to produce 1 mol of Cl*(g)* from Cl_2 *(g)*. This factor of 2 is a major stumbling block in problems. **See problem 11.**

Molar bond enthalpies are on the order of 100 to 1000 kJ mol^{-1}. See Table 13-3, text p. 450. Exact bond enthalpies depend on the details of the surroundings of a chemical bond in a molecule. The bond enthalpy of the first O−H bond in H_2O is 502 kJ mol^{-1}. The bond enthalpy of the O−H bond in the OH molecule left behind by removal of one H from H_2O is only 426 kJ mol^{-1}. The values in Table 13-3 (text) are *average* bond enthalpies.

Tables of bond enthalpies allow convenient estimates of the ΔH° of reactions. One imagines that a reaction proceeds by the complete disruption of all chemical bonds in the reactants to give isolated gaseous atoms followed by the recombination of these atoms to give the products. The total bond enthalpy of the reactants is the amount required to complete the disruption. It is the sum of the bond enthalpies of all the bonds the reactants have. The total bond enthalpy of the products is, similarly, the enthalpy to break all the bonds among the products. The *change* in bond enthalpy is the bond enthalpy of the products minus the bond enthalpy of the reactants. It is the negative of ΔH° of the reaction. See **problems 1 and 39.**

Molecular Geometries

The *valence shell electron pair repulsion* (*VSEPR*) theory holds that valence electron pairs distribute themselves about atoms in such a way as to minimize electron pair repulsions. Electron pairs are either *lone pairs*, not directly involved in bonding, or *bonding pairs*, shared between atoms.

The *steric number* (*SN*) for an atom is the sum of the number of atoms bonded to it and the number of lone pairs on it. The way to get the steric number of an atom in a molecule is first to draw a valid Lewis structure for the molecule and then just to count the atom's neighbors. If the molecule has several resonance Lewis structures, it does not matter which is used. A double or triple bond contributes only 1 to an atom's steric number.

In VSEPR theory the strength of the electron pair repulsions follows the order:

lone pair-lone pair > lone pair-bonding pair > bonding pair-bonding pair

Different geometrical shapes minimize the electron pair repulsions for the different steric numbers. The shapes are shown in text Figure 13-1. Knowing these shapes and employing the repulsion order just quoted, a table of the molecular shapes predicted by VSEPR theory for steric numbers from 2 to 6 is readily constructed. In this table, X is an atom bonded to the central atom A, and E stands for an electron pair.

SN of Atom A	Molecule Type*	Shape of Molecule
2	AX_2	linear
3	AX_3	trigonal planar
3	AX_2E	bent
4	AX_4	tetrahedral
4	AX_3E	trigonal pyramidal
4	AX_2E_2	bent
5	AX_5	trigonal bipyramidal
5	AX_4E	distorted see-saw
5	AX_3E_2	distorted T-shaped
5	AX_2E_3	linear
6	AX_6	octahedral
6	AX_5E	square pyramidal
6	AX_4E_2	square planar
7	AX_7	pentagonal bipyramid

*X is a bonded atom. E is an electron pair

Problems 3, 5, and 7 show the use of VSEPR. The theory also helps in **problems 46 and 62.**

13.2 *IONIC AND COVALENT BONDS*

Ionic bonds arise from the transfer of electrons between atoms and the subsequent Coulomb attractions among the positive and negative ions which are formed.

To remove even one electron from a gaseous atom of any element always requires energy. This is the ionization energy, *IE*. Many atoms release energy, the electron affinity (*EA*), as they accept an electron. For every possible pair-wise combination of atoms, (*IE* − *EA*) is positive. *It always costs energy to transfer an electron from one atom to another.*

Ionic bonds form because of the attraction between charges of unlike sign. The energy of this attraction is:

$$E_{coulomb} = \frac{Z_1 Z_2 e^2}{4 \pi \epsilon_0 R}$$

According to this formula the energy of a pair of unlike ions goes to negative infinity as R goes to zero. But ions are not point charges. At very short distances the electron clouds of the ions start to repel each other and the energy rises steeply. See text Figure 13-4.

The formula applies to a pair-wise interaction between atoms A*(g)* and X*(g)* to form the molecule A^+X^- *(g)*. If the Coulomb energy is converted to a per mole basis then one must imagine a mole of widely separated A^+X^- *(g)* molecules. A collection of such ionic molecules held in the same region of space would, at ordinary temperatures, condense to form an ionic crystal, in which every ion interacts with all others. **Problem 41** shows how to deal with the difference between energy per molecule and energy per mole.

Covalent Bonding

Covalent bonding arises from the sharing of electrons between atoms. Electrons have a negative charge. An electron shared by two nuclei spends most of its time *between* the two. It tends to lower the repulsion between the positively-charged nuclei because it is attracted to both.

One definition of a covalent bond is "a shared pair of electrons." This definition is over-simple (as proved by the covalent bond in the H_2^+ ion, which only *has* one electron). Nevertheless it can sometimes be helpful.

Electronegativity Scales

Chemical bonds are rarely entirely covalent or entirely ionic. Instead they have an intermediate character. *Electronegativities* help estimate just how covalent or ionic a bond is.

The electronegativity of an atom A is a unit-less quantity represented by the symbol χ_A.

- The greater the difference in electronegativity (Δ) between two atoms, then the more ionic is the bond between the two. The less the difference the more covalent is the bond.

Atoms of the same element have the same atomic electronegativity. In compounds like H_2 and F_2 Δ is zero, and the bond is 100 percent covalent.

Linus Pauling set up a scale of atomic electronegativity values using a version of the equation:

$$\Delta^2 = 0.0104 \times [\Delta E_d(AB) - (\Delta E_d(A_2) \times \Delta E_d(B_2))^{1/2}]$$

In this equation, Δ is the difference in electronegativity ($\chi_A - \chi_B$) between the two atoms A and B, and the ΔE_d's are dissociation energies, measured in kJ mol^{-1}, of the three possible diatomic molecules which atoms A and B can form. The constant 0.0104 is the number of electron-volts per molecule that is equivalent to one kJ mol^{-1}. Pauling originally used dissociation energies in eV.

In taking the square root of Δ, the sign that gives the more electronegative element the larger χ is chosen. Fluorine, the most electronegative element, is assigned $\chi = 3.98$ on the Pauling scale. The χ's of other atoms range downward from 3.98. Cesium, a very *electropositive* element, has an electronegativity of only 0.79.

If Δ exceeds about 2.0 for a pair of elements, then the chemical bond between them is mainly ionic. If Δ is less than about 0.4, then the bond is mainly covalent. In the intermediate range, the chemical bonds are covalent with partial ionic character.

The partial ionic character of a bond is approximately:

$$\text{Percent Ionic Character} = 16\,|\Delta| + 3.5\,\Delta^2$$

A Δ of 2.0 corresponds to about 46 percent (about half) ionic character. **Problem 15** illustrates the use of this approximation.

The equation that defines Δ contains three different dissociation energies. If two dissociation energies are given it is possible to work backward from a table of electronegativities to get the missing dissociation energy. **See problems 11 and 21-23.**

Figure 13-6 in the text lists a set of atomic electronegativities. The largest χ's occur at the upper right of the periodic table, and the smallest χ's occur at the lower left of the table.

Dipole Moment

Molecules in which the centers of positive and negative electrical charge do not coincide possess an electric *dipole moment*. In effect they have an electrically-positive end and an electrically-negative end.

If placed between a positively-charged and a negatively-charged plate (that is, in an electric field) a molecule with a dipole moment tends to align itself so that its negative end is nearer the positive plate and its positive end is nearer the negative plate. Molecules not aligned with the field experience a torque, or twist, which tends to align them. For a given electric field the size of the torque depends on the magnitude of the molecule's dipole moment. Numerically the dipole moment is:

$$\mu = Q R$$

where R, is the distance separating charges of opposite sign and equal magnitude, Q.

Electric dipole moments are *vector* quantities. They have both magnitude and direction. Frequently they are represented by arrows with the tail at the positive charge and the head at the negative charge. Larger dipoles are represented by longer arrows.

A difference in electronegativity between two bonded atoms means that there is a charge separation across the bond and consequently a dipole moment associated with the bond. For *diatomic* molecules, which have only one bond, this dipole moment is of course the dipole moment of the molecule.

In molecules which contain more than one bond the overall molecular dipole moment depends both on bond polarity and molecular geometry. It is the *vector sum* of the bond dipole moments. For example, chlorine is more electronegative than carbon. The C−Cl bond is therefore polar. The CCl_4 molecule has four equal C−Cl bond dipoles. According to VSEPR theory the four bonds point from the central C atom to Cl atoms at the corners of a tetrahedron. This symmetrical arrangement means that the vector sum of the four bond dipoles is zero. **See problems 23 and 46** for other examples. In addition, **problem 44** involves the vector summation of bond dipoles.

Dipole moments can be experimentally measured. The magnitude of a dipole moment in a polyatomic molecule provide important information about the molecule's geometry. Absence of a dipole moment argues for some kind of symmetrical molecular structure. **See problem 47.**

Dipole moments are nearly always cited in a non-SI unit, the Debye. One Debye (D) equals 3.336×10^{-30} C m. The Coulomb meter (C m) is the SI unit for dipole moment. The Debye is more convenient for chemists because most molecular dipole moments are between 0 and 10 Debye.

- *The Debye is the dipole moment of a +1 and −1 fundamental charge (charges of magnitude 1.602×10^{-19} C) held 0.2082 Å apart.*

For diatomic molecules (and only for diatomic molecules), experimental dipole moments allow the estimation of the partial ionic character of the bond. The greater the dipole moment for a given bond distance, then the greater is the bond's ionic character. If δ is the fraction of a unit charge on each atom in a diatomic molecule then:

$$\delta = \mu / e R$$

When μ is in Debye, R in Ångstrom then:

$$\delta = 0.2082 \ \mu(\text{in Debye}) / R(\text{in Å})$$

The quantity δ is the fractional ionic character. It ranges from 0.0 to 1.0. The fractional *covalent* character of a bond is $1.0 - \delta$. The equation is straightforward to use in problems. The only complication is the factor of 0.2082 which arises from the use of the non-SI unit for dipole moment. **See problem 13.**

The values for fractional ionic character estimated with the formula correspond fairly well with the values derived from electronegativity differences. **See problem 43.**

13.3 *MOLECULAR ORBITALS FOR DIATOMIC MOLECULES*

Atomic orbitals (abbreviated AO) *localize* the probability of finding their occupying electrons in regions close to the atom's nucleus. For example the average distance of the 1*s* electron from the nucleus in hydrogen is only 0.8 Å. In contrast, *molecular* orbitals *delocalize*, over all the atoms of a molecule, the probability of finding the electrons which occupy them. Like atomic orbitals, molecular orbitals have different shapes, sizes and energies. A molecular orbital (abbreviated MO) is constructed mathematically by mixing together the atomic orbitals of the atoms that make up the molecule. The process is *linear combination of atomic orbitals.*

For example, a linear combination of two 1*s* atomic orbitals on a pair of neighboring atoms A and B is:

$$\psi(\text{bonding}) = C_A\,\psi(1s)^A \;+\; C_B\,\psi(1s)^B$$

$$\psi(\text{antibonding}) = C_A\,\psi(1s)^A \;-\; C_B\,\psi(1s)^B$$

where C_A and C_B are constants telling the degree of mixing, that is, the proportion of each parent's character that the daughter molecular orbital possesses. If the two parent atoms are identical, then $C_A = C_B$ on the basis of symmetry. If not, the constants differ. **See problem 51.** The two atomic orbitals combine in two ways:

- *Bonding.* The first combination places much electron probability density between the nuclei of the atoms. This tends to countervail internuclear repulsions and leads to bonding between the atoms. The energy of a bonding daughter MO is *less* than the energy of either of its AO parents.

- *Antibonding.* In contrast, the second combination (with the minus sign) places a node (a region of zero electron probability density) between the atoms. Its energy is higher than the energy of the parent atomic orbitals. Electrons in an antibonding orbital actively oppose the continued association of the nuclei in a chemical entity.

There are several key points concerning molecular orbitals:

- Molecular orbitals are designated by Greek letters. The letters σ and π are analogous to the designations *s* and *p* used for atomic orbitals. A σ MO has no nodal plane containing the internuclear axis; a π MO has 1 nodal plane containing this axis. See Figure 13-11, text p. 469.

  The parentage of the MO (1*s* or 2*p* etc. atomic orbitals) is indicated by a subscript to the right of the Greek letter.

- Molecular orbitals are bonding, antibonding or non-bonding. Antibonding orbitals are indicated with a superscript * to the right of the Greek letter. Non-bonding orbitals are indicated with a superscript *nb*. Bonding molecular orbitals have no special designation.

- The *bond order* of a molecule is ½ (No. electrons in bonding MO's − No. electrons in antibonding MO's). **See problems 19 and 27.** Determining a bond order requires assigning all electrons to a bonding, an antibonding or a non-bonding orbital. Sharing electrons between atoms does not by itself guarantee bonding since the shared electrons could be in antibonding molecular orbitals.

• The total number of molecular orbitals formed in linear combination always equals the number of atomic orbitals combined. If 10 atomic orbitals are mixed together, then 10 molecular orbitals result. In other words, *in constructing molecular orbitals the number of atomic orbitals is conserved.*

• A *correlation diagram* (text Figures 13-10, 13-12) tells the order of energy and the parentage of a molecule's molecular orbitals. These diagrams are difficult to derive. They should be taken as givens.

• A *molecular electron configuration* is analogous to an atomic electron configuration but uses molecular orbitals instead of atomic orbitals. To write a ground-state molecular electron configuration:

1. Consult the appropriate correlation diagram.
2. Put the available electrons into the MO's starting at lowest energy and going in accord with the Pauli principle and Hund's rule.
3. Write down each MO's designation and show the number of electrons occupying it with a superscript. **See problem 19.**

Like atomic orbitals, MO's hold 0, 1 or 2 electrons. Like atomic orbitals, a single MO designation (example: π_{2p}) can refer to a *set* of two or more orbitals which have the same energy. Therefore, notations like $(\pi_{2p})^4$ are legal. This particular example means that 4 electrons occupy a set of two π MO's of equal energy deriving from the overlap of four $2p$ atomic orbitals. Higher up the correlation diagram in which this symbol is found, the symbol $\pi_{2p}{}^*$ will appear to represent the related set of π^* MO's.

• In writing molecular orbital configurations it is necessary to deal specifically with only the *valence* electrons. Non-valence electrons are always equally distributed between bonding and antibonding orbitals.

13.4 *POLYATOMIC MOLECULES*

The text describes bonding in polyatomic molecules by employing *valence bond theory* for σ bonds and the molecular orbital approach for π bonds.

In the valence bond theory the valence atomic orbitals of atoms with two or more bonds are *hybridized* to form new atomic orbitals. Hybridization involves the same technique of linear combination that was used to construct molecular orbitals except:

• *all of the starting orbitals are on the same atom;*

• *all of the resultant orbitals are on that same atom.*

Hybridization is an attempt to account for the geometrical shape experimentally observed in molecules. Different hybrid combinations give different geometrical shapes. For example, suppose that a $2s$ and 3 $2p$ orbitals ($2p_x$, $2p_y$, $2p_z$) on the same atom are hybridized. The linear combinations are:

$$\psi_1 = \tfrac{1}{2}(s + p_x + p_y + p_z)$$
$$\psi_2 = \tfrac{1}{2}(s + p_x - p_y - p_z)$$
$$\psi_3 = \tfrac{1}{2}(s - p_x + p_y - p_z)$$
$$\psi_4 = \tfrac{1}{2}(s - p_x - p_y + p_z)$$

Taken together, the set of four hybrid daughter wave-functions ($\psi_1 ... \psi_4$) describes the same electron probability density as the four parents. This can be proved by squaring the four members and adding them together:

$$\psi_1^2 + \psi_2^2 + \psi_3^2 + \psi_4^2 = s^2 + p_x^2 + p_y^2 + p_z^2$$

However, unlike the four parents, the daughters in the new hybrid set of orbitals all have the same shape. They are exactly equivalent, except in orientation. *They are sp^3 orbitals and point toward the corners of a tetrahedron.* The superscript in this symbol does not refer to the number of electrons occupying the orbital. Instead it refers to the number of parents of p character. An sp^3 orbital containing two electrons would be denoted by $(sp^3)^2$. This notation is a source of confusion but is deeply entrenched and must be learned.

Valence bond theory invokes sp^3 hybridization when a central atom has a steric number (SN) of 4.

Other important linear combinations are:

$1\ s + 1\ p \Rightarrow 2\ sp$ orbitals	Linear (used when SN = 2)
$1\ s + 2\ p \Rightarrow 3\ sp^2$ orbitals	Trigonal (used when SN = 3)

A σ bond results from the end-to-end overlap of atomic orbitals or hybrid orbitals along a line connecting two atoms. In valence bond theory, a 109.5° angle is predicted between atoms bonded to a central atom using sp^3 hybrids. The predicted angles for sp^2 and sp hybrids are 120° and 180 ° respectively.

Once the framework of a molecule is set up using the valence bond description, orbitals not involved in hybridization are free, within limits imposed by their symmetry, to mix together to form π bonds, The correlation diagrams (such as text Figure 13-20) which result show only π orbitals. The main problem in using these diagrams is making sure that the right number of electrons is used. *All valence electrons do not take part in π bonding.* The number of electrons used in a π-system is the number of valence electrons minus the number used in σ bonding minus the number in lone pairs. **See problems 27 and 53.**

13.5 *BONDING IN ORGANIC COMPOUNDS*

The above bonding theory works well for many organic molecules, molecules that involve carbon and other atoms from the first and second rows of the periodic table. The steps in the analysis of bonding in organic compounds are:

1. Write the Lewis electron dot structure for the molecule. Take particular care to get the number of valence electrons right.

2. Determine the hybridization of every atom. To do this, use VSEPR theory and get each atom's steric number. Read the atom's hybridization from the table:

Steric Number	*Hybridization*
2	sp
3	sp^2
4	sp^3

3. Place electron pairs in each localized (σ) molecular orbital. This constructs the single-bond framework of the molecule. In so doing, it uses two electrons between each pair of bonded atoms.

4. Identify the *p* orbitals that were not used in hybridization. Combine them to form π molecular orbitals.

5. Place the valence electrons that were not used in step 3 into the π molecular orbitals.

13.6 *MOLECULAR SPECTROSCOPY*

Like atoms, molecules possess a series of allowed quantum energy states. Molecular spectroscopy is the study of electromagnetic radiation (light) as it is absorbed or emitted by molecules passing from one of these states to a second. The frequency of the light involved in such transitions is related to the difference in energy between the pair of energy levels (states):

$$|\Delta E| = h\nu$$

The absolute value is taken because ΔE is negative in emission and positive in absorption. The values of the energy in the different levels depends on molecular parameters (bond distances and angles, force constants, moments of inertia, etc.). Molecular spectroscopy thus reveals much about bonding.

The total energy of a molecule is approximately equal to the sum of contributions from three kinds of motion:

$$E(\text{total}) = E(\text{electronic}) + E(\text{vibrational}) + E(\text{rotational})$$

The three contributions are of different orders of magnitude and give rise to three kinds of molecular spectroscopy:

- *Electronic.* Changes in the electronic configuration of molecules generally require from 50 to 500 kJ mol^{-1}. Frequencies of light that supply (or take away) this kind of energy per photon range from 10^{14} to 10^{15} s^{-1}. The corresponding wavelength range is 2400 nm to 240 nm, right across the *visible* region of the spectrum into the ultra-violet. Electronic transitions are studied with ultra-violet and visible molecular spectroscopy.

- *Vibrational.* Changes in the relative positions of the nuclei of a molecule are vibrations. Vibrational transitions range in energy from about 2 to 40 kJ mol^{-1}. These energies correspond to light of frequencies of 5×10^{12} to 10^{14} s^{-1}. Molecules vibrate (oscillate) with frequencies in this range.

  Such frequencies are related to quantities (force constants, atomic masses, etc.) characteristic of the individual molecule. In the case of a diatomic molecule, this relationship has a simple form. The stretching frequency (symbol ν) depends on the force constant of the bond and the molecule's reduced mass:

$$\nu = 1/2\pi\,(k/\mu)^{\frac{1}{2}}$$

where k is the force constant, and the *reduced mass*, μ, of the diatomic molecule is:

$$\mu = m_1 m_2 / (m_1 + m_2)$$

The wavelength of the light involved in transitions between vibrational states ranges from 60,000 nm to 3000 nm. This is in the infra-red region of the electromagnetic spectrum. Thus, in **problem 37** the vibrational transition in question has a wavelength of 28,500 nm.

The allowed vibrational energies of a diatomic molecule are:

$$\text{E(vibrational)} = h\nu\,(v + \tfrac{1}{2}) \qquad v = 0, 1, 2, \ldots$$

where v is the vibrational quantum number. Unlike the electronic energy states, these vibrational energy states are uniformly spaced. See text Figure 13-28. Also, when $v = 0$, the molecule still has some energy, the *zero-point energy.*

• *Rotational.* Rotational transitions involve less energy yet. Typical energy differences range from 0.001 to 0.1 kJ mol^{-1}. The corresponding frequencies are 2.5×10^9 and 2.5×10^{11} s^{-1}. Molecules rotate with frequencies in this range. The corresponding wavelengths are 1200 mm to to 1.2 mm. These wavelengths occur in the microwave and short-wave region of the electromagnetic spectrum.

Rotational energy depends on the molecular moments of inertia. For a diatomic molecule there is only one moment of inertia:

$$I = \mu R_e^2$$

The quantization of the rotational motion of a *linear* molecule is in terms of a new quantum number J:

$$\text{E(rotational)} = [h^2 / 8\pi^2 I]\, J(J + 1) \qquad J = 0, 1, 2, \ldots$$

The rotational energy levels of a linear molecule are thus not uniformly spaced. They get farther apart as J increases. See text Figure 13-28.

In solving problems dealing with molecular spectroscopy it is important to:

¤ Pay proper heed to the units. Work exclusively in SI units and convert at the end as necessary. The units of moment of inertia (kg m^2) and reduced mass (kg) sometimes cause trouble; numbers in amu Å^2 or g Å^2 and in amu or g are used by mistake.

¤ Distinguish between patterns in the *energy levels* of molecules and patterns in the frequencies of the transitions connecting those energy levels. Thus, in **problem 35** the observed spectroscopic lines in the pure rotational spectrum of a diatomic molecule are uniformly spaced. The rotational energy levels of this molecule are *not* uniformly spaced.

DETAILED SOLUTIONS TO ODD-NUMBERED PROBLEMS

1. ***a)*** The point of writing the Lewis structure for propane is to establish the number of bonds of different type in the molecule. Propane has two carbon to carbon single bonds and eight C to H bonds.

b) The problem is to estimate ΔH for the combustion of propane in oxygen to give gaseous carbon dioxide and water vapor. Imagine that this reaction proceeds in two stages: *first*, the complete disruption of the molecules of reactants to give solitary gaseous atoms; *second*, the recombination of those atoms to give the products.

The first stage involves breaking 8 C−H bonds, 2 C−C single bonds and 5 O−O bonds. The enthalpy price for this operation is 8 × 413 kJ plus 2 × 348 kJ plus 5 × 498 kJ for a total of 6490 kJ. The average bond enthalpies come from Table 13-3.

The second stage involves the formation of 6 C=O double bonds (2 in each of 3 molecules) and the formation of 8 H−O bonds. The enthalpy pay-off in this stage is 6 × 728 kJ plus 8 × 463 or 8072 kJ. The difference between the two enthalpy changes is 1582 kJ. Since the formation of the products releases more enthalpy than the disruption of the reactants consumed, the reaction is exothermic. The combustion of a mole of propane gas has a ΔH of −1582 kJ, that is, $\Delta H = -1582 \text{ kJ mol}^{-1}$.

c) According to the balanced equation:

$$C_3H_8\,(g) + 5\,O_2\,(g) \rightarrow 3\,CO_2\,(g) + 4\,H_2O(g)$$

the combustion converts 6 mol of gas to 7 mol of gas. If the gases are ideal, then $\Delta(PV) = \Delta(nRT)$. With the temperature held constant $\Delta(PV) = \Delta n_g RT$. As just shown, Δn_g is +1. Therefore, with T equal to 546 K, $\Delta(PV)$ is $+1 \text{ mol} \times 8.314 \text{ J mol}^{-1}\text{K}^{-1} \times 546 \text{ K}$ or +4.5 kJ for a system which consists of 1 mol of reaction.

d) By definition, $\Delta H = \Delta E + \Delta(PV)$. Substitution of the answers from the previous parts shows that ΔE equals −1582 kJ − 4.5 kJ or −1587 kJ for the combustion of 1 mol of propane. Thus $\Delta E = -1587 \text{ kJ mol}^{-1}$.

---

3. ***a)*** For S in SF_6 the steric number (SN) is six. The geometry is octahedral. ***b)*** For S in $SOCl_2$ the SN is four. This includes 2 Cl's, an O and a lone-pair of electrons. The geometry of the molecule is pyramidal. ***c)*** The I in ICl_3 has an SN of five. The geometry is a distorted T. ***d)*** The C in CBr_4 has an SN of four. The geometry is tetrahedral. ***e)*** The S in SO_3 has an SN of three. The molecule is trigonal planar.

---

5. ***a)*** Use the VSEPR theory. Consult the table on p. 272 of this Guide. CH_3Br has a central C. The four non-carbon atoms are at the corners of a tetrahedron surrounding the C. The C−Br distance exceeds the C−H distance, and the H−C−Br angle is somewhat less than 109.5°

b) ICl_4^- is square planar. The central I has an SN of six. The two lone pairs lie opposite each other on an octahedral pattern, minimizing lone-pair to lone-pair interactions. The four Cl atoms surround the central I atom in a planar square.

c) In OF_2 the central O has an SN of four. The molecule is *bent* to accommodate the steric requirements of the two lone pairs on the O atom. The F−O−F angle is less than 109.5°.

d) In BrO_3^- the central Br has an SN of four. The lone pair on the Br occupies one corner of a tetrahedron around the central Br. The resulting molecule is pyramidal. The presence of the lone pair forces the O's together slightly, so that the O−Br−O angle is less than 109.5°

e) In CS_2 the central C has a SN of 2. The molecule is linear.

---

7. *a)* Planar AB_3: BF_3, BH_3. *b)* Pyramidal AB_3: NH_3. *c)* Bent AB_2^-: ClO_2^-. *d)* Planar AB_3^{2-}: SO_3^{2-}.

---

9. The problem is similar to text Example 13-2. It asks for the gaseous KCl molecule to be treated as two point charges separated by 2.67 Å. In SI units the charges are plus and minus 1.602×10^{-19} C. The Coulomb potential energy of this arrangement is:

$$E_{coulomb} = \frac{Q_1 Q_2}{4 \pi \epsilon_0 R}$$

where Q_1 and Q_2 represent the two charges, R is the distance between the charges, and ϵ_0 is a proportionality factor equal to 8.854×10^{-12} $C^2 J^{-1} m^{-1}$. This equation is equivalent to:

$$E_{coulomb} = \frac{Z_1 Z_2 e^2}{4 \pi \epsilon_0 R}$$

where Z_1 and Z_2 are the charges in units of e. Substitute $R = 2.67 \times 10^{-10}$ m and the various other values, giving $E = -8.64 \times 10^{-19}$ J. This is the potential energy of a *single* K^+ to Cl^- attraction. A *mole* of such attractions has E = −520 kJ mol^{-1}. It would *require* +520 kJ to separate one mole of the KCl molecules, moving Cl^- ions from their initial bond distances near the K^+ ion to an infinite distance.

The result would be a collection of K^+ *(g)* ions and a second collection of Cl^- *(g)* ions. Removing the electrons from the Cl^- *(g)* ions would *consume* 349 kJ mol^{-1}. This is the electron affinity of Cl*(g)*, Table 12-3. Feeding the electrons into the K^+ *(g)* ions would release 419 kJ mol^{-1}. The net energy change accompanying the electron transfer therefore would be a release of 70 kJ mol^{-1} (419 − 349). When this *release* is subtracted from the 520 kJ mol^{-1} consumed in separating the ions the answer is 450 kJ mol^{-1}, the dissociation energy.

---

11. The electronegativities of Cl and F are 3.16 and 3.98 respectively. The *difference* in electronegativity, Δ, is ±0.82. (The sign depends on the order of the values in the subtraction.) Applying the Pauling definition of electronegativity:

$$\Delta^2 = (0.102)^2\ [\Delta E_{dClF} - (\Delta E_{dCl_2} \times \Delta E_{dF_2})^{1/2}]$$

where the ΔE_d's are the dissociation energies in kJ mol⁻¹ of the different molecules. Table 13-1, text p. 448, gives dissociation *enthalpies*, ΔH_d's. Adjust them to ΔE_d's before using them in the electronegativity equation. The dissociation of both F_2 and Cl_2 occasions an increase of 1 in the number of moles of gas, *i.e.*, Δn_g is +1 mol. The adjustment from ΔH to ΔE requires the *subtraction* of Δn_g RT, which is 2.479 kJ (T is 298.15 K and R is 8.314 J $mol^{-1}K^{-1}$), from the ΔH value. The correct ΔE_d values are 240.52 kJ mol^{-1} for Cl_2 *(g)* and 155.52 kJ mol^{-1} for F_2 *(g)*. Then:

$$0.82^2\ /\ (0.102)^2 = \Delta E_{dClF} - [240.52 \times 155.52]^{1/2}$$

$$64.6 = \Delta E_{dClF} - 193.41$$

Hence, the dissociation *energy* of ClF is 258.0 kJ mol^{-1}.

The next step is to convert this dissociation energy to an enthalpy so that it can be combined with other ΔH's. The following reaction is the *reverse* of the dissociation of 1 mol of ClF*(g)*:

$$Cl(g) + F(g) \rightarrow ClF(g)$$

As just shown, the ΔE of this reaction is −258.0 kJ. Its ΔH is ΔE + Δn_g RT. This is equal to −258.0 kJ + (−1 mol) RT. With R equal to 8.314 J $mol^{-1}K^{-1}$ and T 298.15 K, ΔH comes out to be −260.5 kJ.

It takes some enthalpy to make the reactants in this equation from elements in their standard states. Consider the Cl*(g)* first. The ΔH of dissociation of Cl_2 *(g)* is 243 kJ mol^{-1} (Table 13-1 or Appendix D). It therefore takes 121.5 kJ to form 1 mol of Cl*(g)*. Similarly, the enthalpy needed to form 1 mol of F*(g)* is 79.0 kJ. Reaction of 1 mol of F*(g)* and 1 mol of Cl*(g)* *releases* 260.5 kJ as 1 mol of ClF*(g)* forms. Combine these enthalpy changes. 260.5 − 121.5 − 79.0 kJ or 60.0 kJ mol^{-1} is released overall. Chemically, the summation of the three processes amounts to formation of ClF*(g)* from its constituent elements in their standard states. The estimated ΔH°_f of ClF*(g)* is therefore −60.0 kJ mol^{-1}. Compare to the experimental value of −55.65 kJ mol^{-1} (*Handbook of Chemistry and Physics*). The agreement is good, considering the approximate nature of electronegativities.

Since the problem asks for an *estimate* of the ΔH°_f of ClF*(g)*, it is defensible to take a major short-cut in this problem. This is to regard ΔH and ΔE as interchangeable. With this approximation, the ΔE_d of ClF comes out to 260.5 kJ mol^{-1}. Taking this number as the ΔH of dissociation of ClF gives ΔH°_f of ClF equal to −60.0 kJ mol^{-1}. Omitting the two corrections (which are in opposite senses) gives essentially the same final answer as inserting them both.

---

13. In diatomic molecules, the fractional ionic character of the bond, δ:

$$\delta = 0.2082\ (\mu\ /\ R)$$

when μ is in Debyes and R, the bond distance, is in Angstroms. The percent ionic character is just 100 times the fractional ionic character. Applying the formula to the four compounds listed in the problem gives: ClO, 16.4 percent ionic; KI, 73.8 percent ionic; TlCl, 38.0 percent ionic; InCl, 32.8 percent ionic.

---

15. The results can be summarized in a table:

Compound	Δ	$16\|\Delta\| + 3.5\,\Delta^2$	Percent Ionic Character*
HF	1.80	40.1	41
HCl	0.98	19.0	18
HBr	0.178	14.6	12
HI	0.48	8.49	6
CsF	3.19	86.7	70

*Based on Dipole Moment

The point of the problem is the good general agreement between the values calculated from Δ and the dipole moment based values.

---

17. All of the *bonds* in all of the compounds are polar. The symmetry of some of the molecular shapes makes the vector sums of the individual bond dipoles zero, in some cases. Thus, SeF_6 (octahedral), CBr_4 (tetrahedral), and SO_3 (trigonal planar) are *non*-polar. The other two molecules, ICl_3 and $SOCl_2$, are less symmetrical and the vector sums of their bond dipoles are not zero. The VSEPR approach predicts the shapes.

---

19. *a)* Fluorine is a homonuclear diatomic molecule with $Z=9$. The correlation diagram in text Figure 13-12b must be used. The molecular electron configurations are:

F_2 $(\sigma_{2s})^2 (\overset{*}{\sigma}_{2s})^2 (\sigma_{2p})^2 (\pi_{2p})^4 (\overset{*}{\pi}_{2p})^4$

F_2^+ $(\sigma_{2s})^2 (\overset{*}{\sigma}_{2s})^2 (\sigma_{2p})^2 (\pi_{2p})^4 (\overset{*}{\pi}_{2p})^3$

b) F_2 has two more bonding than antibonding electrons. Its bond order is 1. F_2^+ has three more bonding than antibonding electrons. Its bond order is $1\frac{1}{2}$.

c) F_2 has no unpaired electrons and should be diamagnetic. F_2^+ has an odd number of electrons. At least one, which is, according to the molecular electron configuration, a $\overset{*}{\pi}_{2p}$ electron, is unpaired. In any case F_2^+ is paramagnetic.

d) F_2^+ has a larger bond order and therefore requires more energy for dissociation.

---

21. Nitrogen is more electronegative than carbon. The energies of its atomic orbitals are *lowered* in the correlation diagram (text Figure 13-13) relative to the energies of the corresponding orbitals of C. The CN molecule has 9 valence electrons. The molecular electronic configuration of CN is:

$$(\sigma_{2s})^2 (\sigma^*_{2s})^2 (\pi_{2p})^4 (\sigma_{2p})^1$$

The bond order of the molecule is 2½, and it has a single unpaired electron, giving rise to paramagnetism.

---

23. *a)* The central C in CH_4 is sp^3 hybridized and tetrahedral. Although the C–H bonds are slightly polar, their vector sum in the molecule is zero. The molecule is non-polar.

b) The central C in CO_2 is *sp* hybridized. It is linear and non-polar.

c) The central O in OF_2 is sp^3 hybridized. Two of the hybrid orbitals on the O accommodate lone pairs of electrons, and two overlap with orbitals on the fluorines. The molecule is bent and polar.

d) The central C in CH_3^- is sp^3 hybridized. One of the hybrid orbitals contains a lone pair of electrons. The other three overlap with the hydrogen atoms' 1*s* orbitals. The molecule is pyramidal and polar.

e) The central Be in BeH_2 is *sp* hybridized. It is linear and non-polar.

---

25. The central N in NH_2^- is surrounded by the eight valence electrons (5 from the N, 1 each from the H's and 1 for the overall negative charge). The N valence orbitals are sp^3 hybridized. Two of the hybrids overlap in σ bonds with 1*s* orbitals on the two hydrogens, and two contain lone pairs. The H–N–H angle is far less than 180°. In fact, it is 106.7°. The molecule is bent.

---

27. Like carbon dioxide, the azide ion, N_3^-, has 16 valence electrons. Both are triatomic nonhydrides and the correlation diagram of Figure 13-20 in the text applies to both. N_3^- has two N to N σ bonds, resulting from overlap of *sp* hybrid orbitals on the central N and 2*p* orbitals on the outer N atoms. These bonds use 4 electrons. Lone pairs in each of the outer N atoms' 2*s* orbitals use another 4 electrons. The π system (text Figure 13-20) contains the other 8 valence electrons. Four of these are in bonding π orbitals, making two π bonds. The final 4 valence electrons are in *non-bonding* π orbitals. The overall bond order of the molecule is 2 σ bonds plus 2 π bonds equals 4. Each N–N linkage is a double bond. Compare to the Lewis structure of the azide ion on p. 30 of this Guide.

N_3 has 15 valence electrons. It forms as N_3^- loses an electron. Such loss comes from the highest energy molecular orbital which is (see Figure 13-20) a *non*-bonding MO. Thus N_3 has an overall bond order of 4, like N_3^-. Unlike N_3^-, N_3 has a single unpaired electron and is paramagnetic.

N_3^+, a 14 valence electron species, derives from N_3^- by the loss of two non-bonding π electrons. The overall bond order of N_3^+ thus is still 4. There are now *two* unpaired electrons in the set of π^{nb} orbitals so N_3^+ is paramagnetic, too.

---

29. In acetone, the central carbon is sp^2 hybridized and the outer carbons are sp^3 hybridized. There is a C to O double bond. A π orbital system derives from the $2p_z$ orbital of the central C and the $2p_z$ orbital of the O. Two electrons occupy the bonding π orbital; the π^* orbital is vacant. This π interaction accounts for one half of the C to O double bond. The other half is a σ bond.

The three bond angles at the central C $\simeq$ 120°; all six bond angles at each terminal C $\simeq$ 109.5°

---

31. The Lewis structure of methyl acetylene is:

```
   H
   ..
 H:C:C:::C:H
   ..
   H
```

This molecule has 16 valence electrons. The methyl carbon, which has a steric number of 4, is sp^3 hybridized. The other two C's, both with steric numbers of 2, are *sp* hybridized. Localized (σ) bonds between the sp^3 C and 3 H atoms consume 6 of the valence electrons. A σ bond between the outer *sp* hybridized C and an H uses another 2 electrons. Two more σ bonds occur between the sp^3 C and the first *sp* C and between the first and second *sp* C atoms. These use 4 more valence electrons.

Four *p* orbitals, 2 each on the 2 *sp* C atoms, are unused so far. Combine them to make 4 π_{2p} orbitals, 2 bonding and 2 antibonding. Place the remaining 4 valence electrons into these π orbitals in order of ascending energy. The result is the configuration $(\pi_{2p})^4$. The bonding π orbitals contain 2 electrons each, and the antibonding π orbitals are empty.

The molecule is linear (bond angles equal 180°) except the H$-$C$-$H and H$-$C$-$C angles on the sp^3 hybridized C atom $\simeq$ 109.5°.

---

33. The moment of inertia of any diatomic species is:

$$I = \mu R^2$$

where μ is the *reduced* mass of the species (*not* to be confused with its dipole moment for which the same symbol is used) and R is the equilibrium bond length. To compute the reduced mass of $^{12}C^1H$ use the definition:

$$\mu = m_1 m_2 / m_1 + m_2$$

$$\mu = 12.0000\ (1.007825)\ \text{amu} / 13.007825\ \text{amu} = 0.929740\ \text{amu}$$

Dividing this mass in amu by 6.02204×10^{26} amu kg^{-1} gives the reduced mass of CH in kg. It is 1.5439×10^{-27} kg. R is 1.12 Å. Therefore:

$$I = 1.5439 \times 10^{-27}\ \text{kg} \times (1.12 \times 10^{-10}\ \text{m})^2 = 1.94 \times 10^{-47}\ \text{kg m}^2$$

The frequency of radiation exciting the J to $J + 1$ transition is:

$$\nu = (h / 4\pi^2 I)(J_i + 1)$$

where J_i is the initial value of the rotational quantum number. Substitution of $J_i = 0$ gives a ν of 8.67×10^{11} s^{-1}. This frequency corresponds to a wavelength of 0.346 mm, in the far infra-red region of the spectrum.

---

35. ***a)*** The spacing of the absorption lines in the pure rotational spectrum of a diatomic species is *uniform* with a frequency separation of $h / 4\pi^2 I$ where I is the moment of inertia. From the data in the problem the average frequency separation of the lines in the $^{12}C^{16}O$ rotational spectrum is 1.155×10^{11} s^{-1}. Also, the frequency of the *first* rotational line is 1.15×10^{11} s^{-1}, as predicted by:

$$\nu = (h / 4\pi^2 I)(J_i + 1)$$

with $J_i = 0$. The moment of inertia is computed by substitution in the equation:

$$1.155 \times 10^{11}\ \text{s}^{-1} = [h / 4\pi^2 I] \qquad \textit{hence} \qquad I = 1.45 \times 10^{-46}\ \text{kg m}^2$$

b) The *energy* of each rotational state is given by:

$$E_{rot} = [h^2 / 8\pi^2 I]\, J(J + 1) = h/2\, [h / 4\pi^2 I]\, J(J + 1)$$

The term $[h / 4\pi^2 I]$ is 1.155×10^{11} s^{-1}, from part a). Then:

$$E_{rot} = \tfrac{1}{2} h\, (1.155 \times 10^{11})\, J(J + 1)$$

Combining the constants gives a value of 3.826×10^{-23} J. Therefore, the rotational energy is 7.65×10^{-23} J when $J=1$, 2.30×10^{-22} J when $J=2$, and 4.59×10^{-22} J when $J=3$.

c) The moment of inertia of the diatomic molecule is:

$$I = (m_1 m_2 / m_1 + m_2)\, R^2$$

where R is the bond distance and the m's are the two atomic masses. The mass of ^{12}C is 12.0000 amu and the mass of ^{16}O is 15.994915 amu. Dividing these values by 6.02205×10^{26} amu kg^{-1} converts the masses to kg. Substitution of these values together with I (from part a) into the equation gives R equal 1.13×10^{-10} m or 1.13 Å.

---

37. The Raman spectrum involves two photons, the incoming photon with frequency ν_1 and the scattered photon, with frequency ν_2 (text p. 489). The wavelength of 2.85×10^{-5} m must equal this frequency difference divided into c, the speed of light. If so, it is related to the *difference* in energy between two vibrational states of the Li_2 molecule: $h(c/\lambda) = \Delta E_{vib}$.

Strong Raman scattering is seen only for transitions between vibrational states differing by 1 in vibrational quantum number ($\Delta v = \pm 1$). The vibrational energy difference is:

$$\Delta E_{vib} = h[(1/2\pi)(k/\mu)^{1/2}][(v_2 + 1/2) - (v_1 + 1/2)]$$

The second term in brackets is 1 because Δv is 1. Since $hc/\lambda = \Delta E_{vib}$:

$$hc/\lambda = h[1/2\pi\,(k/\mu)^{1/2}]$$

Recall that k is the force constant, μ is the reduced mass (of Li_2 in this case), h, c are Planck's constant and the speed of light, and λ is the quoted wavelength.

Using Table 11-1, text p. 374, the mass of ^7Li is 7.016005 amu; the reduced mass of Li_2 is therefore half of this or 3.508003 amu. Converting this value to kg gives 5.8253×10^{-27} kg. Finally, completing the required arithmetic gives $k = 25.4$ N m^{-1}. Note that h cancels out of the calculation and that the bond distance of Li_2 is not needed.

---

39. ***a)*** The point of drawing the Lewis structures is to learn the number of bonds of the various kinds in the compounds. In O=O there is one O to O double bond; in O=C=O there are two C to O double bonds and in H−O−H there are two H to O single bonds.

b) In CH_4 there are four C−H single bonds. In C_8H_{18} there are 7 C−C single bonds and 18 C−H single bonds. In H_3C-CH_2-O-H there are 5 C−H single bonds, 1 C−C single bond, 1 C−O single bond and 1 O−H single bond.

c) In the combustion of methane:

$$CH_4(g) + 2\,O_2(g) \rightarrow CO_2(g) + 2\,H_2O(g)$$

4 C−H bonds and 2 O=O bonds are broken. This requires (4 × 413) + (4 × 249.2) or 2648.8 kJ. See text Table 13-3. The molar enthalpy of atomization of O_2 is tabulated in kilojoules per mole of O. The atomization of 2 mol of O_2 gives 4 mol of O. This is the reason that 249.2 is multiplied by 4.

In making the products 2 C=O bonds and 4 H−O bonds are formed. This releases (2 × 728) + (4 × 463) = 3308 kJ. The overall reaction releases the difference between the heat freed in formation of the products and the heat absorbed in the atomization of the reactants. This is 659.2 kJ, *i.e.* $\Delta H = -659$ kJ.

d) The calculation of the ΔH for the burning of octane:

$$C_8H_{18}(g) + {}^{25}/_2\,O_2(g) \rightarrow 8\,CO_2(g) + 9\,H_2O(g)$$

is exactly similar. The enthalpy investment is:

Enthalpy to break bonds = 7(348) + 18(413) + 25(249.2) = 16100 kJ

The enthalpy released upon formation of CO_2 and H_2O is:

Enthalpy released = 16(728) + 18(463) = 19982 kJ.

The ΔH of the reaction is the first figure minus the second: −3882 kJ. Hence, ΔH° for the combustion of octane is −3882 kJ mol^{-1}.

e) The ΔH° for the reaction:

$$C_2H_5OH(g) + 3\,O_2(g) \rightarrow 3\,CO_2(g) + 3\,H_2O(g)$$

is −2424 kJ by the same method. Therefore, ΔH° for the combustion of ethanol is −2424 kJ mol^{-1}.

f) Take 1 cm^3 as the volume, V, of liquid fuel to compare. Given the densities of the liquids, the weights and then the chemical amounts in 1 cm^3 of each fuel can be computed. The following table answers the question.

Fuel	*V*	*Weight*	*Chemical amount*	*ΔH° / V*
C_2H_5OH	1.00 cm^3	0.789 g	0.0172 mol	−41.7 kJ cm^{-3}
CH_4	1.00	0.415	0.0259	−17.1
C_8H_{18}	1.00	0.704	0.00617	−24.0

In the table, the chemical amount of each fuel is the weight of it which is present in 1.00 cm^3 divided by its molar weight. The molar weights come from the chemical formulas of the fuels.

---

41. *a)* At the critical distance, which is R_c, the Coulomb potential energy between the $M^+(g)$ and $X^-(g)$ ions just compensates for the energy required to extract an electron from M(*g*) and place it on X(*g*). This energy is the difference between the ionization energy (*IE*) of M(*g*) and the electron affinity (*EA*) of X(*g*):

$$-\frac{Z_1 Z_2 e^2}{4\pi\epsilon_0 R_c} = (IE - EA)$$

In the MX case, Z_1 is +1, Z_2 is −1, ϵ_0 is 8.854×10^{-12} C^{2}J^{-1}m^{-1}, and *e* is -1.602×10^{-19} C. Substitution and solution for R_c give:

$$R_c = (2.3066 \times 10^{-28} \text{ J m}) / (IE - EA)$$

Values of *IE* and *EA* are generally given on a per mole basis. Multiplying 2.3066×10^{-28} J m by Avogadro's number converts it to a per mole basis, too:

$$R_c = (1.3894 \times 10^{-4} \text{ J m mol}^{-1}) / (IE - EA)$$

b) For LiF, (*IE* − *EA*) is 520 − 328 or 192×10^3 J mol^{-1}. The numbers come from text Table 12-3. Substituting in the expression derived in part a) for R_c gives R_c (LiF) as 7.23×10^{-10} m. Similarly, for KBr (*IE* − *EA*) is 93×10^3 J mol^{-1} and R_c is 14.9×10^{-10} m. For NaCl (*IE* − *EA*) is 147 kJ mol^{-1} and R_c is 9.45×10^{-10} m.

---

43. The electronegativity of Au is 2.54, and the electronegativity of Cs is 0.79. This gives Δ equal to 1.75. Substitution into the formula:

$$\text{Percent Ionic Character} = 16\,|\Delta| + 3.5\,\Delta^2$$

gives 39 percent as the ionic character of the CsAu bond. This means a build-up of 0.39 of the charge on the electron on each end of the CsAu molecule. This fractional build-up, δ, is related to the bond distance and dipole moment:

$$\delta = 0.2082\,\mu\,/\,R$$

where R, the bond length, must be in Å and μ, the dipole moment, must be in Debye to cancel the units of the constant. Substitution of R = 3.69 Å gives μ = 6.9 Debye.

---

45. *a)* The Lewis structure of N=S−F places a −1 formal charge on the nitrogen atom and a +1 formal charge on the sulfur atom. The fluorine has a zero formal charge. In S=N−F all three atoms have zero formal charges. In S−F=N, the S has a −1 , the F a +2, and the N a −1 formal charge. Notice that a structure featuring the three atoms bonded in a triangle is possible. Such a structure however eliminates the distinction between terminal and central atoms which is the point of the problem.

b) On the basis of the preferment of structures with least separation of formal charge the second structure, S=N−F, would be predicted. This is *not* the observed structure (N=S−F).

c) The electronegativity of F exceeds that of S by 1.40. The F to S bond has substantial ionic character with the S positive. The N to S bond (Δ = 0.46) is also polar with the S positive. An arrangement of atoms which makes the S formally positive is thus understandable.

---

47. *a)* The observation of a non-zero dipole moment for ozone, O_3, rules out symmetrical structures like a *linear* or a triangular structure (three equivalent O's at the corners of an equilateral triangle). The molecule must be bent.

b) The best Lewis structures are a canonical pair with one O=O double bond and one O−O single bond in each, differing only in which side has the double bond and which the single. See the diagram on p. &ozone. of this Guide. The central O has a lone pair in both structures.

c) The VSEPR model gives the central O in the bent molecule a steric number of 3 and thereby predicts approximately a 120° bond at the central O.

---

49. A ground-state H_2 molecule would lose an electron from a σ bonding molecular orbital. Such an electron is harder to extract from H_2 than H. It is lower in energy in the molecule than in the atom. See the correlation diagram in text Figure 13-10.

An O_2 molecule would lose its highest energy electron, which is in a π^* (antibonding) orbital. The highest energy electron is thus easier to extract from O_2 than O. It is higher in energy in the molecule than in the atom.

In F_2 the highest energy electron is in a π^* orbital, just as in O_2. The correct prediction is that the ionization energy of F exceeds the ionization energy of F_2.

---

51. The molecular orbital for the ground state of the heteronuclear molecule is:

$$\psi = C_A\psi_A + C_B\psi_B$$

The *square* of the wavefunction is the quantity that is related to the probability of finding the electron. The square of the above is:

$$\Psi^2 = C_A^2\,\Psi_A^2 + 2\,C_A\,C_B\,\Psi_A\,\Psi_B + C_B^2\,\Psi_B^2$$

Neglecting the overlap of the two orbitals means neglecting the cross-term in the above:

$$\Psi^2 \simeq C_A^2\,\Psi_A^2 + C_B^2\,\Psi_B^2$$

If the electron spends 90 percent of its time in orbital Ψ_A then $C_A^2 = 9\,C_B^2$. Also, the electron is on either atom A or atom B so $C_A^2 + C_B^2 = 1$. Solution of the two simultaneous equations gives $C_A = 0.9486$ and $C_B = 0.3162$.

---

53. Carbon dioxide and sulfur dioxide differ substantially in their bonding. CO_2 has 16 valence electrons. The molecule is linear: :Ö=C=Ö: There are 4 valence electrons in localized σ bonds between the O's and the C atom. Another 4 valence electrons are localized in 2*s* orbitals, one pair each on the 2 O atoms. These are oxygen lone pairs.. The remaining 8 valence electrons join in the π bonding system. The correlation diagram is in text Figure 13-20. The π configuration is $\pi_x^2\,\pi_y^2\,(\pi_x^{nb})^2\,(\pi_y^{nb})^2$. The two pairs of electrons in π^{nb} orbitals (non-bonding π orbitals) correspond to the second lone pair that appears on each O in the Lewis structure of CO_2.

Sulfur dioxide has 18 valence electrons and is *bent.* The Lewis structure is :Ö–S̈=Ö: plus the resonance hybrid in which the double bond is on the left side. The central S has three sp^2 hybrid orbitals. One contains the lone pair on the sulfur and the other two overlap in σ bonds with *p* orbitals on the oxygens. This system accounts for 6 valence electrons.

Another 8 valence electrons are localized, 4 each on the two oxygen atoms. These electrons are the 2 lone pairs on each of the O's.

The remaining 4 electrons comprise the π system, constructed from the overlap of a p_z orbital from each atom. (See text Figure 13-21). The π electronic configuration is $(\pi_z)^2\,(\pi_z^{nb})^2$. The first of these π electrons accounts for the additional ½ bond between each O and the S. The second pair, the non-bonding π pair, is the final lone pair that appears on one O or the other in the resonance Lewis diagrams.

55. The standard enthalpy change for the reaction:

$$6\,C(g) + 6\,H(g) \rightarrow C_6H_6\,(g)$$

is equal to the standard enthalpy change of formation of $C_6H_6\,(g)$ (82.93 kJ mole^{-1}) minus six times the standard enthalpy change of formation of $H(g)$ minus six times the standard enthalpy change of formation of $C(g)$. Neither of the latter two quantities is zero because gaseous monatomic H and gaseous C are *not* the standard states of these two elements. Instead these two ΔH°_f's are 716.68 kJ mole^{-1} and 217.96 kJ mol^{-1}. See Appendix D. The ΔH° for the reaction is 82.93 kJ − 4300.1 kJ − 5607.84 kJ or −5524.9 kJ.

Formation of one of the Kekulé structures of benzene means the formation of 3 C to C double bonds, 3 C to C single bonds and 6 C to H bonds. Using the average bond enthalpies in Table 13-3 reveals that the bond enthalpy of $C_6H_6\,(g)$ is (6 × 413 kJ) + (3 × 348 kJ) + (3 × 615 kJ) or +5367 kJ. This positive answer means that the enthalpy of 1 mole of $C_6H_6\,(g)$ is 5367 kJ less than the enthalpy of 6 mol of uncombined $C(g)$ plus 6 mol of uncombined $H(g)$. The bond enthalpy calculation thus predicts a ΔH° for the reaction of −5367 kJ, 158 kJ less negative than the ΔH° calculated the other way. The difference is important. It is the *resonance stabilization energy.*

- -

57. The six π molecular orbitals of pyridine arise as combinations of the six p_z orbitals of the ring atoms. Number these atoms 1 through 6 as in the following figure:

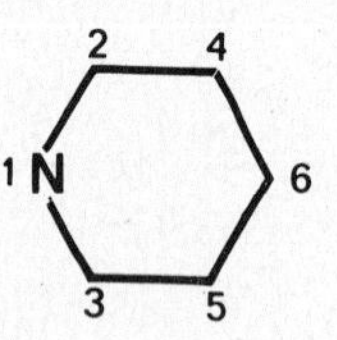

According to the problem, molecular orbitals that apportion electron density onto the N atom will be lower in energy in pyridine than in benzene. Thus, molecular orbitals which have the N atom p_z atomic orbital among their parents will be lower in energy. Bear in mind that the N atom occupies position 1 in the numbering scheme and contrast the above diagram with Figure 13-25, text p. 484, in which the six π MO's of benzene are sketched. The strongly bonding and strongly antibonding (the highest and lowest) MO's in Figure 13-25 both have parentage that includes the p_z orbital on atom 1. These two molecular orbitals are therefore both *lowered* in energy in pyridine relative to benzene. One of the two weakly bonding molecular orbitals in benzene has p_z(atom 1) parentage, but the other does not. Instead, its parentage includes $p_z(2)$, $p_z(3)$, $p_z(4)$, and $p_z(5)$. The first of the two weakly bonding MO's (located on the left in Figure 13-25) is therefore lowered in energy in pyridine relative to benzene, and the other is not affected. Similarly, the two weakly antibonding MO's in benzene are split in energy. The one that has some $p_z(1)$ parentage is lowered, and the other is unchanged. The final result is an pyridine energy level diagram with six different π orbital energies, four lower than the corresponding benzene π orbitals, and two unchanged in relative energy.

- -

59. The moment of inertia of a diatomic molecule is:

$$I = \mu R_e^2$$

where R is the equilibrium bond distance and μ is the reduced mass;

$$\mu = m_1 m_2 / (m_1 + m_2)$$

The two m's are the atomic masses of the two atoms making up the molecule.

Using Table 11-1 as a source for the m's, μ for $^1H-^{19}F$ is 0.957055 amu and μ for $^1H-^{81}Br$ is 0.99543 amu. Division by 6.0220×10^{26} amu kg^{-1} converts these masses to kg. They are 1.5893×10^{-27} kg for HF and 1.6530×10^{-27} kg for HBr. The equilibrium bond distances are 0.926×10^{-10} m for HF and 1.424×10^{-10} m for HBr.

Completing the substitution in the formula for I gives the moments of inertia of the two molecules: 1.363×10^{-47} kg m^2 for HF and 3.352×10^{-47} kg m^2 for HBr.

The rotational spectra of these molecules consist of series of equally spaced lines separated by the frequency $h / 4\pi^2 I$. This is the meaning of the formula:

$$\nu = \Delta E / h = (h / 4\pi^2 I)(J_i + 1)$$

The frequency spacing for HF is 5.001×10^{11} s^{-1}. For HBr it is 12.32×10^{11} s^{-1}.

The large change in mass between HF and HBr causes only a small change in the reduced mass of the diatomic molecule. Note the behavior of the reduced mass as a mathematical quantity. If either of the atoms is much heavier than the other, then μ approachs the mass of the *lighter* atom. In these cases the reduced mass is close to the mass of ^1H for HF and does not change much more going to HBr.

---

61. The electron in ethylene is excited from a π to a π^* orbital. The transfer reduces the overall bond order of the molecule by 1. The C to C bond in the molecule in the excited state should be *longer* than it was in the ground state. Because the bond is weaker, its force constant, k, is diminished, and the vibrational frequency of the C to C stretching mode is reduced.

Chapter 14

INTERMOLECULAR FORCES AND THE LIQUID STATE

14.1 *INTERMOLECULAR FORCES*

Electromagnetic forces alone are responsible for the interactions among molecules. If two electrical charges are at rest, then *Coulomb's law* gives the electrostatic force between them:

$$F = \frac{-\,Q_1\;Q_2}{4\pi\epsilon_0 R^2}$$

where Q_1 and Q_2 are the charges, R is their separation and ϵ_0 is a constant (8.854×10^{-12} $C^2 J^{-1} m^{-1}$) to make the units come out right. The Q's are signed quantities. Charges of opposite sign attract each other. With the negative sign in the above equation a positive Coulomb force means an attraction between the charges; a negative force is a repulsion.

A deceptively similar but distinct formula gives the Coulomb *energy* of interaction between two charges. In this equation R is not squared and the negative sign does not appear:

$$V = \frac{Q_1\;Q_2}{4\pi\epsilon_0 R}$$

When the charges have opposite signs they attract each other, and the energy of interaction is negative.

Dipole-Dipole Interactions

Electrostatic forces occur between electrically neutral molecules, as well as ions, as long as the molecules have a non-spherical distribution of electrical charge. Such molecules are *electric dipoles.* They have a positive end and a negative end separated by a distance R.

There can be considerable complexity when the Coulomb energy of a system of two or more dipolar molecules must be computed. The Coulomb energy depends on the relative *orientation* of the dipoles are well as the magnitude of the charges on their ends and the distances between them. A good procedure is:

1. Count all of the pair-wise intermolecular combinations of charge centers. Every dipole has two centers of charge. Two molecules have 4 combinations. An arrangement of three molecules would have 8. Combinations *within* molecules are not counted.

2. Figure out the distances between the centers of charge in each pair. Calculating these distances in an arbitrary arrangement of dipoles can involve much drawing of triangles and geometrical constructions which are subject to error. **Problem 13** illustrates the assignment of Cartesian coordinates to each charge to facilitate the calculation of the distance between them.

3. Determine the magnitude and sign of every charge. This information may be concealed in the form of the dipole moments of the molecules. Recall that the dipole moment, μ, is QR, the product of the distance separating charges of opposite sign and magnitude Q.

4. Compute the total Coulomb energy, the sum of all of the pair-wise contributions. **See problem 1.** If the sum is negative, then the interaction is an attractive one. If it is positive then the interaction is a repulsive one.

Dispersion Forces

Dispersion forces between molecules arise from the correlation of the motions of electrons. They are induced-dipole to induced-dipole attractions. The electron probability density on one molecule experiences a momentary fluctuation. This induces a correlated fluctuation in the probability density on the second. A build-up of negative charge on one side of the first atom induces a build-up of positive charge on the near side of the second atom. The result is an attraction between the molecules.

Pauli Repulsive Forces

Pauli repulsive forces operate *in addition to* electrostatic repulsive forces as two atoms are pushed close together. The electron probability densities of the two distort to maintain the Pauli principle in the new, two-atom system. This distortion is a rearrangement of the electron probability density of the atoms to a situation of higher energy.

Potential Energy Curves

As two atoms (or molecules) approach each other they are at first attracted. The potential energy of the system drops. This continues until short-range repulsive forces become important. Then the potential starts to rise again. The result is a characteristic hook-shaped potential energy curve. (See Figure 14-3, text p. 504.) The minimum of the curve is the bottom of the system's *potential energy well.* The depth of the well varies enormously. When two atoms form a chemical bond, the well is hundreds of kJ mol^{-1} deep. When atoms experience only dipole-dipole attractions or dispersion forces it may be only a few tenths of a kJ mol^{-1} deep.

The *Lennard-Jones* function approximates the shape of the potential energy curve for non-bonding interactions between two identical atoms and molecules. It consists of a repulsion term (the term involving the 12-th power) and an attraction term:

$$V = 4\,\epsilon\,[\,(\sigma/R)^{12} - (\sigma/R)^{6}\,]$$

where ϵ and σ are constants that differ for various atoms and molecules (see Table 14-1, text p. 505). It is instructive to graph the intermolecular potential energy curve for one or two different gases, given ϵ and σ. **Problem 15** shows how quickly the repulsive part of the Lennard-Jones function builds into importance at small R. **Problem 17** lends additional physical meaning to the Lennard-Jones parameters by relating them to the van der Waals constants.

14.2 *LIQUIDS*

In a gas, even at high temperatures, attractive intermolecular forces temporarily hold molecules together in pairs or in small clusters. These associations quickly break apart under the random jostling of collisions from other molecules. As the temperature is lowered, the collisions from outside become weaker. Instead of breaking a small cluster apart, impinging molecules tend to join it. Finally, when the temperature gets down to the boiling point, a single large cluster forms and grows rapidly. The result is a *liquid*.

Structure of Liquids

In liquids there is short-range, local order in the arrangement of the molecules but no long-range order. A predictable, repeating pattern holds temporarily within a cluster of molecules. The pattern is frayed at its edges by holes or defects. With time it disintegrates completely. Another region of order pops up. The relative positions of the molecules constantly change as regions of order form and decay. From another point of view, ordered regions migrate from place to place, changing their membership as they go.

The average structure of a liquid is given by a *radial distribution function*, abbreviated $g(R)$. To understand this function, consider the surroundings of a *test molecule* first in an ideal gas and then in a liquid. Let ρ represent the average density over time of the gas or liquid. This average density has the units of molecules per unit volume. In the ideal gas there is no order, either at short or long range, among the molecules, which are points. Therefore, ρ is the same up close to the test molecule as it is far away from it. In the liquid, the test molecule influences its surroundings. It exerts intermolecular attractions and also has a small volume of its own. As a result, the time-average density near the test molecule depends on R, the distance from the test molecule:

$$\text{Average Density} = \rho\ g(R)$$

where the dependence on R has been separated off into the function $g(R)$, the radial distribution function. This function is a unitless multiplier that modifies ρ. For an ideal gas, $g(R) = 1.0$. In a liquid, the function $g(R)$ is very small at short distances because the intermolecular forces are strongly repulsive at short distances. At intermediate distances, $g(R)$ has one or more maxima (see Figure 14-5, text p. 507). It then goes to 1.0 at distances around 10 to 15 Å because the order in the liquid is only short-range.

To find the average number of neighbors that the test molecule has at distance R, imagine a spherical shell of radius R englobing the test molecule and having thickness ΔR. The volume of this shell is $4\pi R^2 \Delta R$. Multiplying this volume by $\rho g(R)$, which tells the time-average density of molecules at the distance R, gives the desired result.

Example. Estimate the average number of neighbors at a distance of 3.1±0.6 Å from a typical molecule in liquid argon (density 1.4 g cm^{-3}), if the liquid has the radial distribution function graphed in Figure 14-5, text p. 507.

Solution. The number of neighbors within the specified range is:

$$\text{N(neighbors)} = 4\pi^2 R^2 \rho g(R) \Delta R$$

ΔR is 1.2 Å, and R is 3.1 Å. The graph of $g(R)$ uses Ångstroms as the unit of R so it is sensible to make everything involve Å by converting the density of the liquid from g cm^{-3} to molecules $Å^{-3}$. The molar weight of argon is 39.95 g mol^{-1} so its density becomes 0.021 $Å^{-3}$. This is the *number density* (see Chapter 3) of the liquid. Referring to Figure 14-5, as R goes from 2.5 to 3.7 Å, g(R) goes from 1.0 to a maximum near 2.5 and then returns to about 1.0. It has an average value of around 1.3 (estimated by eye). Substituting these values gives:

$$\text{N(neighbors)} = 4\pi^2 (3.1\ \text{Å})^2 (0.021\ \text{Å}^{-3}) (1.3) (1.2\ \text{Å}) \simeq 12$$

Thermodynamics of Liquids

As a gas condenses to a liquid at its boiling temperature, T_b, the enthalpy of the system *decreases.* The internal potential energy in the liquid is much less than in the vapor. The kinetic energy is about the same as long as the temperature is still T_b. The total energy of the liquid is therefore less and the total enthalpy is closely related. **See problem 21.**

On the other hand, vaporizing the liquid requires enthalpy from outside to pry the molecules away from each other, overcoming the intermolecular attractions. In this reverse process, the enthalpy of the system increases.

During condensation of a gas to a liquid, the entropy of the system decreases because there are fewer microstates available to the molecules in the product liquid than in the gas (see Chapter 8). Obviously, in the reverse process the entropy of the system increases.

Whether approached by lowering the temperature, to condense gas, or by raising the temperature, to vaporize liquid, T_b is the temperature at which liquid and gas phases are in equilibrium. At equilibrium:

$$\Delta G = \Delta H_{vap} - T_b \Delta S_{vap} = 0$$

$$\Delta S_{vap} = \Delta H_{vap} / T_b$$

The subscripts could as well indicate condensation.

The magnitude of the ΔH of vaporization depends on the strength of the intermolecular forces in the liquid. If the forces are strong, then ΔH is large. Stronger attractive forces also make for a higher boiling temperature. The upshot is that ΔS_{vap} is roughly constant for a wide variety of liquids. This is *Trouton's rule:*

$$\Delta S_{vap} \simeq 88\ \text{J mol}^{-1}\text{K}^{-1}$$

14.3 *DYNAMICS OF LIQUIDS*

Molecules in a liquid diffuse according to the same law given for gases:

$$r_{rms} = (6\,Dt)^{1/2}$$

where r_{rms} is the-root-mean-square displacement of a molecule after time *t*, and *D* is the diffusion coefficient of the liquid. An important difference between diffusion in liquids and gases is that the diffusion coefficient for typical liquids is about 10,000 times smaller than for gases at STP, about 10^{-9} m^2 s^{-1}. Compare **problem 25** to **problem 3-55b.** Diffusion in gases is already slow. In liquids it is orders of magnitude slower. Note in problem 25 that the diffusion coefficient is given as 2.3×10^{-5} cm^2 s^{-1} which converts to 2.3×10^{-9} m^2 s^{-1}.

The motion of a molecule in a liquid consists of random-walk diffusion upon which is superimposed a rattling motion as the molecule is trapped by intermolecular attractions in a temporary cage of its neighbors. An *encounter* between two molecules consists of a series of collisions within such a cage over a long period of time. Liquid phase encounters are much more productive of chemical reaction than simple gas phase collisions because the candidates for reaction have many more chances to interact just right. Moreover, there are plenty of near neighbors to carry away any excess energy.

For *diffusion-controlled reactions* in liquids the slow step in reaction is the step of diffusing togther. Again, **see problem 25.**

14.4 *HYDROGEN BONDING AND WATER*

Hydrogen bonds are the strongest type of intermolecular forces. They are found only in compounds containing hydrogen chemically bonded to nitrogen, oxygen or fluorine. The latter three elements are all strongly electronegative. **Problem 11** treats the dependence of hydrogen bond strengths on the identity of the H-*donor*, the atom chemically bonded to the hydrogen, and the H-*acceptor*, the atom with lone-pairs of electrons which interact with the hydrogen. Hydrogen bonds range in strength from 5 to 25 kJ mol^{-1}. This is roughly 10 percent of the strength of regular chemical bonds.

DETAILED SOLUTIONS TO ODD-NUMBERED PROBLEMS

1. The H to F bond in hydrogen fluoride is polar. The charge distribution approximates that of a point charge, $-\delta e$, at the fluorine and another point charge, $+\delta e$, at the hydrogen where δ is 0.41, and e is the charge on the electron. When two HF molecules approach each other there are four interactions: two attractive (H to F) and two repulsive (H to H and F to F). The pairs *within* the molecule are not counted.

Let R_1 and R_2 represent the H to H and F to F distances respectively, and let R_3 and R_4 represent the two H to F distances. The energy of the interaction between any two HF molecules is the sum of the pair-wise Coulomb energies of interaction:

$$V = (e^2 / 4\pi \epsilon_0) [\delta^2 / R_1 + \delta^2 / R_2 - \delta^2 / R_3 - \delta^2 / R_4]$$

Attractions *lower* V and repulsions *raise* it. This explains the signs of the four terms in the brackets. Substitution of $e = 1.602 \times 10^{-19}$ C, $\epsilon_0 = 8.854 \times 10^{-12}$ C^{2}J^{-1}m^{-1} and $\delta = 0.41$ gives:

$$V = 3.8774 \times 10^{-29} \text{ J m} [1 / R_1 + 1 / R_2 - 1 / R_3 - 1 / R_4]$$

In the geometry of Figure 14-1a, text p. 501, the H to H and F to F distances (R_1 and R_2) are both 3.00 Å (3.00×10^{-10} m). The two H to F distances (R_3 and R_4) are 3.138×10^{-10} m. The latter result comes by applying the Pythagorean theorem to the right triangle defined by three of the four atoms. Substituting these R values gives V equal to 1.1×10^{-20} J.

When the two HF molecules have the relative orientation shown in text Figure 14-1c, R_1 is 1.16×10^{-10} m, R_2 is 3.00×10^{-10} m, and R_3 and R_4 are both 2.08×10^{-10} m. The four distances are easily read from a sketch of the geometry. The term in brackets is $+0.23386 \times 10^{10}$ m^{-1}, and V is 9.1×10^{-20} J.

The major pitfall in this problem is getting the wrong values of R. This is especially true because Figure 14-1, which shows the required orientations, has a *5.00* Å distance (not 3.00) indicated between *bond centers* (and not atoms). Draw a careful sketch.

---

3. Trouton's rule states that the enthalpy change of vaporization of a liquid divided by the normal boiling temperature is a constant:

$$\Delta H_{vap} / T_b = 88 \text{ J mol}^{-1}\text{K}^{-1}$$

The constant has the units of entropy and in fact is the entropy change of vaporization. For toluene T_b is 273.15 + 110.6 or 383.75 K. Therefore, ΔH_{vap} for toluene is 34 kJ mol^{-1}.

The van't Hoff equation relates the equilibrium vapor pressure of a liquid to its temperature and enthalpy of vaporization:

$$\ln P_2 / P_1 = (-\Delta H_{vap} / R) [1 / T_2 - 1 / T_1]$$

Let T_1 be the normal boiling point of toluene. The corresponding vapor pressure, P_1, is 1.00 atm. This is the pressure used in the definition of normal boiling temperatures. T_2, as given in the problem, is 50° C (323.15 K). R is 8.314 J mol^{-1}K^{-1} (or 8.314×10^{-3} kJ mol^{-1}K^{-1}). Substitution gives:

$$\ln P_2 = -2.0 \quad \textit{and} \quad P_2 = 0.13 \text{ atm}$$

---

5. Identification of the type of intermolecular forces in these cases depends upon evaluation of the polarity of the bonds and the shape of the molecules. Electronegativity values allow the first and VSEPR theory helps with the second.

a) In CCl_4 the C−Cl bonds are somewhat polar, but the symmetrical tetrahedral shape of the molecule means that it has no net dipole moment. There are *dispersion* interactions between the molecules in liquid and solid CCl_4.

b) The major attractive forces in solid and liquid KBr are *ionic* (electrostatic, Coulomb) interactions. Dispersion forces make a minor contribution.

c) The $POCl_3$ molecule has a permanent dipole moment with its negative end positioned at the more electronegative atom attached to the central P. This is the O atom. The major contributors to the stability of solid and liquid $POCl_3$ are *dipole-dipole* interactions. As always, dispersion forces are present and make a minor contribution.

d) In xenon the only intermolecular attractions are *dispersion* forces.

e) The HF molecule is a permanent dipole. The high electronegativity of the F and the presence of H allow particularly strong dipole-dipole interactions: hydrogen bonds. Dispersion forces make a small contribution.

---

7. Substances with the strongest intermolecular forces require the highest temperature to make them boil. Liquid RbCl has strong electrostatic forces holding its ions together. It has the highest boiling point. NH_3 has dipole-dipole attractions as does NO. In NH_3 these are particularly strong; they are hydrogen bonds. In NO the bond is less polar than the N−H bond so the dipole-dipole attractions are weaker. NH_3 boils at a higher temperature than NO. Weak dispersion (van der Waals) forces are the only intermolecular attraction in liquid neon. It has the lowest boiling point of all.

---

9. ***a)*** The root-mean-square displacement by diffusion after time *t* is:

$$r_{rms} = (6\,Dt)^{1/2}$$

D for water is given as 2.3×10^{-9} m^2 s^{-1}. After 1 s, r_{rms} is 1.2×10^{-4} m.

b) After 1 day (8.64×10^4 s), r_{rms} is 3.5×10^{-2} m. The calculation shows just how *slow* diffusion is in liquids. After a day at room temperature an average water molecule moves less than 1½ inches from its starting position.

---

11. The problem is to rank the strength of four hydrogen bond interactions. Number the four: **1.** O−H...F; **2.** N−H...S; **3.** F−H...F; **4.** F−H...O. The atom that is covalently bonded (solid line) to the H is the *donor atom.* The other atom is the *acceptor atom.*

On the basis of the criteria summarized in the problem, the hydrogen bonds in **3** and **4** are both stronger than in **1.**

Comparing **1** and **2,** we see that, in the latter, sulfur is the acceptor atom. Its large size makes make it inferior to fluorine as an acceptor even though its ionization energy is less than fluorine's. Furthermore, the donor atom in **2** is nitrogen which forms weaker hydrogen bonds than the donor in **1** which is oxygen. Therefore the hydrogen bond is stronger in **1** than in **2.**

Next compare **3** and **4.** The donor is the same in the two and O is a better acceptor than F. Hence **4** has stronger hydrogen bonds than **3.** The predicted order of strength is:

$$2 < 1 < 3 < 4$$

The ΔH° values (that is, the strengths) of hydrogen bonds between a given donor and different classes of acceptors fall into rather broad ranges that overlap each other.

---

13. For a diatomic molecule:

$$\mu = (R\ /\ 0.2082)\ \delta$$

where μ is the dipole moment in Debye, R is the internuclear distance in Å, and δ is the fraction of the charge of the electron on each atom. For HBr R is 1.414 Å and μ is 0.78 D. Hence, δ is 0.13.

The problem describes two HBr molecules in a geometry derived from the one shown in text Figure 14-1a by having the two molecules cocked at a 30° angle. Call the atoms of the two molecules H_1, Br_1, H_2 and Br_2. The distance between the centers of the molecules is 5.000 Å. The H_1 to H_2 distance exceeds 5.000 Å and the Br_1 to Br_2 distance is less than 5.000 Å by an equal amount.

There are many ways to get the four intermolecular distances. One is to construct a Cartesian coordinate system with its origin midway between the molecules' centers and with its x axis parallel to one molecule's bond. The coordinates (in Å) of the atoms in such a system are: H_1 (−0.7070, 2.500); Br_1 (0.7070, 2.500); H_2 (−0.6123, −2.8535); Br_2 (0.6123, −2.1465). Note that 0.6123 is the sine of 30° times one-half the HBr bond distance and that the average of −2.1465 and −2.8535 is −2.5000. The coordinates reflect the symmetry of the atom layout and assure that no errors have been made.

The distance between any two atoms is:

$$R = [(x_2 - x_1)^2 + (y_2 - y_1)^2]^{\frac{1}{2}}$$

The distances (R_1 through R_4) come out: H_1 to Br_2, 4.830 Å; H_1 to H_2, 5.354 Å; Br_2 to Br_2, 4.647 Å; Br_1 to H_2, 5.514 Å.

The Coulomb energy of interaction of the two HBr molecules is:

$$V = (e^2\,\delta^2 / 4\pi\epsilon_0)\,[-1/R_1 + 1/R_2 + 1/R_3 - 1/R_4]$$

This is equation 14.1 from the text specialized to this case. δ is 0.13; e and ϵ_0 are 1.602×10^{-19} C and 8.854×10^{-12} $C^2\,J^{-1}m^{-1}$ respectively. Converting the four R's to meters and completing the arithmetic give $V = 3.1 \times 10^{-20}$ J. The positive result means that in this geometry the two molecules repel each other.

If the second molecule were cocked by 30° the *other* way, so that the H atoms of the molecules were close, V would be the same. Note however that the problem would be perceptibly more difficult if it were vague about the sense of the 30° tilt.

---

15. ***a)*** Two argon atoms move toward a head-on collision. The speed of each atom is the root-mean-square speed of argon atoms at 300 K. The kinetic energy of each is $^3/_2\, k_BT$, where k_B is the Boltzmann constant. See Chapter 3. The kinetic energy of the two-particle system is twice this, or 3 k_BT. For a mole of such atom-pairs the kinetic energy is 3 RT. At 300 K, 3 RT equals 7.482 kJ mol^{-1}.

As the members of the pair draw closer together, the *total* energy remains unchanged as kinetic energy (*K.E.*) is converted to potential energy, *V*. At the distance of closest approach the two atoms come to a dead stop. All of their kinetic energy has been converted to potential energy. The Lennard-Jones function gives this potential energy:

$$V = 4\epsilon\,[(\sigma/R)^{12} - (\sigma/R)^6]$$

Substituting for V according to the above discussion:

$$7.482 \text{ kJ mol}^{-1} = 4\epsilon\,[(\sigma/R)^{12} - (\sigma/R)^6]$$

For argon ϵ is 0.996 kJ mol^{-1} and σ is 3.40 Å (text Table 14-1). With R in Å the equation becomes:

$$1.878 = (3.40 / R)^{12} - (3.40 / R)^6$$

When R is 3.40 Å, the right-hand side of this equation is zero. When R is *less* than 3.40 Å, the right-hand side is positive. With this in mind it is fairly easy (with a calculator) to find an R to satisfy the equation by successive approximation. R equals 3.04 Å.

b) The total energy, E, of the pair of Ar atoms remains constant during their interaction. It is the sum of the kinetic and potential energy of the system.

$$E = K.E. + V$$

When R is 3.80 Å, the distance given in the problem, the quantity (σ / R) is 0.8947. Under those conditions:

$$V = 4(0.996 \text{ kJ mol}^{-1})[0.8947^{12} - 0.8947^{6}] = -0.995 \text{ kJ mol}^{-1}$$

The kinetic energy of a pair of argon atoms at R = 3.80 Å is the total energy minus V:

$$K.E. = E - V = 3\,RT - (-0.995 \text{ kJ mol}^{-1}) = 7.482 + 0.995 = 8.477 \text{ kJ mol}^{-1}$$

The kinetic energy of a single argon atom is *half* of this value or 4.239×10^{3} J mol^{-1}. The mass of argon is 39.948 g mol^{-1} or 39.938×10^{-3} kg mol^{-1}. Kinetic energy and speed are related by the equation:

$$K.E. = \tfrac{1}{2} mu^{2}$$

Substitution gives:

$$4.239 \times 10^{3} \text{ J mol}^{-1} = \tfrac{1}{2}(39.938 \times 10^{-3} \text{ kg mol}^{-1})\, u^{2}$$

The units are chosen so that u^2 comes out in m^2 s^{-2}. Completing the arithmetic gives u = 461 m s^{-1}

---

17. ***a)*** The data for the plot are in the following table:

Gas	b (L mol^{-1})	$N_0 \sigma^3$ (L mol^{-1})
Ar	0.03219	0.0237
H_2	0.02661	0.0151
CH_4	0.04278	0.0336
N_2	0.03913	0.0305
O_2	0.03183	0.0273

There is a strong correlation between b and $N_0 \sigma^3$. The ratio of b to $N_0 \sigma^3$ confines itself to a range between 1.17 to 1.36 except for the case of H_2 (where it is 1.76).

b) The units of the van der Waals constant a are atm L^2 mol^{-2}. Both the joule and the liter·atmosphere are units of energy. 1 L atm equals 101.325 J. Also, 1 L equals 10^{-3} m^3. Thus, 1 atm L^2 mol^{-2} equals 0.101325 J m^3 mol^{-2}. The combination of Lennard-Jones constants, $\epsilon\sigma^3$, has units of J m^3 mol^{-1}. Multiplying it by Avogadro's number, which has units mol^{-1}, gives the combination $\epsilon\sigma^3 N_0$ with the same dimensions as a. Taking data from Table 3-2 (text p. 95) and Table 14-1 gives:

Gas	a (J m^3 mol^{-2})	$\epsilon\sigma^3 N_0$ (J m^3 mol^{-2})
Ar	0.13628	0.0236
H_2	0.02476	0.00466
CH_4	0.22829	0.0414
N_2	0.14084	0.0241
O_2	0.13780	0.0270

There is a strong correlation between a and $\epsilon\sigma^3 N_0$. The ratio of the two varies within a fairly narrow range (from 5.1 to 5.8) for the six gases.

---

19. The two high-boiling liquids (LiF, BeF_2) have electrostatic intermolecular forces which require much thermal energy to overcome. The remaining six compounds all have much lower boiling points. The intermolecular forces are dipole-dipole or van der Waals attractions between covalent molecules. In the series of fluorides, as the electronegativity difference (between F and the other atom) decreases, the strength of the attractions decreases and the boiling point also decreases.

---

21. Consider the liquid-to-gas phase transition, at the normal boiling temperature, T_b, of a quantity of the monatomic liquid mentioned in the problem. Assume that the gas in contact with the liquid is ideal. According to Trouton's rule, the ΔH of this process is 88 T_b J mol^{-1}. In vaporization, Δn_g is +1. Use these facts in the definition of ΔH in terms of ΔE:

$$\Delta H = \Delta E + \Delta n_g RT \qquad \textit{hence} \quad \Delta E = 88\ T_b - RT_b$$

Refer to p. 145 of this Guide for a review of this relationship. ΔE is the total energy of the liquid subtracted from the total energy of the gas. For each phase, the total energy is the sum of the kinetic energy, *K.E.*, and the potential energy, *V*. Therefore:

$$\Delta E = [K.E.(g) + V(g)] - [K.E.(l) + V(l)]$$

The kinetic energies of both the gas and the liquid are equal to $^3/_2$ RT_b, since the Maxwell-Boltzmann distribution of molecular kinetic energies applies to both. Inserting this fact in the previous equation and taking the potential energy of the gas, *V(g)*, to be zero give:

$$-V(l) = \Delta E = 88\ T_b - RT_b$$

The *K.E.(l)* is, as just stated, equal to $^3/_2$ RT_b. The ratio of the potential energy of the liquid to its kinetic energy therefore is:

$$V(l)\ /\ K.E.(l) = -(88\ T_b - RT_b)\ /\ ^3/_2\ RT_b$$

The temperature cancels out on the right-hand side. Taking R in J mol^{-1}K^{-1}, so that it is commensurate with the units of the Trouton constant, gives a numerical value for the ratio of −6.4

---

23. The velocity distribution of the molecules in a liquid is the Maxwell-Boltzmann distribution. This means that the average kinetic energy per molecule of liquid is $^3/_2$ k_BT. The kinetic energy is $\frac{1}{2}\ m\ (u_{rms})^2$. See Chapter 3, p. 42, of this Guide. The procedure for determining the temperature of the computer-simulated liquid is to determine the mean of the squares of the speeds of the molecules, multiply it by the molecular mass and divide by $3k_B$. This procedure assumes all the molecules have the same mass.

---

25. Suppose that the concentrations of H_3O^+ *(aq)* and of OH^- *(aq)* are 1.0 M. Then there would be 6.022×10^{23} H_3O^+ ions and a like number of OH^- ions in a liter of solution. A liter is 10^{-3} m^3 so there would be $2 \times 6.022 \times 10^{26}$ ions per cubic meter of solution. Hence there would be, on the average, 8.303×10^{-28} m^3 available for every ion.

If this volume is spherical the radius of the sphere is 5.831×10^{-10} m (computed using $V = {}^4/_3\, \pi r^3$). The mean separation of H_3O^+ and OH^- ions is then the sum of the radii of the spheres surrounding each or 11.7×10^{-10} m. This is 11.7×10^{-8} cm.

Make the approximation that this distance is the root mean square distance, r_{rms}, which the ions traverse by diffusion to collide. It is known that:

$$r_{rms} = (6\, D\, t)^{1/2}$$

Moreover, D is 2.3×10^{-5} $cm^2 s^{-1}$. Therefore, the average time initially required for diffusion before collision is 9.9×10^{-11} s.

In the instant that the reaction starts, the number of collisions per second by an average pair of ions is the reciprocal of the diffusion time or 1.0×10^{10} s^{-1}. There is a mole of pairs in the solution. Assume that every collision is a productive reaction event. Then, as the reaction starts, H_3O^+ and OH^- pairs diffuse toward each other throughout the bulk of the 1 L of solution, producing water at the initial rate of 1.0×10^{10} mol $L^{-1}s^{-1}$.

Since the reaction is second-order:

$$\text{rate} = k\,[H_3O^+][OH^-]$$

Substitution of the initial values of the rate and of the two concentrations gives $k = 1.0 \times 10^{10}$ L $mol^{-1}s^{-1}$.

---

27. ***a)*** Among the Group IV hydrides the enthalpy change of sublimation, ΔH_{subl}, increases by about 4.3 kJ mol^{-1} with each step going down the periodic table:

CH_4, 8.4; SiH_4 ?; GeH_4 17; SnH_4 21 kJ mol^{-1}

The expected value for SiH_4 would be about 12.7 kJ mol^{-1}. The two Group VI hydrides H_2Te and H_2Se have ΔH_{subl} 5 to 7 kJ mol^{-1} larger than the Group IV elements in the same row. In Group VI, ΔH_{subl} increases going down the table by 6 kJ mol^{-1} from H_2Se to H_2Te. Extrapolating this trend upward would give a ΔH_{subl} of 10 kJ mol^{-1} for H_2O. Extrapolating across the chart (from CH_4) would give ΔH_{subl} of 14 kJ mol^{-1}. Estimate water's hypothetical ΔH_{subl} as a compromise between these two: 12 kJ mol^{-1}.

b) The experimental ΔH_{subl} of ice is 51 kJ mol^{-1}. Presumably 39 kJ mol^{-1} of this is due to hydrogen bonds. Each O atom in a mole of ice is hydrogen-bonded to 2 other O atoms through its two hydrogens. The ratio is 2 mol of H-bonds per mole of ice. The estimated energy of an H-bond in ice is half of 39 or 20 kJ mol^{-1}.

---

29. Polar solutes will be more soluble in H_2O, the more polar solvent, and have higher concentration in it than in CCl_4. Non-polar solutes will have higher concentrations in the CCl_4 phase.

a) CH_3OH -- water ***b)*** C_2Cl_6 -- carbon tetrachloride ***c)*** Br_2 -- carbon tetrachloride ***d)*** NaCl -- water ***e)*** $HClO_4$ -- water

Chapter 15

THE SOLID STATE

15.1 *CRYSTAL STRUCTURE AND SYMMETRY*

Crystals are classified according to the number and kind of their *symmetry elements.* Elements of symmetry include:

- *Rotational axes.* If the rotation of a crystal about an axis leaves it superimposed upon its original appearance then this is an axis of symmetry. A C_2 symmetry axis has a 180° rotation. A C_4 axis has a 90° rotation. A C_n axis has a 360°/n rotation.

- *Mirror planes.* Imagine a plane slicing through a crystal. If reflection across the plane, exchanging left for right and right for left, leaves the crystal unchanged in appearance then the plane is a mirror plane.

- *Centers of inversion.* A crystal has a center of inversion if an imagined reflection through its central *point* leaves it unchanged in appearance.

Empirical and theoretical studies show that there are only seven different possible combinations of symmetry elements in crystals, the seven *crystal systems.* Each has a descriptive name:

cubic, tetragonal, orthorhombic, monoclinic, triclinic, hexagonal & rhombohedral.

The text identifies the symmetry elements for the cubic system. **Problem 1** does the same for the tetragonal system. An obvious exercise is to make such a listing for the other five crystal systems.

On the microscopic level a crystal can be thought of as built from blocks of a unit structure, its *unit cell,* stacked side by side in three dimensional space. Unit cells always have three pairs of parallel faces so that they can stack snugly top to bottom, side to side, and front to back. There are of course no little cells walls inside actual crystals because unit cells are constructions of the mind. As such, unit cells are chosen as *the smallest units that retain all the symmetry of the crystal system.*

Usually, the walls of a unit cell will slice through one or more atoms. When portions of atoms are excluded just outside one wall, the loss is compensated for inside the unit cell since a similar portion is automatically included just *inside* the opposite wall.

There are seven types of unit cell, one for each of the crystal systems. Each type has the same symmetry as one of the seven crystal systems. A set of six parameters, three edge lengths (axial lengths) and three angles, specify the size and shape of a unit cell. The angles are the *interaxial angles*, the angles between the edges, and are not necessarily 90°. Unit cells (and crystal systems) with higher symmetry have more conditions relating the values of a, b and c, the edge lengths, and restricting α, β, and γ, the angles, to certain values. See Table 15-1, text p. 523. Unit cells have axial lengths starting at about 4 Å. **See problem 7.** The structure of the entire crystal can be understood in terms of the unit cell.

Real crystals are huge compared to unit cells. Most unit cells are buried deep in the crystalline interior. In the interior, every unit cell is just like the next only moved over in one direction or another by the length of a cell edge. Therefore, the surroundings of the eight corners of every unit cell are identical. These corners are *lattice points.* They are at the intersections of a regularly-spaced three dimensional lattice, or grid, of lines running along the edges of the unit cells.

- **Every lattice point has identical surroundings.**

In many problems and applications it is important to know how many lattice points there are per unit cell. A unit cell has eight corners but shares each one with seven other cells. Its corners alone contribute $1/8 \times 8 = 1$ lattice point.

Some lattices have more than one lattice point per unit cell and still conform to the symmetry of one of the seven crystal systems. The additional lattice points appear:

- At the interior center of the unit cell. This is *body-centering.* A body-centered cell has 2 lattice points. One stems from the 8 corners and the second is the point at the body center.

- At the centers of all six faces. This is *face-centering.* A face-centered unit cell has 4 lattice points. Understanding the face-centered structure is essential in many problems such as **problem 3.**

- At the centers of just one of the three sets of parallel faces. This is *end-centering.* An end-centered unit cell has 2 lattice points.

Bravais showed that there are exactly seven simple lattices, one in each crystal system, and an additional seven centered lattices for a total of 14 *Bravais space lattices.* The 14 Bravais lattices are drawn in Figure 15-4, text p. 525. In studying this figure, try to draw new lattices and then see how they reduce to one of the 14. For example, a hypothetical end-centered tetragonal lattice is really just a simple tetragonal lattice in a different orientation (and having smaller lattice parameters). For this reason Bravais excluded it from his list.

Unit cell parameters are also called *lattice parameters* because they tell the spacing of the lattice points.

Crystal Structure of Simple Substances

The cubic crystal system is important in the structures of many simple substances. Three Bravais lattices have cubic symmetry: the simple cubic (s.c.), the face-centered cubic (f.c.c.) and the body-centered cubic (b.c.c.). Some geometrical facts about the cube are useful:

1. One parameter, the length of an edge a, fully characterizes a cube. By the same token one parameter, the length of a unit cell edge, fully characterizes a cubic lattice.

2. The volume of a cube is equal to its edge cubed (a^3). Similarly the volume of a cubic unit cell is a^3.

3. A line drawn diagonally across one face of a cube is the *face diagonal, f.* The length of f is $\sqrt{2}\,a$.

4. A line connecting the opposite corners of a cube is the *body diagonal, b.* The length of b is $\sqrt{3}\,a$

The density of a crystal is equal to its mass divided by its volume. In terms of the unit cell, the density of a crystal is the mass of the contents of the cell divided by the volume of the cell:

$$\text{density} = \rho = m_{\text{contents of cell}} / V_{\text{cell}}$$

This leads to the useful expression:

$$\rho = n\,(\text{MW}) / (N_0\ V)$$

where n, which is always an integer, is the number of atoms (or molecules) per unit cell, MW is their molar weight, N_0 is Avogadro's number, and V is the volume of the unit cell. If MW is in g mol^{-1} and V is in cm^3, the density comes out in g cm^{-3}. This equation is the basis for **problems 3, 5, 7, 9, and 11.** One error in its use is failure to compute V properly. If lattice parameters are given in Ångstroms then cell volume will come out in Å^3. Remember that 1 cm^3 is 10^{24} Å^3. Another error in computing densities is to omit N_0. One view is on the presence of N_0 is that MW has units of g mol^{-1} and some way has to be found to get rid of the mol^{-1} part because densities do not have anything to do with moles in their units. **Problem 3** gives an additional perspective on why N_0 appears.

Even in simple substances, all of the atoms in a unit cell do *not* automatically sit at lattice points. Atoms can be anywhere in the cell. Their locations are given in terms of a set of three *fractional coordinates.* One of the corners of the unit cell serves as the origin and the three cell edges which intersect at the origin serve as axes. Any point inside the cell has coordinates x,y,z where x tells how far away from the origin the point is parallel to the first edge in units of a, that edge's length and y and z do the same for the second and third edges. For example, an atom in a cubic cell might have the fractional coordinates (0.40, 0.15, 0.35). If the three cell edges were all equal to 4.0 Å, this atom would lie 1.60 Å out from the origin along the a cell edge, 0.60 Å out along b, and 1.40 Å out along c.

In terms of fractional coordinates the eight corners of the unit cell have the coordinates:

(0,0,0) (1,0,0) (0,1,0) (0,0,1)
(1,1,1) (0,1,1) (1,0,1) (1,1,0)

These eight corners are exactly equivalent. In fractional coordinates, adding or subtracting 1 to any member of the triple does not create new or unique positions but instead refers to the *same* location in the adjoining unit cell.

Example. What are the fractional coordinates of atoms located at the face-centers of a cubic unit cell?

Solution. The face centers all have one fractional coordinate equal to 0 and the other two equal to ½. The correct answer:

$$(\tfrac{1}{2},\tfrac{1}{2},0)\ (\tfrac{1}{2},0,\tfrac{1}{2})\ (0,\tfrac{1}{2},\tfrac{1}{2})$$

The trouble with this answer is that everybody knows that a cube has six face centers because it has six faces. The centers of the other three faces are:

$$(\tfrac{1}{2},\tfrac{1}{2},1)\ (\tfrac{1}{2},1,\tfrac{1}{2})\ (1,\tfrac{1}{2},\tfrac{1}{2})$$

These coordinates do *not* specify new points, but points that are equivalent to the first three translated by the length of a cell edge along z, y and x respectively. This is a mathematical way of showing that a unit cell has just three face centers because each of its six faces is shared 50:50 with a neighboring unit cell.

Fractional coordinates are used to advantage in **problem 23.**

Atomic Packing in Crystals

In elemental metals there is only one kind of atom in the crystal. This makes for simplicity in the crystal structure, and many metals crystallize in cubic lattices in which the atoms are in contact with their neighbors and are located at each point of the Bravais lattice:

- In a simple cubic lattice the distance between neighboring lattice points is a. The radius of metal atoms sitting at these points and in contact is $\frac{1}{2}a$.

In a face-centered cubic lattice nearest-neighbor atoms at the lattice points touch along a face diagonal. The distance between their centers is $\sqrt{2}\ a / 2$, and the radius of the metal atoms is $\sqrt{2}\ a / 4$.

- In a body-centered cubic lattice nearest-neighbor lattice points touch along a body diagonal. The distance between their centers is $\sqrt{3}\ a / 2$, and the radius of the metal atoms is $\sqrt{3}\ a / 4$.

If atoms are hard spheres, then it is impossible not to leave gaps between them when packing them together in a crystal. A *close-packed* structure is a structure in which identical spheres (atoms) occupy the greatest possible fraction of the total space. The two most efficient methods for packing identical spheres are *cubic close-packing* and *hexagonal close-packing.* In both, the atoms occupy 74.0 percent of the volume. The other 26.0 percent of the volume of the unit cell is empty space. Cubic close-packing corresponds to setting down layers of identical spheres in the sequence *abcabcabc...*, where the letters refer to the offset in position of the higher layers relative to the first. The recurrence of *a* in the fourth position means that atoms in the fourth layer are positioned exactly above atoms in the first layer. Cubic close-packing gives rise to a face-centered cubic lattice. Hexagonal close-packing corresponds to the sequence *ababab....*

The gaps between close-packed atoms are *interstitial sites.* Octahedral interstitial sites are surrounded by 6 atoms. Tetrahedral interstitial sites are smaller and surrounded by only 4 atoms. **See problem 25.** In the face-centered cubic lattice there are 4 octahedral sites and 8 tetrahedral sites per unit cell.

In other kinds of lattices there are other kinds of interstitial sites. An examination problem might refer to the geometry of a *cubic* site (surrounded by 8 atoms) to test understanding. **See problem 29.**

Scattering of X-rays by Crystals

X-rays have wavelengths, λ, on the order of the distances between layers of atoms in crystals. When a beam of X-rays strikes a layer of atoms in a crystal, it is *scattered* in all directions by interaction with the electrons of the atoms in the layer. The atoms lie in a regular array. Some of the scattered X-rays from one layer *constructively interfere* with some of the scattered X-rays from another layer. The result is a large number of scattered beamlets shooting out from the crystal in all directions. If X-rays were visible the effect might resemble what is seen at dance-halls when a spotlight is played on a large sphere which is tiled with small flat mirrors. The reason there are so many beamlets is that many different sets of layers of atoms exist simultaneously in the crystal.

Each scattered beamlet is called a *reflection.* The *Bragg law* is the criterion for scattering:

$$n\lambda = 2\, d \sin\theta$$

In this important equation, n is an integer, λ is the wavelength of the X-rays, d is the perpendicular distance between the layers of atoms, and θ is one-half of the angle between the incident beam of X-rays and the reflected beamlet in question. θ is the *Bragg angle.*

In common applications, λ is known and 2θ (and thus θ) is measured. The order of the reflection is often given explicitly. **See problems 9 and 11.**

Example. Suppose that X-rays of wavelength 1.54 Å are scattered from a set of evenly-spaced layers of atoms in a crystal with a spacing of 4.62 Å. Compute the Bragg angle of all the resultant reflections.

Solution. From the Bragg equation with λ = 1.54 Å and d = 4.62 Å:

$$\theta = \sin^{-1} n\lambda/2d = \sin^{-1} (n / 6.00)$$

Systematic substitution for the integer n gives:

$n = 1$	$\theta = 9.59°$	$n = 4$	$\theta = 41.81°$
$n = 2$	$\theta = 19.47$	$n = 5$	$\theta = 56.44$
$n = 3$	$\theta = 30.00$	$n = 6$	$\theta = 90.00$

The listing ends at n=6 because higher n's give $\sin\theta > 1.0$, and numbers exceeding 1.0 do not have an inverse sine. Every reflected beamlet indexed in the list is scattered at the angle 2θ relative to the incident beam. For n=6 this angle is 180°. The n=6 reflection back-scatters from the crystal right into the incident beam.

X-rays are not unique in being scattered by crystals. Both neutrons and electrons have wave-like properties, as predicted by the DeBroglie equation (Chapter 12):

$$\lambda = h / p = h / mv$$

where p is the particle's momentum. When beams of neutrons or electrons of the proper momentum impinge on crystals, they are scattered according to the Bragg equation. **See problem 26,** which is like **problem 9** except that the proper λ has to be calculated (using the DeBroglie equation) before applying the Bragg equation.

15.2 *BONDING IN CRYSTALS*

The strength of the forces that hold crystalline substances together varies enormously. Crystals are classified according to these forces. The categories are: *molecular, ionic, metallic and covalent.*

Molecular crystals

Molecular crystals are held together by van der Waals or dipole forces. These forces are weak. The building-block molecules in the crystal are only slightly affected in their internal geometry by the interaction with their neighbors.

The lattice energy of a molecular crystal is the energy decrease that results when the crystal forms from a gas at 0 K. Lattice energies of molecular crystals can be estimated from the Lennard-Jones parameters of the molecules. They are much smaller than the lattice energies of other types of crystals. For example, compare the answer to **problem 27a** with the answer to **problem 13a.**

Ionic crystals

Compounds composed of elements which differ considerably in electronegativity occur as ionic crystals. Such crystals behave to a good approximation as if constructed from of hard, charged spheres held together *in contact* by electrostatic attractions. Such attractions are strong but *non-directional* in character.

To maximize the attractions (between ions of unlike charge) and to minimize repulsions (between ions of like charge), the lattice of an ionic crystal has every positive ion surrounded by as many negative ions as possible and vice-versa. Since fixed bond angles play no part in electrostatic attractions, the ionic crystal favors densely packed structures. Each positive ion occupies an interstitial site in a lattice of negative ions and each negative ion occupies an interstitial site in a lattice of positive ions. The shape (octahedral, tetrahedral or other) and proportion of occupied sites depend on the relative size and charge of the ions.

In ionic crystals, the negative ions are nearly always larger than the positive ions. The cation/anion *radius ratio,* a number less than 1, determines the structure which charged hard spheres adopt in a lattice. As the radius ratio gets larger, more anions can surround the cation and just touch it without bumping into each other. **See problem 25.**

Electrostatic forces are the predominant type of interaction in ionic lattices, and the major contributor to the *lattice energy* of the ionic crystal. The lattice energy is the energy necessary to break up the crystal and create a collection of widely-separated, non-interacting positive ions and negative ions.

The electrostatic (or Coulomb) contribution to an ionic crystal's lattice energy is:

$$V = \frac{MN_0 Z_1 Z_2 e^2}{4 \pi \epsilon_0 R_0}$$

where the two Z's are the charges on the two ions in the lattice in units of e, (e is 1.6021×10^{-19} C, the electron charge), ϵ_0 is 8.854×10^{-12} $C^2\ J^{-1} m^{-1}$, a factor to make the units work out properly, R_0 is the *minimum* distance between the two ions, and M is a special constant, the *Madelung constant.* The Madelung constant accounts for the fact that the electrostatic energy of a crystal is the sum of all the pair-wise attractions and repulsions in the crystal. The value of the Madelung constant depends on the type of crystal structure. In problems, the Madelung constant is usually a given.

Calculations using the above formula give *theoretical* estimates of lattice energies. Lattice energies are measured *experimentally* using the *Born-Haber cycle,* an application of the first law of thermodynamics. It is impossible directly to decompose an ionic crystal into its constituent ions in their gaseous states. Fortunately, ΔE for this process, which defines the lattice energy, can be determined as the sum of the ΔE's of several measurable steps which add together to this reaction. **See problem 13a** for a listing of the steps and an example of their use. Form a rough idea of the size of the different energies involved in the Born-Haber cycle. This will allow a common-sense approach to checking lattice energy calculations. **See problem 31.** Difficulties in working with the energetics of an ionic crystal come from:

- Carelessness with the sign. Either Z_1 and Z_2 in the above formula must be negative. All of the other quantities are positive. Consequently the Coulomb energy of a crystal is always negative.

- Confusion between lattice energy and Coulomb energy. The lattice energy of an ionic crystal is an experimental number. It is positive, since it is the energy required to take a crystal apart. It approximates the Coulomb energy in magnitude but is always 10 to 15 percent smaller. The Coulomb energy (multiplied by −1) errs as an estimate of the lattice energy because the hard-sphere picture of the lattice is an over-simplification. **See problem 31.**

- Confusion between ΔH and ΔE. As defined, the lattice energy is a ΔE, an energy change, not a ΔH, an enthalpy change. The Born-Haber cycle must use ΔE values for all of its steps to compute a lattice energy properly. In practice, the difference between ΔH and ΔE is usually small. **See problem 13a.**

Many examination problems require some creative use of a Born-Haber cycle or a variant.

Example. Estimate the enthalpy of formation of the hypothetical compound CaCl*(s)*. Why does $CaCl_2$ *(s)* always form when Ca*(s)* and Cl_2 *(g)* react?

Solution. The enthalpy of formation of CaCl*(s)* must be estimated with a Born-Haber cycle because the compound does not exist, and no direct measurements are possible. The creative insight is to use the lattice energy of KCl to approximate the lattice energy of the hypothetical compound. First, either look up or calculate this lattice energy. Then apply the usual cycle to the formation of the hypothetical CaCl*(s)*. The answer is a ΔH°_f of about −150 kJ mol^{-1}. The compound is favored thermodynamically relative to the elements. Why does it not form? The answer is in the second part of the question. The real compound,

$CaCl_2$ *(s)*, has a ΔH_f° of -795.8 kJ mol^{-1} (Appendix D). It is even more favored.

Metallic Crystals

In a metallic crystal, the attractive force holding the crystal together is the metallic bond. Metallic bonds are *nondirectional* in character and involve wide-spread *delocalization* of electrons. They involve the sharing of electrons among all the atoms in the crystal. The electrons occupy molecular orbitals arising from the overlap of atomic orbitals of similar energy on all of the atoms. For example, in 1 mol of sodium metal, there are 6.02×10^{23} $3s$ orbitals. These orbitals overlap to make a *band* of molecular orbitals. The molecular orbitals comprise a band because they are so numerous and their energies are so nearly the same. The band has room for $2 \times 6.02 \times 10^{23}$ electrons and therefore is only half-full, under the Pauli principle.

The excellent electrical and thermal conductivities of metals are explained by the great mobility of electrons at the top of the sea of occupied levels. The uppermost occupied or half-occupied molecular orbital in the band is the *Fermi level.* Electrons at or near the Fermi level require only slight amounts of energy to excite them to occupy levels lying above.

Covalent Crystals

In covalent crystals, the atoms are linked by covalent bonds. These bonds are *strong* and *directional* in character. Diamond, discussed in **problem 23** and presented structurally in Figure 15-16, text p. 544, is a typical case. Its crystal structure consists of a face-centered cubic array of C atoms with additional C atoms occupying every other tetrahedral interstitial site. In this arrangement, each C atom is surrounded by 4 nearest neighbors situated perfectly for σ overlap between sp^3 hybrid orbitals.

There are just enough valence electrons to complete the requirements of this bonding scheme. The gap between the energy of these electrons and the next unoccupied orbitals is large. Hence, diamond is a poor electrical conductor. **See problem 17.**

Silicon and germanium both have the diamond structure, and both are covalent crystals. The energy gap for excitation of an electron to a conduction band is however much smaller in these substances than in diamond. Both are *semiconductors.* The electrical conductivity of silicon is too low for use in solid-state electronic devices. Doping pure silicon with phosphorus or gallium gives a crystal in which some of the Si sites are occupied by impurity atoms with either one more or one fewer valence electron. The result is enhanced electrical conductivity either as an *n-type* or *p-type* semiconductor. The first type of semiconductor conducts through the motion of the extra valence electrons. The second type conducts through the motion of holes, apparent positive charges, in the valence band.

15.3 *DEFECTS AND AMORPHOUS SOLIDS*

Real crystals are imperfect. In a real crystal some lattice sites are unoccupied. Sometimes the missing atom is entirely lost, and sometimes it is just displaced to near-by interstitial site. The first case is a *Schottky defect.* This second is a *Frenkel defect.*

The existence of defects explains why many solid compounds have stoichiometries differing from strict whole number ratios. If some proportion of the positive ions in an ionic lattice is oxidized (meaning that some ions lose electrons), then there can be Schottky defects at some of the positive ion sites. The result is a ratio of positive ion to negative ion that is slightly *less* than stoichiometric. **See problems 21 and 35.**

DETAILED SOLUTIONS TO ODD-NUMBERED PROBLEMS

1. The tetragonal crystal system has $a = b \neq c$ and all angles equal to 90°. It derives from the cubic system by the elongation (of shortening) of one parallel set of the cube's edges. The resulting shape is a rectangular prism.

A cube has 3 C_4 axes. (See text, Section 15-1.) Only one of these (the axis perpendicular to the square face) remains after the distortion. Use Figure 15-3, text p. 524, to verify this. The other two turn into C_2 axes. All of the cube's C_3 axes vanish. The 6 C_2 axes passing through the centers of the edges of the cube are reduced to 2 C_2 axes in the tetragonal case. The 9 mirror planes of the cube are reduced by the distortion to 5 planes. Since two pairs of opposing faces become rectangles during the distortion, the planes perpendicular to these faces and cutting them on the diagonal are no longer mirror planes. The 1 center of symmetry persists.

Text Figure 15-3 shows 23 distinct elements of symmetry for the cubic system. Suppose that just leaving an object alone (the identity operation) is counted as an element of symmetry. Then the cubic system has 24 symmetry elements, and the tetragonal system has 12 symmetry elements (the 11 just listed and the identity element). The tetragonal system is one-half as symmetrical as the cubic system.

---

3. AgBr crystallizes in the rock salt structure. Therefore there are four Ag^+ and four Br^- ions per unit cell. The mass of the contents of one unit cell is 751.116 amu, four times the molecular weight of AgBr expressed in amu. The edge of the cubic unit cell is twice the distance between adjacent Ag^+ and Br^- ions. These ions alternate along the edge. This distance is 5.78 Å or 5.78×10^{-8} cm. The volume of the unit cell is the cube of this value or 1.93×10^{-24} cm^3. The density, ρ, of the AgBr crystal is its mass divided by its volume:

$$\rho = m / V = 751.116 \text{ amu} / 1.93 \times 10^{-24} \text{ cm}^3 = 3.89 \times 10^{24} \text{ amu cm}^{-3}$$

The density is also, from the problem, 6.473 g cm^{-3}. The apparent difference between the given density and the calculated density stems from a difference in their units. The *ratio* of the second to the first is 6.01×10^{23} amu g^{-1}. Since there are Avogadro's number of amu in a gram, 6.01×10^{23} is the estimate of N_0.

---

5. ***a)*** The body-centered cubic structure implies 2 Fe atoms per unit cell, one in the center of the cell and 1 at each of the 8 corners of the cell. Each corner atom is shared by 7 neighboring cells. The Fe atoms are in contact along the body diagonal of the cell. They do *not* touch along the cell edges.

The volume of the unit cell is the mass of its contents divided by its density. The density, as given in the problem, is 7.86 g cm^{-3}. The mass of the contents is 2×55.847 amu or 111.694 amu. Converting this mass from amu to grams requires division by the factor 6.02205×10^{23} amu g^{-1} and gives 1.8548×10^{-22} g. The volume is this mass divided by the density. The volume is 2.36×10^{-23} cm^3. The edge, a, of the cubic unit cell is the cube root of the volume. It is 2.87×10^{-8} cm which is 2.87 Å. The nearest-neighbor distance is one-half the body diagonal b of the unit cell. Because:

$$b = \sqrt{3}\, a$$

b is 4.97 Å and the nearest neighbors are 2.48 Å apart.

b) The lattice parameter is 2.87 Å, the cubic cell's edge. See above.

---

7. ***a)*** In a body-centered cubic structure there are two lattice points per cell. In metallic sodium a single Na atom is associated with each lattice point so there are two Na atoms per cell.

b) Let r equal the radius of the Na atoms. These atoms touch along the body diagonal, b, of the cubic cell, which has its corners exactly at the centers of Na atoms. This means:

$$4\, r = b$$

But b is $\sqrt{3}$ times the edge of the cell:

$$4\, r = \sqrt{3}\, a$$

Cubing the equation gives:

$$64\, r^3 = 3\sqrt{3}\, a^3$$

The volume of the cell is a^3. The volume of a single Na atom is $^4/_3\, \pi r^3$. 2 Na's have twice this volume:

$$V_{cell} = a^3 \qquad and \qquad V_{2Na} = {}^8/_3\, \pi r^3$$

Substituting these relationships in the previous equation gives:

$$8\, V_{2Na} = \pi \sqrt{3}\, V_{cell}$$

Solving for the ratio of the two volumes and completion of the arithmetic gives V_{2Na} / V_{cell} of 0.6802. It was not necessary to have an actual value for r.

---

9. ***a)*** The Bragg equation is:

$$n\lambda = 2d \sin \theta$$

The problem gives λ (1.539 Å), n (2), and θ (15.84°). Substitution gives a d of 5.638 Å. Note that the units of d are always the same as the units of λ. This reflection is from the planar faces of the unit cell so d, the interplanar spacing, is equal to a.

b) In the NaCl structure, the Na^+ and Cl^- ions lie in contact along a cell edge. The distance between their centers is the smallest Na^+ to Cl^- distance. The length of the edge is the distance from one Na^+ to the next Na^+ or from one Cl^- to the next Cl^-. The Na^+ to Cl^- distance is one-half of a cell edge or 2.819 Å.

---

11. *a)* The Bragg equation for this case is $2\lambda = 2d \sin 27.35°$, hence d is 3.613 Å. This is also a because the reflection is specified to come from the parallel faces of the unit cell. These faces are separated by the distance a.

b) The density, ρ, of a cubic crystal is related to the molar weight (MW) of its molecules and its cell edge, a by:

$$MW / \rho = N_0 a^3 / n$$

where n is the number of molecules per unit cell and N_0 is Avogadro's number. For copper metal the molar weight is 63.54 g mol^{-1}. From part a), a is 3.613×10^{-8} cm. The density is 8.92 g cm^{-3}. When the quantities are expressed this way and substituted in the equation all units cancel out, and n is a pure number. It is 3.99. The number of molecules per cell must be a whole number so $n = 4$.

---

13. *a)* The lattice energy of LiF is the negative of the energy change of the reaction:

$$Li^+(g) + Cl^-(g) \rightarrow LiF(s) \qquad \text{lattice energy} = -\Delta E$$

Experimental measurement of this ΔE is not possible. The Born-Haber cycle replaces the direct combination of the ions with a circuitous series of steps, taking place at 25° C, for all of which ΔE can be measured. The steps are: *step 1*, the transfer of electrons from $Cl^-(g)$ ions to $Li^+(g)$ to give neutral gaseous atoms: *step 2*, the condensation of $Li(g)$ to $Li(s)$ and association of $F^-(g)$ to $F_2(g)$; *step 3*, the reaction of $Li(s)$ and $F_2(g)$ in their standard states to give $LiF(s)$.

Applying the first law of thermodynamics to this cycle:

$$\Delta E = \Delta E_1 + \Delta E_2 + \Delta E_3$$

ΔE_1 is the electron affinity of $F(g)$ *minus* the first ionization energy of $Li(g)$. This is 328 kJ mol^{-1} minus 520 kJ mol^{-1} or -192 kJ mol^{-1}. The negative sign means that energy is released to the surroundings by the electron transfer.

ΔE_2 is the energy change accompanying the condensation of $Li(g)$ to $Li(s)$ plus the energy change accompanying the association of $F(g)$ to $F_2(g)$. The $\Delta H°$'s for these processes are in text Appendix D. They are -159.37 kJ mol^{-1} and -78.99 kJ mol^{-1}. Although these $\Delta H°$ values are not equal to the $\Delta E°$'s, they are related. The relationship is:

$$\Delta E° = \Delta H° - RT\,\Delta n_g$$

For the condensation of $Li(g)$ to $Li(s)$, Δn_g is -1, so $\Delta E° = -159.37$ kJ mol^{-1} + 1 RT. RT is 2.8 kJ mol^{-1}. For the association of $F(g)$ to $F_2(g)$, Δn_g is $-\frac{1}{2}$ so $\Delta E°$ for this portion of the step is $-78.99 + 1.4$ kJ mol^{-1}:

$$\Delta E_2 = -159.37 + 2.8 - 78.99 + 1.4 = -234.2 \text{ kJ mol}^{-1}$$

The third step is the reaction of $Li(s)$ and $F_2(g)$ to give $LiF(s)$. From Appendix D, the ΔH° for this reaction is -615.97 kJ mol^{-1}. Again, there is a small correction to get a ΔE°:

$$\Delta E^\circ = \Delta H^\circ - RT\,\Delta n_g$$

This time Δn_g is $-\frac{1}{2}$. Hence:

$$\Delta E_3 = -615.97 \text{ kJ mol}^{-1} + 1.4 \text{ kJ mol}^{-1} = -614.6 \text{ kJ mol}^{-1}$$

Now ΔE values for all three steps in the Born-Haber cycle are available.

$$\Delta E_{overall} = -192 - 234.2 - 614.6 = -1041 \text{ kJ mol}^{-1}$$

If the $\Delta n_g RT$ corrections in step 2 and 3 are omitted, the answer is 2RT more negative, that is, -1046 kJ mol^{-1}. The omission causes less than a one percent change. Considering the uncertainty of some of the ΔE values (mainly electron affinities), taking ΔE to equal ΔH is probably defensible. Compare to **problem 13-11.** The lattice energy is $+1041$ kJ mol^{-1}.

b) The computation of the Coulomb energy for LiF follows the pattern of Example 15-4 in the text. The Coulomb energy is:

$$V = \frac{MN_0 Z_1 Z_2 e^2}{4\pi \epsilon_0 R_0}$$

Since LiF crystallizes in the rock salt structure, the Madelung constant, M, is 1.748. Z_1 and Z_2 are $+1$ and -1. The charge on the electron is 1.602×10^{-19} C, R_0 is 2.014×10^{-10} m (2.014 Å), and the constants ϵ_0 and N_0 are 8.854×10^{-12} C^{2}J^{-1}m^{-1} and 6.022×10^{23} mol^{-1} respectively. Substitution and the usual care with units gives a Coulomb energy of -1.206×10^6 J mol^{-1} or, equivalently -1206 kJ mol^{-1}.

The experimental lattice energy of LiF has the opposite sign from this Coulomb energy because lattice energy is defined as the energy required to take the lattice apart. One might expect the two to have the same absolute value. Instead the Coulomb energy is about 15 percent bigger than the experimental (Born-Haber) lattice energy. The discrepancy arises because the Coulomb calculation ignores non-electrostatic interactions.

---

15. The fact that indium phosphide absorbs only those photons with wavelengths, λ, *less* than 920 nm means that only photons with energies greater than the energy of 920 nm photons exceed the band gap. The energy of a photon of light is related to its wavelength by:

$$E = hc/\lambda$$

where h is Planck's constant and c is the speed of light. Substitution gives E equal to 2.16×10^{-19} J. This is the band gap in joules. Dividing it by 1.602×10^{-19} J eV^{-1} converts it to eV. It is 1.35 eV.

---

17. The problem requires substitution in the formula for the number of electrons excited and some care with the units.

The gap energy E_g is 5.4 eV which is 8.65×10^{-19} J or 5.21×10^5 J mol^{-1}. With T equal to 300 K and R = 8.314 J $mol^{-1}K^{-1}$ the exponential term is:

$$\exp(-E_g / 2RT) = e^{-E_g /2RT} = 4.42 \times 10^{-46}$$

Combination with the rest of the formula gives n_e equal to 1.11×10^{-26} cm^{-3}. In 1 cm^3 of diamond at room temperature only 1.11×10^{-26} electrons are excited to the conduction band. It is no wonder that diamond is a poor electrical conductor.

---

19. In making the classification, consider the electronegativities of the elements and their locations in the periodic table.

a) ionic ***b)*** covalent ***c)*** molecular ***d)*** metallic

---

21. ***a)*** A sample of, say, 100 g of this sample of iron(II) oxide contains 77.45 g of Fe and 22.55 g of O. These values correspond to 1.3868 mol of Fe and 1.4094 mol of O. This is equivalent to $FeO_{1.0163}$ and also to $Fe_{0.9839}O$. It is improper to round off to the stoichiometric formula FeO. The experimental analysis is precise to four significant figures and the chemical formula should have the same precision.

b) Only 98.39 percent of the stoichiometrically required Fe is present. 1.61 percent of it is absent which means that 0.0161 of the Fe sites are vacant.

c) Let a equal the fraction of sites occupied by Fe^{3+} ions and b equal the fraction of sites occupied by Fe^{2+} ions. The Fe^{3+} ions occurring in Fe^{2+} sites make up for missing Fe^{2+} ions elsewhere and make the compound as a whole electrically neutral. The average positive charge per site must be 2. Therefore:

$$3a + 2b = 2$$

Also the *sum* of a and b is 0.9839. Combination of these two relationships gives a equal to 0.0322. This is the fraction of Fe^{2+} sites occupied by Fe^{3+} ions.

d) From the same equations, b equals 0.9517. This is the fraction of Fe^{2+} sites occupied by their proper occupants, Fe^{2+} ions.

---

23. In diamond the C−C bond distance will be the distance between the two most closely neighboring atoms. Reviewing the list of coordinates given in the problem shows these are the C at (0,0,0) and the one at (¼,¼,¼). This is also clear in Figure 15-16 in the text. Other pairs of carbons are equally close but none is closer. These carbons are separated by one-fourth of the body diagonal of the unit cell. The body diagonal is $\sqrt{3}$ times a, the edge of the cell, 3.57 Å. The bond distance is $(\sqrt{3} \times 3.57\ \text{Å}) / 4$ or 1.55 Å.

---

25. Any tetrahedral interstitial site can be viewed as occupying the center of a cube which has an alternate four of its eight corners occupied by spherical atoms of radius r_1. Let the edge, a, of such a cube have length 1. Then the face diagonal, f, has length $\sqrt{2}$, and the body diagonal, b, has length $\sqrt{3}$. The four atoms at the alternate corners surround the center and touch each other along the face diagonals of the cube. Therefore:

$$2\, r_1 = \sqrt{2}$$

Let r_2 be the radius of a spherical atom placed *at* the interstitial site, the center of the cube. The largest such atom will just touch all four atoms at the corners. One half of the body diagonal in that case equals r_1 added to r_2:

$$r_1 + r_2 = \sqrt{3} / 2$$

Dividing the second equation by the first gives:

$$(r_1 + r_2) / r_1 = \sqrt{3} / \sqrt{2} \qquad \textit{hence} \qquad 1 + r_2 / r_1 = 1.225$$

Since r_2 / r_1 is 0.225, the largest value for r_2 is 0.225 of r_1. A larger atom at the site would push the four surrounding atoms out of contact with each other.

---

27. ***a)*** According to the equations developed in Section 15-2, text p. 536, for a face-centered cubic molecular crystal:

$$R_0 \simeq 1.09\, \sigma$$

where σ is the Lennard-Jones parameter and R_0 is the equilibrium atomic spacing (at T = 0 K). For N_2, σ is 3.70 Å. (Table 14-1). Therefore R_0 is about 4.03 Å.

The potential energy of the molecular lattice is related to the other Lennard-Jones parameter, ϵ:

$$V_t = -8.61\, \epsilon$$

From Table 14-1, ϵ is 0.790 kJ mol^{-1}. The potential energy of the lattice is −6.80 kJ mol^{-1}. The lattice energy is the energy required to break up the lattice. It is +6.80 kJ mol^{-1}.

b) The density of $N_2\,(s)$ is 1.026 g cm^{-3}. The crystal has four N_2 molecules per unit cell for a total mass of 112.054 amu or 1.8607×10^{-22} g per unit cell. The volume of the cell is this mass divided by the density of the crystal or 181.36×10^{-24} cm^3. The edge of the cubic cell is the cube root of its volume and equals 5.660×10^{-8} cm (5.660 Å).

In a face-centered cubic lattice there is a nitrogen molecule at the center of every face of the unit cell and at every corner. The face diagonal is $\sqrt{2} \times 5.660$

Å or 8.005 Å long. One-half of this is the distance from an N_2 at the face's center to one at a corner. This is the intermolecular distance: 4.002 Å. This result is only about 0.7 percent less than the distance computed using the Lennard-Jones parameter. The agreement tends to confirm the analysis in text section 15-2.

---

29. In the NaCl structure each Na^+ is surrounded by a set of 6 nearest-neighbor Cl^-'s. It is at an octahedral site in a face-centered cubic array of chloride ions. The 6 neighbors are $\frac{1}{2}a$ distant where a is the lattice parameter of the f.c.c. cell. Then a set of 12 Na^+ ions is found at the distance $(\sqrt{2}/2)\ a$ and a set of 8 Cl's occurs at the distance $(\sqrt{3}/2)\ a$. Further sets of neighbors are found at the distances $(\sqrt{4}/2)\ a$, $(\sqrt{5}/2)\ a$, etc.

In the simple cubic CsCl lattice, the positive ion is larger than in NaCl and is surrounded by more Cl^- ions. It has 8 nearest neighbor Cl^- ions. Then comes a set of 6 more distant Cs^+'s and then a set of 12 yet more distant Cs^+ ions.

Get this answer by imagining a Cs^+ at the center of a cube with Cl^-'s on its eight corners. These are the nearest-neighbor Cl^-'s. The cube has 6 faces and 12 edges. The 6 second-nearest neighbors are the Cs^+ ions at the centers of the 6 face-adjoining cubes in the lattice. The 12 Cs^+ ions are at the centers of the 12 edge-adjoining cubes.

The danger in this problem is getting bogged down using messy sketches to count neighbors and decide which neighbors are nearer. A better way uses the formalism of *fractional coordinates.* Take the case of CsCl. Define an origin at the Cs^+ ion. Then express the location of any neighbor in terms of its fractional coordinates. For instance there is an ion (a Cl^-) at $(\frac{1}{2},\frac{1}{2},\frac{1}{2})$. The cubic symmetry means that the x, y and z coordinates are equivalent and that the plus and minus directions on each coordinate are equivalent, too. Therefore *permuting* the fractional coordinates and *changing the signs* of the fractional coordinates generate equivalent locations. For $(\frac{1}{2},\frac{1}{2},\frac{1}{2})$ the following unique sets of coordinates result from these operations:

$$(+\tfrac{1}{2},+\tfrac{1}{2},+\tfrac{1}{2})\ (-\tfrac{1}{2},+\tfrac{1}{2},+\tfrac{1}{2})\ (+\tfrac{1}{2},-\tfrac{1}{2},+\tfrac{1}{2})\ (+\tfrac{1}{2},+\tfrac{1}{2},-\tfrac{1}{2})$$
$$(-\tfrac{1}{2},-\tfrac{1}{2},-\tfrac{1}{2})\ (+\tfrac{1}{2},-\tfrac{1}{2},-\tfrac{1}{2})\ (-\tfrac{1}{2},+\tfrac{1}{2},-\tfrac{1}{2})\ (-\tfrac{1}{2},-\tfrac{1}{2},+\tfrac{1}{2})$$

The eight locations are the eight equivalent nearest-neighbors of the Cs^+ previously identified. Their distance from the central Cs^+ is the square root of the sum of the squares of the three coordinates. In this case the distance is $(^3/_4)^{\frac{1}{2}}\ a$.

This method is even more helpful farther away from the center. For example, there is a neighbor of the central Cs^+ at $(^3/_2,\frac{1}{2},\frac{1}{2})$. Permuting these fractional coordinates gives:

$$(+^3/_2,+\tfrac{1}{2},+\tfrac{1}{2})\ (+\tfrac{1}{2},+^3/_2,+\tfrac{1}{2})\ (+\tfrac{1}{2},+\tfrac{1}{2},+^3/_2)$$

Each of these permutations has eight distinct combinations of plus and minus signs. Therefore there are 3 × 8 or 24 equivalent Cl ions all at the distance:

$$d = [(^3/_2)^2 + (\tfrac{1}{2})^2 + (\tfrac{1}{2})^2]^{\frac{1}{2}} = (^{11}/_4)^{\frac{1}{2}}$$

The form of these fractional coordinates even tells the identify of this set of neighbors. The site $(+^3/_2,+\frac{1}{2},+\frac{1}{2})$ is just the $(+\frac{1}{2},+\frac{1}{2},+\frac{1}{2})$ site moved over by 1 cell edge in the x direction. Unit cells have identical contents, so the new site must be occupied by the same kind of ion as the old, a Cl^-.

---

31. Compare the Born-Haber steps for the making of CuX_2 *(s)*, the copper(II) halides, with those for the making of CuX*(s)*, the copper(I) halides, from the constituent elements.

The two processes start out even by requiring 745 kJ mol^{-1} to ionize Cu*(g)* to Cu^+ *(g)*. But making Cu(II) halides then consumes much *more* ionization energy (1958 kJ mol^{-1}) to remove the second electron. See text Table 12-3. In the second step, the Cu(II) halides also require more enthalpy (and energy) for the vaporization and dissociation of the X_2 molecules because they use twice as many X atoms. The combined enthalpy change for vaporization plus dissociation varies among the halogens from 31 to 122 kJ mol^{-1}. (See Appendix D and Table 13-3). There is some small compensation in the third step. Transferring electrons *to* the X*(g)* atoms releases more energy in making CuX_2 *(s)* than CuX*(s)* because the electron affinity of twice as many X*(g)* atoms is involved. The electron affinities for the halogens are however all less than 350 kJ mol^{-1}.

Despite the last, favorable contribution, the formation of CuX_2 *(s)* is, up to this point, energetically *dis*favored relative to CuX*(s)* by roughly 1600 kJ mol^{-1}. The final step in the Born-Haber cycle involves the lattice energy. The lattice energies of Cu(II) halides are *far* larger than those of Cu(I) halides because a 2+ ion (not just a 1+ ion) is involved in Coulomb interactions. The following table tells the story.

Lattice Energies of Copper Halides

Substance	ΔE(lattice)	Substance	ΔE(lattice)
CuF_2	3082 kJ mol^{-1}	CuF	– kJ mol^{-1}
$CuCl_2$	2811	CuCl	996
$CuBr_2$	2763	CuBr	979
CuI_2	2640	CuI	966

This is the reason most Cu(II) halides are stable relative to the corresponding Cu(I) halide.

What about the case of the copper iodides? The iodides have the smallest lattice energies both for Cu(I) and Cu(II). Iodide (I^-) ion is the biggest of the halide ions. A large ion is polarizable and behaves less like a point charge than does a small ion. It lends itself well to non-Coulomb, repulsive interactions, effects which are greater in the presence of a 2+ ion. Going to iodide reduces the lattice energy in both kinds of halides, but does so proportionately more for the copper(II) halides. The difference in lattice energy between CuI_2 *(s)* and CuI*(s)* is just not enough to compensate for the huge investment in removing the second electron from copper.

---

33. NaCl is an ionic solid. If there are Schottky defects, a fraction of the Na^+ sites is vacant. To maintain electrical neutrality an equal fraction of the Cl^- sites must be vacant. The density of defect-free NaCl is 2.165 g cm^{-3}. Introducing 0.0015 mole fraction of Schottky defects reduces the chemical amount of NaCl per cm^3 to 0.9985 of what had been. Therefore, the mass of NaCl per cm^3 is 0.9985 of what it had been. The density is 2.162 g cm^{-3}.

Frenkel defects involve displacement from a regular lattice site to an interstitial site. No mass is removed from the crystal, so the density stays at 2.165 g cm^{-3}.

---

35. ***a)*** The compound is 28.31 percent O and therefore 71.69 percent Ti. In 100 g of the compound there is 1.497 mol of Ti and 1.769 mol of O. The formula is $Ti_{0.8458}O$.

b) Only 0.8458 of the stoichiometric quantity of Ti is present. 0.1542 of the Ti sites then must be vacant. Let a equal the fraction of Ti^{2+} sites with a Ti^{3+} occupying them, and b the fraction of sites with a Ti^{2+}. Then:

$$a + b = 0.8458$$

The net positive charge per oxygen must be +2. Each Ti^{3+} contributes +3 and each Ti^{2+} contributes +2:

$$3a + 2b = 2$$

Solution of the equations in a and b gives b equal to 0.5375 and a equal to 0.3083. Just under 31 percent of the Ti^{2+} sites contain a Ti^{3+}.

Chapter 16

CHEMISTRY OF THE PERIODIC TABLE AND ITS APPLICATIONS

Chapter 16 begins the text's coverage of *descriptive chemistry*, the application of theoretical concepts to problems of chemical interest.

16.1 *PERIODIC PROPERTIES OF THE ELEMENTS*

The periodic table of the elements is the central organizing vehicle in the study of descriptive chemistry. Its structure arises from the systematic filling of atomic orbitals by electrons according to the Pauli principle and Hund's rule, as detailed in Chapter 12. Trends among important physical and chemical properties arise from repeating patterns of electron configuration among the elements:

• Compounds of *similar stoichiometry* are found among elements in the same group (column) in the periodic table. Valence electron configurations govern stoichiometry. The electron configurations of elements in the same group are the same type. For example, the valence electron configuration of the Group VII elements all have the form $ns^2\ np^5$.

• *Ionization energy* (the minimum energy necessary for the removal of an electron from an atom) *increases* going to right across the periodic table and *decreases* going down the table. Exceptions occur and are explained by the special stability of half-filled subshells. A half-filled subshell is an electron configuration like p^3 or d^5.

Study this trend by sketching graphs of the *IE* data in text Table 12-3. Spot exceptions (*e.g.* the *IE* of O is *less* that the *IE* of N) and explain them.

• *Electronegativities* in general *increase* to the right across the periodic table and *decrease* going down the table. There are exceptions. For example, among the transition metal groups headed by Ni, Co and Zn the electronegativity decreases going to the right and increases going down the table. High electronegativity in an atom is favored by small size and a large effective charge felt by the outer electrons. The first reversal just cited occurs because size increases going to the right (instead of decreasing) in the Ni, Co and Zn groups. The second occurs because the size increases only slightly going down the table. This allows the outer electrons to feel a stronger effective nuclear charge in the elements with larger Z. In other areas of the periodic table the size effect dominates.

• *Atomic size* (the radii of ions and of metallic atoms) *decreases* with increasing atomic number across a period and *increases* going down a group in the periodic table. **See problem 1.** The *rate* of increase in size going down a group diminishes sharply once the 3*d* orbitals start to be filled. As *Z* increases across the transition metals, any one 3*d* electron is only imperfectly shielded from the attraction of the larger nuclear charge by the other electrons in the same subshell. Atomic size therefore contracts. The same effect occurs with *p* subshells and is responsible for the general contraction in size going across the table. The intervention of the first block of 10 elements in which *d* orbitals are filled makes immediately subsequent elements smaller than they otherwise would have been.

Isoelectronic species have the same electron configuration but different nuclear charge. S^{2-}, Cl^-, Ar, and K^+ are isoelectronic. The smallest species among an isoelectronic group is the one with the highest *Z*.

• *Oxidation states* that give the elements a closed shell outer electron configuration of $ns^2 np^6$ or nd^{10} are the most common, although other oxidation states occur. Heavier elements in Groups III through VII show oxidation states that are lower (in steps of 2) than required by this rule. Many transition elements show multiple oxidation states. Among these elements, the highest oxidation states occur at the center of the period. In addition, higher oxidation states are favored going down the table. It is best to *memorize* the common oxidation states of the transition elements.

The following diagram helps in predicting how properties vary with position in the periodic chart:

small atomic radius
high *IE*
high electronegativity

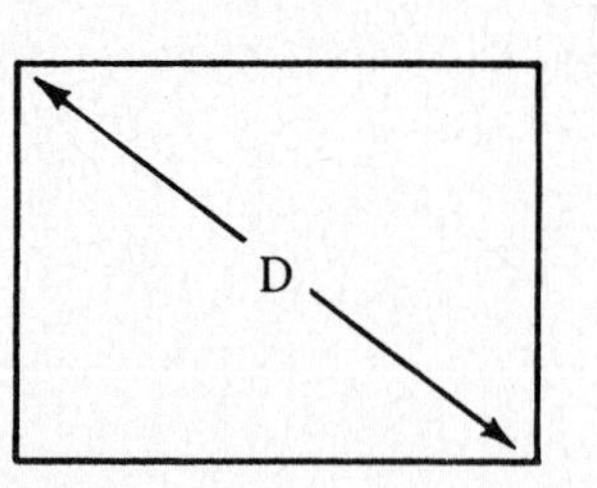

large atomic radius
low *IE*
low electronegativity

Properties are roughly constant along the diagonal line D.

16.2 *HYDROGEN AND THE HYDRIDES*

The descriptive chemistries of the individual elements are often studied on the basis of the following outline:

- Natural Occurrence and Preparation of the Element
- Characteristic Chemical and Physical Properties
- Important Compounds and Types of Compounds
- Uses in Society

In this section the text discusses hydrogen. In studying the material, note how the theoretical concepts of previous chapters are put to work in systematizing great quantities of chemical information. For example, hydrogen forms compounds with nearly every other element. It might be possible to memorize the ΔH_f° of all of these compounds. But linking the data to the periodic table allows useful *general* conclusions (the hydrides of the strongly electropositive and electronegative elements are the most stable; the hydrides of heavier elements from the center of the table are less stable). Look for valid generalizations and avoid memorizing minutiae.

The term *hydride* is the class name of all binary compounds of hydrogen. Thus, HCl and NaH are both hydrides. The specific name of a hydrogen-containing compound includes the word hydride only if hydrogen is in the -1 oxidation state.

The problems recall the theoretical concepts of earlier chapters and apply them with emphasis on hydrogen and its compounds. Thus, **problem 7** treats the ionic bonding of sodium hydride and resembles **problem 15-13. Problem 11** uses concepts developed in Chapter 2. **Problem 3** is like **problem 9-7b.**

16.3 *OXYGEN AND THE OXIDES*

This section introduces a particularly useful tool in studying the descriptive chemistry of the elements. It is the *reduction potential diagram.* Such diagrams consist of a list of species containing a particular element either by itself or in in combination with hydrogen, oxygen or both. The diagrams always concern reactions taking place in aqueous solution. The species are written in order of decreasing oxidation state of the subject element from left to right across the page. Lines connecting pairs of species stand for the properly balanced half-equation for the reduction of the form on the left to the one on the right. The corresponding standard reduction potentials appear above the lines.

Since the reduction potentials are for half-reactions going on in water, the pH of the solution sometimes affects reduction potentials sharply. Reduction potential diagrams always include a statement of whether the solution is acidic (pH 0) or basic (pH 14). The balanced half-reactions must take the availability of H^+ *(aq)* or OH^- *(aq)* into account, too.

It is essential to be able to determine the oxidation numbers of all the elements appearing in a reduction potential diagram. This skill is a preliminary to computing even *more* reduction potentials. Thus, the compactly displayed information in a reduction potential diagram can be expanded into many balanced chemical half-equations and equations.

Example. The reduction potential diagram for the element oxygen in *base* is:

$$O_3 \overset{1.24\text{ V}}{\text{―――――}} O_2 \overset{-0.086\text{ V}}{\text{―――――}} HO_2^- \overset{0.87\text{ V}}{\text{―――――}} OH^- \quad \text{(base)}$$

Write out and balance half-equations for *all* of the half-reactions represented either explicitly or implicitly in the diagram and state their standard reduction potentials.

Solution. Four different species appear in this reduction potential diagram. Therefore, there are six possible reduction half-reactions. In base, H^+ *(aq)* is scarce. Only $H_2O(l)$ and OH^- *(aq)* should appear explicitly in the balanced half-reactions. Balance the three half-reactions for which the reduction potential appears in the diagram:

$$O_3(g) + H_2O(l) + 2\,e^- \rightarrow O_2(g) + 2\,OH^-(aq) \qquad \mathcal{E}^\circ = 1.24\text{ V}$$
$$O_2(g) + H_2O(l) + 2\,e^- \rightarrow HO_2^-(aq) + OH^-(aq) \qquad \mathcal{E}^\circ = -0.086\text{ V}$$
$$HO_2^-(aq) + H_2O(l) + 2\,e^- \rightarrow 3\,OH^-(aq) \qquad \mathcal{E}^\circ = 0.87\text{ V}$$

The first reduction consists of one of the oxygen atoms in $O_3(g)$ (ozone) gaining two electrons and simultaneously abstracting a H^+ from water to give the product hydroxide. Thus, some of the product hydroxide comes from the reduction, and some comes from the solvent water. The same thing happens in the last half-equation.

Three more half-equations come from all the possible pair-wise combinations of the first three:

$$O_3(g) + 3\,H_2O(l) + 6\,e^- \rightarrow 6\,OH^-(aq) \qquad \mathcal{E}^\circ = 0.68\text{ V}$$
$$O_2(g) + 2\,H_2O(l) + 4\,e^- \rightarrow 4\,OH^-(aq) \qquad \mathcal{E}^\circ = 0.40\text{ V}$$
$$O_3(g) + 2\,H_2O(l) + 4\,e^- \rightarrow HO_2^-(aq) + 3\,OH^-(aq) \qquad \mathcal{E}^\circ = 0.528\text{ V}$$

The reduction potentials were computed using the rules of Section 9-2. For instance, the reduction potential in the first half-reaction is:

$$\mathcal{E} = [2(1.24) + 2(-0.086) + 2(0.87)]\,/\,6 = 0.68\text{ V}$$

This half-reaction involves the gaining of 6 electrons and is the sum of the three half-reactions given in the diagram, each of which involves a 2 electron gain. **See problem 3.** Also, **see problem 20-7.**

Members of the second group of three half-equations and reduction potentials could be represented explicitly in the reduction potential diagram by lines directly connecting the species which react. This is seen in the diagram on text p. 571.

The half-reactions represented in a reduction potential diagram can also be combined to give *whole* reactions. The $\Delta\mathcal{E}^\circ$ for a whole reaction is:

$$\Delta\mathcal{E} = \mathcal{E}(\text{cathode}) - \mathcal{E}(\text{anode})$$

Thus, in base, $O_3(g)$ reacts with $HO_2^-(aq)$ to give $O_2(g)$ with a standard cell potential of $1.24 - (-0.086) = 1.34$ V. Moreover, $OH_2^-(aq)$ reacts with itself with a cell potential, $\Delta\mathcal{E}^\circ$, of $0.87 - (-0.086) = 0.96$ V. The products are $OH^-(aq)$ and $O_2(g)$. This is a *disproportionation.*

• *A species in a reduction potential diagram disproportionates spontaneously whenever a reduction potential on a line leading from it to the right exceeds a reduction potential on a line leading from it to the left.*

The reduction potential diagram for oxygen in base (previous page of this Guide) lists different voltages than the diagram for oxygen in acid (text p. 571). The differences are not random. For example, the reduction potential for the reduction of ozone in *acid:*

$$O_3(g) + 2\,H^+(aq) + 2\,e^- \rightarrow O_2(g) + 2\,H_2O(l)$$

is 2.07 V, substantially higher than the 1.24 V reduction potential for the same gain of electrons taking place in base:

$$O_3(g) + H_2O(l) + 2\,e^- \rightarrow O_2(g) + 2\,OH^-(aq)$$

It is much easier to reduce O_3 *(g)* in acid than in base. In base, the half-reaction has to generate OH^- in a solution with an OH^- concentration of 1 M. Making the solution acidic reduces the OH^- concentration and so favors the products (LeChatelier's principle). The larger voltage in acid is no surprise.

Eventually, as a basic solution is acidified, the OH^- disappears from among the products in the equation, and the best representation becomes the acid half-reaction, which has H^+ *(aq)* (with activity 1), on the left-hand side.

The voltage for the O_3 *(g)* to O_2 *(g)* reduction at pH 14 is computed by putting $[H^+] = 10^{-14}$ M in the Nernst equation for the half-reaction in acid. The voltage at pH 0 is computed by putting $[OH^-] = 10^{-14}$ M in the Nernst equation for the half-reaction in base.

In reduction potential diagrams care is taken to represent the correct dominant species according to the pH. Thus, at pH 1, H_2O_2 is the major species containing oxygen in the -1 oxidation state. HO_2^-, the conjugate base of H_2O_2, is the major species at pH 14.

Reduction potential diagrams are a compact means of communicating great quantities of chemical information. They can replace paragraphs of descriptive text. Often, reduction potential diagrams of members of a group in the periodic table are displayed one after another to highlight trends. In subsequent chapters of the text, the reduction potential diagram is regularly given as a summary of the solution chemistry of the various elements. **See problem 20-31** for a case where a reduction potential diagram allows prediction of the products of an unknown reaction.

Acid-Base Properties

Hydrogen forms compounds with nearly every other element. The element is intermediate in its electronegativity. The *ionic* hydrides, which hydrogen forms with elements of either low electronegativity or high electronegativity, are basic (in water) when the element comes from the left side of the periodic table and acidic when the element comes from the right side of the table. The *covalent* hydrides, the compounds hydrogen forms with elements of intermediate electronegativity, have little acid or base character.

Oxygen, like hydrogen, forms compounds with nearly every other element. In aqueous solution, strongly ionic oxides dissolve (or react) to give basic solutions. The ionic oxides are *basic anhydrides.* In writing equations to represent these processes, remember that no oxidation numbers change, but that new bonds form between the element and oxygen. *Covalent* oxides react with water to give *oxyacids.* They are acid anhydrides. **See problem 2-35** A few elements have oxides on the border between these classes. These are *amphoteric* oxides.

The acid strength of the *oxyacids* increases with:

- Increasing oxidation state of the element (which is usually a non-metal).

- Increasing number of oxygen atoms which are bonded to the element but *not* at the same time bonded to hydrogen.

These facts are the basis for **problem 11, 15 and 29.** Study Table 16-2 (text p. 574).

16.4 *THE CHEMICAL INDUSTRY*

The text traces the methods and development of chemical industry. Two industrial processes typify the considerations involved in the design of a chemical industrial process. These are the *LeBlanc process* **(see problem 17)** for making sodium carbonate and the *Solvay process* which has replaced the LeBlanc process **(see problem 1-51).**

The products of a practical process must be obtained at a sufficient yield, at a sufficient rate and either pure enough to use immediately or readily purified. Thermodynamics imposes fundamental limitations on the yield of chemical reactions. Pressure and temperature must be selected accordingly. But practical processes may not proceed arbitrarily slowly. Compromises sometimes must be made between the best thermodynamic conditions and conditions which the process go at an acceptable rate. Much effort goes into designing catalysts to speed up thermodynamically favored processes enough to make them practical. Finally, if side-reactions contaminate the desired product with by-products that are dangerous to handle or hard to remove or both, the conditions of the process may again have to be modified to eliminate the troublesome by-products.

DETAILED ANSWERS TO ODD-NUMBERED PROBLEMS

1. Use periodic trends in ionic size to select the larger ion in each pair. Thus, the larger ion is located to the *right* when the candidates are in the same row of the table and have the same charge, toward the *bottom* when they are in the same column and have the same charge. Between ions of the same element the one with greater positive charge is smaller.

a) S^{2-} *b)* Ti^{2+} *c)* Mn^{2+} *d)* Sr^{2+}.

3. A reduction diagram shows the compounds arranged from left to right in order of decreasing oxidation number on the common element. For example, for the three lead species in this problem:

	$PbO_2(s)$	$PbO(s)$	$Pb(s)$
oxidation state	IV	II	0

Lines connecting pairs of species stand for the half-cell reaction, involving the two, written as a reduction. The reduction potential for each half-cell reaction appears above the line:

$$PbO_2(s) \overset{0.28\text{ V}}{\text{———}} PbO(s) \overset{-0.576\text{ V}}{\text{———}} Pb(s) \quad \text{(in base)}$$

$PbO(s)$ disproportionates to $PbO_2(s)$ and $Pb(s)$ only if the voltage on the right exceeds the voltage on the left. In this case $-0.576 < 0.28$ so there is no tendency to disproportionate. In other terms, the $\Delta\mathcal{E}°$ for the disproportionation is -0.856, and the reaction is non-spontaneous.

---

5. *a)* Na_2O is more stable with respect to its elements than Cs_2O because the Na^+ ion is smaller than Cs^+. There will be stronger Coulomb interactions in the Na_2O ionic lattice.

b) OF_2 is more stable than OBr_2 because it is more ionic.

c) CaO is more stable with respect to its elements than SrO because the Ca^{2+} ion is smaller.

d) SrO is more stable than Rb_2O (with respect to their elements) because the Sr^{2+} ion is smaller and doubly charged.

---

7. The goal is to estimate the enthalpy change of formation ΔH_f°, of $NaH(s)$. The reaction for the formation of $NaH(s)$ from its constituent elements in their standard states is:

$$Na(s) + \tfrac{1}{2} H_2(g) \rightarrow NaH(s)$$

It is the *sum* of the following reactions:

$Na(s) \rightarrow Na(g)$ vaporization of Na
$Na(g) \rightarrow Na^+(g) + e^-$ ionization of Na
$\frac{1}{2} H_2(g) \rightarrow H(g)$ dissociation of H_2
$H(g) + e^- \rightarrow H^-(g)$ formation of hydride ion
$Na^+(g) + H^-(g) \rightarrow NaH(s)$ lattice formation

Table 12-3, text p. 439, gives ΔE° for the ionization of Na*(g)* (496 kJ mol^{-1}) and the formation of $H^-(g)$ ion (-73 kJ mol^{-1}). The latter is the *negative* of hydrogen's electron affinity. Appendix D gives ΔH° for the vaporization of Na*(s)* (107.32 kJ mol^{-1}) and ΔH° for the dissociation of H_2. Only the energy change in the last equation requires more than reference to a table. The formation of the NaH crystal lattice releases energy. The electrostatic potential energy of the crystal is:

$$V = \frac{Z_1 Z_2 e^2 M N_0}{4\pi \epsilon_0 R_0}$$

where M is the Madelung constant, Z_1 and Z_2 are the charges on the ions, R_0 is the interionic distance, and the other quantities are the charge on the electron (e), Avogadro's number (N_0) and the proportionality factor ϵ_0 (8.854×10^{-12} C^2 J^{-1} m^{-1}).

For the rock salt structure, M is 1.7476 (Table 15-5, text p. 539). The value of R_0 is 2.52×10^{-10} m. It is the sum of the ionic radii of Na^+ and H^-. Substitution in the electrostatic potential energy formula gives $V = -1000$ kJ mol^{-1} (See text Example 15-4). The lattice energy is $+1000$ kJ mol^{-1}. Therefore, ΔE of the lattice-forming reaction is -1000 kJ mol^{-1}.

The small differences between ΔE's and ΔH's are negligible especially since non-electrostatic repulsions in the crystal are not being considered. Treat the ΔE of each reaction as approximately equal to its ΔH. Then:

$$\Delta H_f^\circ = (496 - 73 + 107.32 + 217.96 - 1000) \text{ kJ mol}^{-1}$$

$$\Delta H_f^\circ = -252 \text{ kJ mol}^{-1}$$

A more elaborate calculation, taking repulsions into account, gives V equal to -782 kJ mol^{-1}, from which derives a ΔH_f° equal to -34 kJ mol^{-1}. This is still not quite right since the experimental ΔH_f° of NaH*(s)* is -57.32 kJ mol^{-1}. The true lattice energy of NaH is approximately 806 kJ mol^{-1}. Really precise calculated lattice energies are elusive.

The ΔH_f° of metal hydrides will become more negative for smaller cohesive energy of the metal, for smaller ionization energy of the metal, and for smaller metal ion radius. In other words, the compound forms more easily if the metal is easier to take apart, if the metal loses an electron or electrons more easily or if the metal ion is closer to the hydride ion in the metal hydride lattice.

---

9. The classification is made on the basis of position in the periodic table. PoO_3, Cl_2O_7, SiO_2 and OF_2 are, to varying degrees, covalent compounds and hence acidic in water. SrO and Li_2O are ionic compounds and are basic oxides.

---

11. Lewis structures that satisfy the octet rule place formal charges of +1 on the N in HNO_3, +2 on the S in H_2SO_4 and +3 on the Cl in $HClO_4$. The acid strength of these acids increases in the same order.

---

13. The formula of formic acid, HCOOH, can be written $C(OH)_2$. This is, however, misleading in terms of the structure-to-acidity correlations presented in Table 16-2. In fact, formic acid has one H bonded to the central C atom. Its structure is OCH(OH). Formic acid should therefore be a weak acid like $SO(OH)_2$, etc., and its pK_a should be about 3. At high pH (pH 14) the ion $HCOO^-$ *(aq)* will predominate in an aqueous solution of HCOOH. The second H is *not* removed in base. It is bonded to the C atom. Note the similarity of HCOOH to $HPO(OH)_2$ (text p. 575), which also has a non-acidic hydrogen atom bonded directly to the central atom.

---

15. The four acids to be compared are $H_2B_4O_7$, H_3BO_3, $H_5B_3O_7$ and $H_6B_4O_9$. Rewrite their formulas as, respectively:

$$B_4O_5(OH)_2 \quad B(OH)_3 \quad B_3O_2(OH)_5 \quad B_4O_3(OH)_6$$

The acid with the largest ratio of lone oxygens to B should be the strongest. This ratio is the four cases is 1.25, 0, 0.67, 0.75. The predicted order of acid strength is therefore:

$$B(OH)_3 < B_3O_2(OH)_5 < B_4O_3(OH)_6 < B_4O_5(OH)_2$$

---

17. **a)** The following are the steps in the Leblanc process for Na_2CO_3 together with the ΔH°_f of each substance in kJ mol^{-1}:

$$\begin{array}{ccccccc} 2\,NaCl(s) & + & H_2SO_4(l) & \rightarrow & Na_2SO_4(s) & + & 2\,HCl(g) \\ -411.15 & & -813.99 & & -1387.08 & & -92.31 \end{array}$$

$$\begin{array}{ccccccc} Na_2SO_4(s) & + & 2\,C(s) & \rightarrow & Na_2S(s) & + & 2CO_2(g) \\ -1387.08 & & 0 & & -373.21 & & -393.51 \end{array}$$

$$\begin{array}{ccccccc} Na_2S(s) & + & CaCO_3(s) & \rightarrow & Na_2CO_3(s) & + & CaS(s) \\ -373.21 & & -1206.92 & & -1130.68 & & -482.4 \end{array}$$

$$\begin{array}{ccccccc} CaCO_3(s) & + & C(s) & \rightarrow & CaO(s) & + & 2\,CO(g) \\ -1206.92 & & 0 & & -635.09 & & -110.52 \end{array}$$

All of the numbers come from Appendix D with the exception of the ΔH°_f of Na_2S*(s)* which does not appear in that compilation and is taken from the *Handbook of Chemistry and Physics* (1979). The heading of Appendix D mentions this source. The standard enthalpy change for each step is computed by application of the equation:

$$\Delta H^\circ = \Sigma\, \Delta H^\circ_f \text{ (products)} - \Sigma\, \Delta H^\circ_f \text{ (reactants)}$$

The answers are: ΔH° (step 1) = 64.59 kJ mol^{-1}; ΔH° (step 2) = 226.85 kJ mol^{-1}; ΔH° (step 3) = −32.95 kJ mol^{-1}; ΔH° (step 4) = 350.79 kJ mol^{-1}.

b) The sum of the first three equations in part a) is:

$$NaCl(s) + H_2SO_4(l) + 2\,C(s) + CaCO_3(s) \rightarrow 2\,HCl(g) + 2\,CO_2(g) + Na_2CO_3(s) + CaS(s)$$

This reaction summarizes the actual process leading to sodium carbonate. The fourth reaction:

$$CaCO_3(s) + C(s) \rightarrow CaO(s) + 2\,CO(g)$$

is an unavoidable side reaction that occurs between two of the reactants in the main sequence of the process.

c) The ΔH° of the main reaction is the sum of the ΔH°'s of its three steps. It is +258.49 kJ mol^{-1}. The LeBlanc process is endothermic. Heating the reactants favors the products. As actually carried out, at about 1000° C, the endothermic fourth reaction also occurs.

---

19. The production of synthesis gas is:

$$CH_4(g) + H_2O(g) \rightarrow CO(g) + 3\,H_2(g)$$

For this (and for any) reaction at constant temperature and pressure:

$$\Delta G = \Delta H - T\Delta S$$

Get ΔH° and ΔS° using the values in Appendix D. Then assume the two values are sufficiently independent of temperature to be used for ΔH and ΔS across a temperature range. The equilibrium constant of the reaction is exactly 1 when $\Delta G = 0$. Set ΔG in the above expression equal to 0 and solve for T. Following this plan gives $\Delta H^\circ = -110.52 - (-241.82) - (-74.81) = 206.11$ kJ, then $\Delta S^\circ = 197.56 + 3(130.57) - 186.15 - 188.72 = 214.4$ J K^{-1}. Substitution and the usual care with the units give:

$$\Delta G = 0 = 206.11 \text{ kJ} - T\,(0.2144 \text{ kJ K}^{-1}) \quad \textit{and} \quad T = 961.3 \text{ K}$$

---

21. The reaction of calcium hydride with water is:

$$CaH_2(s) + H_2O(l) \rightarrow Ca(OH)_2(aq) + H_2(g)$$

This key fact from the balanced equation is that 1 mol of CaH_2 produces 1 mol of H_2. The molar weight of CaH_2 is 42.096 g mol^{-1}. The 12.21 g of CaH_2 specified in the problem therefore is 0.2900 mol. Assuming that the $H_2(g)$ is ideal, its volume at P = 1 atm and T = 298.15 K (which is 25° C) is derived readily from PV = nRT with R = 0.082057 L atm mol^{-1}K^{-1}. The volume is 7.096 L.

---

23. The reaction of Na_2O and SiO_2 to give Na_2SiO_3 can indeed be considered a Lewis acid-base reaction:

$$Na_2O + SiO_2 \rightarrow Na_2SiO_3$$

Na_2O is an ionic compound and a basic oxide. The oxygen in Na_2O is in the -2 oxidation state and has a closed shell electron configuration. It donates an electron pair to the SiO_2, which accepts the pair. A new bond is formed between the two. The Lewis base is the O^{2-} and the Lewis acid is SiO_2.

---

25. The problem is reminiscent of problem 1-37. 5.00 g of Na reacts with oxygen to give some sodium peroxide, Na_2O_2, and some sodium superoxide, NaO_2. The total mass of these products is 9.25 g. Let x equal the mass of the oxygen in the Na_2O_2 and y the mass of the oxygen in the NaO_2. Then:

$$x + y = 4.25$$

Na_2O_2 has a molar weight of 77.99 g mol^{-1}. Based on its chemical formula, it is 58.96 percent Na and 41.04 percent O by weight. Similarly, NaO_2 (molar weight 54.99) is 41.81 percent Na and 58.19 percent O by weight. If there is x g of O in the Na_2O_2, then there is $(58.96 / 41.04)x$ g of Na, according to the fixed composition of the compound. If there is y g of O in the NaO_2, then there is $(41.81 / 58.19)y$ g of Na. The total weight of Na is 5.00 g. Hence:

$$(58.96 / 41.04)x + (41.81 / 58.19)y = 5.00$$

The two equations in x and y are easily solved. x is 2.71 g and y is 1.54 g. The peroxide oxygen is $(2.71 / 4.25) \times 100$ or 63.8 percent, and the superoxide oxygen is 36.2 percent.

---

27. The equations for the formation of the two copper oxides are:

$$Cu(s) + \tfrac{1}{2}\,O_2(g) \rightarrow CuO(s) \text{ (black)}$$

$$2\,Cu(s) + \tfrac{1}{2}\,O_2(g) \rightarrow Cu_2O(s) \text{ (red)}$$

The problem is a recapitulation of problem 8-17. For each reaction compute ΔH° and ΔS°, using data from Appendix D. Then apply the equation:

$$\Delta G = \Delta H^\circ - T\Delta S^\circ$$

For each reaction the required temperature range is the one that makes its ΔG negative.

From Appendix D, ΔH° of the first reaction is -157.3 kJ and ΔH° of the second is -168.6 kJ. These are just the enthalpies of formation of 1 mole each of the two oxides. ΔS° for the first reaction is $42.63 - 33.15 - \tfrac{1}{2}(205.03) = -93.03$ J K^{-1}. Similarly, ΔS° for the second reaction is $93.14 - 2(33.15) - \tfrac{1}{2}(205.03) = -75.67$ J K^{-1}. For the first reaction:

$$\Delta G = -157.3 \text{ kJ} - T(-0.09303 \text{ kJ K}^{-1})$$

Assuming that ΔH° and ΔS° are *not* dependent on the temperature, the ΔG for this reaction goes to zero at a temperature of 1691 K. At lower temperatures it is *less* than zero. For the second reaction:

$$\Delta G = -168.6 \text{ kJ} - T(-0.07567 \text{ kJ K}^{-1})$$

The ΔG of this reaction becomes zero at 2228 K. It is negative at all lower temperatures. Now, subtract the second equation from the first, inserting a subscript on the ΔG's to keep them straight:

$$\Delta G_1 - \Delta G_2 = 11.3 \text{ kJ} - T(-0.01736 \text{ kJ K}^{-1})$$

When the first reaction is favored over the second, $\Delta G_1 < \Delta G_2$ and $\Delta G_1 - \Delta G_2$ is negative. This never happens because the right-hand side of the equation for the ΔG difference is positive as long as T is positive. Therefore the second reaction is favored over the first at all temperatures.

$\Delta G_1 - \Delta G_2$ is really the ΔG of the reaction:

$$Cu_2O(s)_{red} \rightarrow Cu(s) + CuO(s)_{black}$$

which is the chemical equation that results when the equation for the formation of red copper oxide is subtracted from the equation for the formation of black copper oxide. Red $Cu_2O(s)$ is thermodynamically unstable with respect to black $CuO(s)$ plus Cu metal at all temperatures.

---

29. The four oxyacids of the transition metals can be formulated:

$$Zr(OH)_4 \qquad OMn(OH)_3 \qquad O_2Mn(OH)_2 \qquad O_3Mn(OH)$$

By comparison to Table 16-2, the $Zr(OH)_4$, having no lone oxygens on the central metal, should be "very weak" and have K_a on the order of $(10^{-4} \times 10^{-10}) = 10^{-14}$. The $OMn(OH)_3$ should be "weak" $K_a \sim 10^{-7}$. The $O_2Mn(OH)_2$ should be "strong" $K_a \sim 10^{-2}$. The $O_3Mn(OH)$ should be "very strong", having 3 lone oxygens on the Mn, $K_a \sim 10^3$.

Chapter 17

METALLURGY AND THE PROPERTIES OF METALS

17.1 *THE PRODUCTION AND REFINING OF METALS*

About three-fourths of the naturally-occurring elements are metals. Extractive metallurgy deals with the recovery of these metals from their sources in the earth.

The Production and Refining of Metals

Most metals occur in *minerals. Ores* are minerals that are economically viable sources of metals. The extraction of a metal from an ore involves three major operations:

- Concentration of the ore
- Extraction of the crude metal
- Refining

The winning of a metal from an ore requires chemical reduction, since the metal is in some positive oxidation state. *Smelting* is the heating of an ore with a reducing agent, usually coke, to *reduce* it and *melt* it. Smelting of many ores is carried out in a blast furnace.

The choice of carbon (coke) as a reducing agent in smelting metals involves some neat applications of thermodynamic principles. The ΔG°'s for the conversion of metals to their oxides are nearly all negative at ordinary conditions. This makes sense in connection with the fact that so many metals occur as oxides. Over geological time, the oxidation of the metals has had a good opportunity to progress toward equilibrium. ΔG values for the metal to metal oxide conversions are strongly dependent on temperature:

$$\Delta G = \Delta H^\circ - T\Delta S^\circ$$

The nature of the dependence is important. As a rule, the ΔG of oxide formation *becomes less negative* (increases algebraically) as the temperature increases. The rule exists because ΔS° is nearly always negative for the formation of oxides. The negative sign reflects the fact that the reactions all involve the consumption of $O_2(g)$ to make a solid. The dependence on the temperature of ΔG_f of carbon monoxide:

$$2\,C(s) + O_2(g) \rightarrow 2\,CO(g)$$

is different. This ΔG becomes *more negative* as the temperature increases. ΔS° is positive for $CO(g)$ formation because the reaction is a net producer of 1 mol of gas. **See**

problem 1. It follows that at high enough temperatures, carbon is thermodynamically capable of reducing almost all metal oxides. This thermodynamic point is the theme of **problem 29.** Moreover, carbon (from wood and coal) is cheap and widely-available.

The more electropositive metals are produced by the methods of *electrometallurgy* because of the difficulty of producing and maintaining the very high temperatures that reduction with coke would entail.

17.2 *THE ALKALI AND ALKALI EARTH METALS*

The alkali metals and alkali earth metals occupy Groups I and II in the periodic table respectively. The alkali metals have:

- Low densities and large atomic radii.

- Low ionization energies and electronegativities.

- Large *negative* standard reduction potentials. This means that they are hard to reduce but easily oxidized. They serve as reducing agents in many reactions.

- The +1 oxidation number exclusively in their compounds.

- Salts (like NaCl or Li_2SO_4, etc.) that are mostly *soluble* in water.

The *alkali earths* all exhibit exclusively the +2 oxidation number in their compounds. They are all active metals (like their neighbors in Group I) but to a substantially lesser extent.

Lithium, the leading member of the alkali metals, is the most atypical of that group. Similarly, beryllium, the first of the alkali earths, deviates the most from the general pattern of chemical and physical properties of the other alkali earths. In many ways it resembles Al more than Mg, its nearest group-mate. *In the chemistry of the groups in the periodic table, trend-breaking, atypical behavior is most prevalent with the lightest members of the groups.*

17.3 *PROPERTIES OF THE METALS*

The text reviews salient facts on the chemical and physical properties of the metals. The organizing theme is the periodic table of the elements. Productive study will concentrate on the trends in the physical and chemical properties of the metals. Refer to text Appendix F for data on the metals. Draw graphs of values of various properties versus group number or atomic number. **See problem 16.** Problems on metallurgy and the metals include:

- Examination of the *thermodynamics* of prominent processes for the recovery and refining of metals. For example, the Hall-Héroult process is the electrolytic reduction of a solution of aluminum oxide in Na_3AlF_6 *(l)*. Another electrochemical process is the *Downs* process. In **Problem 15** the operating voltage of this process is estimated. The thermodynamics of the *thermite* reaction, by which many metal oxides can be reduced, is treated in **problem 21.**

- Evaluation of the economics of recovery processes. **See problem 22.**

- Prediction of necessary conditions for the success of a proposed process involving the production of metals. **See problem 20 and 24.**

- Computation of yields of reactions using the stoichiometric principles of Chapter 1. **See problem 27.**

DETAILED SOLUTIONS TO ODD-NUMBERED PROBLEMS

1. This problem applies the ideas of Chapter 8 to the important equilibrium:

$$C(s) + CO_2(g) \rightleftarrows 2\,CO(g)$$

Using the data of Appendix D, ΔH° for the reaction is:

$$\Delta H^\circ = 2(-110.52) - (-393.51) = 172.47 \text{ kJ}$$

$$\Delta S^\circ = 2(197.56) - (213.63) - 5.74 = 175.75 \text{ J K}^{-1}$$

The reaction is endothermic and goes with an *increase* in entropy. (There are two moles of gas on the right versus only one on the left). If ΔH° and ΔS° do not depend on the temperature:

$$-RT \ln K = \Delta H^\circ - T\Delta S^\circ$$

At 298 K substitution and numerical evaluation give $K = 8.9 \times 10^{-22}$. At 2150 K substitution gives $K = 9.8 \times 10^{4}$. High temperatures favor the products in this case. Indeed, high temperatures *always* favor the products of reactions which have a positive ΔS.

---

3. The balanced equation for the reduction of pyrolusite with aluminum is:

$$3\,MnO_2(s) + 4\,Al(s) \rightarrow 3\,Mn(s) + 2\,Al_2O_3(s)$$

The ΔH° of the reaction is:

$$\Delta H^\circ = 2(-1675.7) - 3(-520.03) = -1791.31 \text{ kJ}$$

The numbers come from Appendix D. Recall that the ΔH°_f's of $Al(s)$ and $Mn(s)$ are zero. The reaction involves 3 mol of $MnO_2(s)$ so the standard enthalpy change per mole is -597.1 kJ mol^{-1}. Compare this problem to **problem 21.**

The standard free energy change per mole of $MnO_2(s)$ is calculated using the same method as for ΔH°, but taking the appropriate ΔG°_f instead of ΔH°_f values from Appendix D. It is -589.7 kJ mol^{-1}.

---

5. The details of the chemical steps used to precipitate $K^+(aq)$ as $K_2PtCl_6(s)$ are not important as long as *all* the $K^+(aq)$ is captured in the precipitate. Start with the last sentence of the problem. The molar weight of K_2PtCl_6 is 486.012 g mol^{-1}. Since K contributes 78.204 g mol^{-1} of this, the weight percentage of K in K_2PtCl_6 is 16.091 percent. The 0.2026 g of K_2PtCl_6 contains 0.03260 g of K. The clay sample weighs 0.8255 g. Dividing 0.03260 by 0.8255 and multiplying by 100 percent gives 3.949 percent K in the clay.

---

7. The sodium serves as a reducing agent to bring the Ti(IV) down in oxidation state to Ti(0):

$$TiCl_4(l) + 4\ Na(l) \rightarrow Ti(s) + 4\ NaCl(s)$$

Express the quantity of Ti (100 kg) in mol by dividing by the molar weight (which is 47.90 g mol^{-1} or 0.04790 kg mol^{-1}. The answer is 2087.68 mol. The equation shows that 4 times this chemical amount of Na(l) is consumed. This is 8350.73 mol of Na(l) or (at 22.9898 g mol^{-1} for Na) 192 kg of Na.

---

9. The equations for the reduction of iron(III) oxide with coke and air are:

$$2\ C(s) + O_2(g) \rightarrow 2\ CO(s)$$

$$Fe_2O_3(s) + 3\ CO(g) \rightarrow 2\ Fe(s) + 3\ CO_2(g)$$

In this process, CO(g) is the effective reducing agent.

The overall reaction for the formation of Fe(s) from its ore is the sum of the above two reactions adjusted so that CO(g) cancels out:

$$6\ C(s) + 3\ O_2(g) + 2\ Fe_2O_3(s) \rightarrow 4\ Fe(s) + 6\ CO_2(g)$$

Using the data in Appendix D, ΔH° for this reaction is -712.66 kJ. The tactic of combining the reactions *first* saves the labor of adding in and then subtracting out a term for ΔH_f° of CO(g). The reaction produces 4 mol of Fe(s) so the molar ΔH° is -178.16 kJ mol^{-1}. A metric ton of Fe is 10^6 g which is (10^6 g / 55.847 g mol^{-1}) or 1.791×10^4 mol. The standard enthalpy change for the production of a metric ton of Fe(s) is thus 1.791×10^4 mol $\times -178.16$ kJ mol^{-1} or -3.19×10^6 kJ.

---

11. The reduction potential diagram for manganese at pH 0 shows that manganese(III) disproportionates to $Mn^{2+}(aq)$ and $MnO_2(s)$. As $Mn_2O_3(s)$ dissolves in acid at pH 0 the reaction is:

$$Mn_2O_3(s) + 2\ H_3O^+(aq) \rightarrow MnO_2(s) + Mn^{2+}(aq) + 3\ H_2O(l)$$

b) The half-equation for the reduction of $MnO_2(s)$ to Mn(s) shows a 4 electron transfer:

$$MnO_2(s) + 4\ H_3O^+(aq) + 4\ e^- \rightarrow Mn(s) + 6\ H_2O(l)$$

The reduction potential for this half-reaction is *not* the simple sum of the reduction potentials in the three steps on the reduction potential diagram:

$$MnO_2(s) \rightarrow Mn^{3+}(aq) \qquad \mathcal{E}^\circ = 0.95\ V$$

$$Mn^{3+}(aq) \rightarrow Mn^{2+}(aq) \qquad \mathcal{E}^\circ = 1.51\ V$$

$$Mn^{2+}(aq) \rightarrow Mn(s) \qquad \mathcal{E}^\circ = -1.03\ V$$

That is, the answer is *not* 1.43 V. It is the ΔG°'s of reactions that are truly additive. The ΔG° of a half-reaction depends on $\mathcal{E}$ *multiplied by n, the number of electrons transferred.* In this case:

$$4\,\xi^\circ\,(\text{overall}) = 1(0.95) + 1(1.51) + 2(-1.03)\text{ V}$$

where the multipliers of the voltages are the numbers of electrons in the respective half-equations. The answer is 0.10 V.

---

13. Metals in higher oxidation states tend to be more covalent in their bonding than in lower oxidation states. If $PbCl_4$ is more covalent than $PbCl_2$ it should have a lower boiling point. The boiling point of $PbCl_2$ is 950° C at 1 atm pressure. The boiling point of $PbCl_4$ is less than 100° C.

---

15. This problem may present difficulties because ΔH_f° and S° of Na(*l*) and NaCl(*l*) are not in Appendix D. However, the enthalpy of fusion of Na(*s*) is tabulated in Appendix F. Remember also that only an *estimated* minimum voltage must be calculated. The reaction is:

$$NaCl_{\text{dissolved in } CaCl_2(l)} \rightarrow Na(l) + \tfrac{1}{2}\,Cl_2(g)$$

It is the sum of the three steps:

$$NaCl(s) \rightarrow Na(s) + \tfrac{1}{2}\,Cl_2(g)$$

$$Na(s) \rightarrow Na(l)$$

$$NaCl_{(\text{dissolved in } CaCl_2(l))} \rightarrow NaCl(s)$$

The three steps are the main chemical change, with reactant and products all in their standard states, followed by two adjustments which cope with the fact that neither the sodium nor the sodium chloride is solid at 600° C. The ΔG for the overall reaction is the *sum* of the ΔG's for the three steps.

Compute ΔG for the first step at 600° C (873.15 K). It is:

$$\Delta G = \Delta H - T\Delta S$$

If ΔH_{873} and ΔS_{873} are assumed to equal ΔH° and ΔS°, then:

$$\Delta G_{873} = \Delta H^\circ - 873.15\,\Delta S^\circ$$

ΔH° for the first step is just the negative of the ΔH_f° of NaCl(*s*). It is +411.15 kJ. The absolute entropy data in Appendix D allow calculation of ΔS° of the first step:

$$\Delta S^\circ = \tfrac{1}{2}\,(222.96) + 51.21 - 72.13 = 90.56\text{ J K}^{-1}$$

Substitution of ΔH° and ΔS° gives ΔG_{873} *for the first step* as +332.08 kJ.

Let ΔG_2 and ΔG_3 equal the free energy changes accompanying the fusion of Na(*s*) at 873.15 K and the solidification of NaCl from the $CaCl_2$ melt at 873.15 K, the second and third steps, respectively. Then:

$$\Delta G(\text{overall}) = 332.1\text{ kJ} + \Delta G_2 + \Delta G_3$$

Estimate ΔG_2 as follows. The temperature exceeds the melting point of sodium (97.81° C, see Appendix F) so ΔG_2 is *negative.* The enthalpy of fusion of Na*(s)* is 2.602 kJ mol^{-1} (Appendix F). The $\Delta S°$ of fusion of Na*(s)* is not in Appendix F, but can reasonably be expected to be small. If so, then ΔG_2 is a small negative number. [In fact, ΔS of fusion of Na*(s)*, according to other references, is 7.02 J mol^{-1} and the ΔG_{873} of fusion of 1 mol of sodium is 2.602 − 873.15 (0.00702) = −3.53 kJ.]

Solid NaCl is in equilibrium with NaCl dissolved in molten $CaCl_2$ at 873.15 K so $\Delta G_3 = 0$. Combine the three ΔG's:

$$\Delta G(\text{overall}) = \Delta G_1 + \Delta G_2 + \Delta G_3 = 332.1 - 3.53 + 0 = 328.6 \text{ kJ}$$

The voltage of the overall reaction is related to its free energy change:

$$\Delta G(\text{overall}) = -nF\Delta\mathcal{E}(\text{overall})$$

In this electrolysis n is 1 mol because the balanced equation involves the transfer of 1 electron. The constant F is 96485 C mol^{-1}. Substituting and rearranging:

$$\Delta\mathcal{E}(\text{overall}) = -3.40 \text{ V}$$

Because the contribution of ΔG_2 is only about 2 percent of ΔG_1 it could have been omitted in making the estimate. The estimated minimum voltage for the electrolysis is 3.4 V.

---

17. Suppose for the sake of concreteness that the strong acid HCl has acidified the lake. If the pH is 3.75 then $[H_3O^+]$ is 1.78×10^{-4} M. Before any limestone ($CaCO_3$) is added:

$$[H_3O^+] = [Cl^-] = 1.78 \times 10^{-4} \text{ M}$$

The $CaCO_3$*(s)* is added. The pH rises to 6.50, so $[H_3O^+]$ is 3.16×10^{-7} M. $CaCO_3$*(s)* reacts with acid to give Ca^{2+}*(aq)*, CO_3^{2-}*(aq)*, HCO_3^-*(aq)* and H_2CO_3*(aq)*. The following equilibria are in effect:

$$H_2O(l) + H_2CO_3(aq) \rightleftarrows HCO_3^-(aq) + H_3O^+(aq) \qquad K_{a1} = 4.3 \times 10^{-7}$$

$$H_2O(l) + HCO_3^-(aq) \rightleftarrows CO_3^{2-}(aq) + H_3O^+(aq) \qquad K_{a2} = 4.8 \times 10^{-11}$$

Substitution of $[H_3O^+] = 1.78 \times 10^{-4}$ M into the mass-action expressions based on these two equilibria gives the following two relationships at a pH of 6.50:

$$[HCO_3^-]/[H_2CO_3] = 1.36 \quad \textit{and} \quad [CO_3^{2-}]/[HCO_3^-] = 1.51 \times 10^{-4}$$

Meanwhile the principle of electrical neutrality (p. 110 of this Guide) for the lake water requires that:

$$2\,[Ca^{2+}] + [H_3O^+] = [Cl^-] + [OH^-] + [HCO_3^-] + 2\,[CO_3^{2-}]$$

The equilibrium concentration of calcium ion, $[Ca^{2+}]$, is of prime interest because it tells directly how much $CaCO_3$ has dissolved. For every $Ca^{2+}(aq)$ ion in the lake there is one carbon-containing ion in some form or other. This fact is a material balance. Expressed mathematically:

$$[Ca^{2+}] = [HCO_3^-] + [CO_3^{2-}] + [H_2CO_3]$$

The $[H_3O^+]$, $[OH^-]$ and $[Cl^-]$ are *knowns*, at 3.16×10^{-7} M, 3.16×10^{-8}, and 1.78×10^{-4} M respectively. There are thus 4 equations relating 4 unknowns. The $[CO_3^{2-}]$ is four orders of magnitude *less* than $[HCO_3^-]$. Neglect it in the electrical neutrality equation. Also neglect $[H_3O^+]$ and $[OH^-]$ in this equation as they are much smaller than $[Cl^-]$:

$$2\,[Ca^{2+}] \simeq [Cl^-] + [HCO_3^-] \simeq 1.78 \times 10^{-4} + [HCO_3^-]$$

Using the same considerations, the material balance equation becomes:

$$[Ca^{2+}] \simeq [HCO_3^-] + [H_2CO_3]$$

But $[H_2CO_3]$ is known in terms of $[HCO_3^-]$:

$$[H_2CO_3] = [HCO_3^-] / 1.36$$

Eliminate $[H_2CO_3]$ between the last two equations and combine with the electrical neutrality relationship. Solving the resulting equation shows that to make the pH equal 6.50 the concentration of $Ca^{2+}(aq)$ in the lake must be brought to 1.25×10^{-4} M. The assumptions can be easily justified. In particular, taking the acidified lake to contain HCl*(aq)* was just an easy way of working the total concentration of acid into the equations.

Knowing the dimensions of the lake allows computation of its volume. The diameter ($2\,r$) is converted from miles to meters (1 mi = 1609 m) and the average depth (h) is converted from feet to meters (1 foot = 0.3048 m). The values for r and h are substituted in the formula for the volume of a cylinder:

$$V = \pi r^2 h = \pi (2414 \text{ m})^2 \ (9.14 \text{ m}) \ = \ 1.67 \times 10^8 \text{ m}^3$$

This volume equals 1.67×10^{11} L. Raising the pH requires dissolution of 1.25×10^{-4} mol L^{-1} of $CaCO_3(s)$. One mole of $Ca^{2+}(aq)$ comes from every mole of $CaCO_3(s)$ placed in the lake (assuming no $CaCO_3(s)$ collects on the bottom). This means the presence of 2.09×10^7 mol of $Ca^{2+}(aq)$ in the lake which implies that 2.09×10^7 mol of $CaCO_3(s)$ does the job. The molar weight of $CaCO_3$ is 100 g mol^{-1}, hence 2.09×10^9 g of $CaCO_3(s)$ or 2.09×10^3 metric ton of $CaCO_3(s)$ is needed.

---

19. The reaction of Mg*(s)* with water is:

$$Mg(s) + 2\,H_2O(l) \rightarrow Mg^{2+}(aq) + 2OH^-(aq) + H_2(g)$$

To compute its ΔG° at 25° C, combine the ΔG°_f values of its products and reactants in the usual way:

$$\Delta G^\circ = \Sigma\,\Delta G^\circ_f \text{ (products)} - \Sigma\,\Delta G^\circ_f \text{ (reactants)}$$

In Appendix D, ΔG_f° of $OH^-(aq)$ is -157.24 kJ mol^{-1}; of $Mg^{2+}(aq)$ -454.8 kJ mol^{-1}; of $H_2O(l)$ -237.18 kJ mol^{-1}. The two elements, $Mg(s)$ and $H_2(g)$, are in their standard states and have ΔG_f°'s of zero. ΔG° of the reaction is:

$$\Delta G^\circ = 2(-157.24) - 454.8 - 2(-237.18) \text{ kJ} = -294.92 \text{ kJ.}$$

Using the data in Appendix E, the above reaction is constructed as the combination of the half-reactions:

$$2\,H_2O(l) + 2\,e^- \rightarrow H_2(g) + 2\,OH^-(aq) \qquad \mathcal{E}^\circ = -0.828 \text{ V}$$
$$Mg^{2+}(aq) + 2\,e^- \rightarrow Mg(s) \qquad \mathcal{E}^\circ = -2.375 \text{ V}$$

The $\Delta\mathcal{E}^\circ$ is $-0.828 - (-2.375) = 1.547$ V. The free energy charge is related to $\Delta\mathcal{E}^\circ$ by:

$$\Delta G^\circ = -n F \Delta\mathcal{E}^\circ$$

Taking F as 96,485 C mol^{-1} and n as 2 mol (of electrons transferred), ΔG° comes out to -298.52 kJ.

Although this value is about 1 percent more negative than the ΔG° from strictly calorimetric data, its message is the same. $Mg(s)$ and water react spontaneously at room conditions. In practice, this reaction is very slow because of the formation of an oxide coating that protects the $Mg(s)$ beneath it.

---

21. ***a)*** The problem recalls principles covered in Chapter 7. The thermite reaction is:

$$2\,Al(l) + Fe_2O_3(s) \rightarrow 2\,Fe(l) + Al_2O_3(s)$$

The standard enthalpy change of this reaction is:

$$\Delta H^\circ = 2\,\Delta H_f^\circ\,(Fe(l)) + (-1675.7 \text{ kJ}) - 2\,\Delta H_f^\circ\,(Al(l)) - (-824.2 \text{ kJ})$$

The two unknown quantities on the right of this equation, the enthalpies of formation of liquid Fe and Al, are in fact the two metals' enthalpies of fusion. If the two enthalpies of fusion are equal, as stated in the problem, they cancel and:

$$\Delta H^\circ = -851.5 \text{ kJ}$$

The ΔG° of the reaction is:

$$\Delta G^\circ = 2\,\Delta G_f^\circ\,(Fe(l)) + (-1582.3) - 2\,\Delta G_f^\circ\,(Al(l)) - (-742.2) \text{ kJ}$$

$$\Delta G^\circ = -840.1 \text{ kJ}$$

The cancellation of the heat capacities of the reactants and products means that ΔH° and ΔS° are independent of temperature.

b) Imagine that the reaction proceeds at constant pressure to give 2 mol of $Fe(l)$ and 1 mol of $Al_2O_3(s)$. It releases 851.5 kJ of heat. If none of this heat is lost to the surroundings, it all stays with the products which thereby attain their maximum temperature. For the system consisting of the products:

$$\Delta H = n\ C_p \Delta T$$

where:

$$n\ C_p = 1\ \text{mol} \times 124\ \text{J mol}^{-1}\text{K}^{-1} + 2\ \text{mol} \times 68\ \text{J mol}^{-1}\text{K}^{-1} = 260\ \text{J K}^{-1}$$

The products absorb the heat freed by the reaction so their ΔH is +851.5 kJ. ΔT is therefore 3275 K. If the reaction started at 298.15 K, the maximum temperature is 3573 K or 3300° C.

---

23. The equation for the reduction of $WO_3\,(s)$ to $W(s)$ with hydrogen gas is:

$$WO_3\,(s) + 3\ H_2\,(g) \rightarrow W(s) + 3\ H_2O(g)$$

The reaction is endothermic. Its $\Delta H°$, calculated from the data in Appendix D, is 117.41 kJ. Its $\Delta S°$ is 131.19 J K^{-1}, using data from the same source. High temperatures favor the products in such a case. See problem 8-17. Assuming that $\Delta H°$ and $\Delta S°$ are independent of temperature, ΔG becomes equal to 0 at the temperature fulfilling the equation:

$$\Delta G = 0 = \Delta H° - T\Delta S°$$

T comes out to 894.96 K. When ΔG is zero, K is 1 (from the relationship: $\Delta G = -RT \ln K$). The answer is 895 K.

---

25. In fire-assaying, the gold and silver-containing ore is heated with lead and borax. Some of the lead is oxidized:

$$Pb(s) + \tfrac{1}{2}\ O_2\,(g) \rightarrow PbO(s)$$

and the product reacts with the borax:

$$Na_2B_4O_7 \cdot 10\ H_2O + PbO(s) \rightarrow Pb(BO_2)_2\,(l) + 2\ NaBO_2 + 10\ H_2O(g)$$

The $Pb(BO_2)_2\,(l)$ dissolves impurities from the ore; the Au and Ag form an alloy with Pb and sink to the bottom of the vessel. After cooling, the solid alloy is separated and then re-melted. The molten lead reacts:

$$Pb(l) + \tfrac{1}{2}\ O_2\,(g) \rightarrow PbO(s).$$

and the product is removed into the walls of the clay crucible. After another cooling, the mixture of Au*(s)* and Ag*(s)* is treated with nitric acid. Only the Ag*(s)* reacts:

$$3\ Ag(s) + 4\ H_3O^+\,(aq) + NO_3^-\,(aq) \rightarrow 3\ Ag^+\,(aq) + NO(g) + 6\ H_2O(l)$$

---

27. Air containing some CO*(g)* passes through 100.0 cm^3 of 0.0100 M $PdCl_2$. For every mole of CO*(g)* present in the air one mole of $Pd^{2+}\,(aq)$ is removed from solution. After the passage of the air, addition of dimethylglyoxime anion ($DMGH^-$) precipitates 0.2850 g of $Pd(DMGH)_2\,(s)$.

The molar weight of this product is $(2 \times 115.1) + 106.4 = 336.6$ g mol^{-1}. (The 115.1 g mol^{-1} is the molar weight of $DMGH^-$. It is 1.01 g mol^{-1} *less* than the molecular weight of $DMGH_2$.) The chemical amount of solid $Pd(DMGH)_2$ is its mass divided by its molar weight or 0.8466×10^{-3} mol. Based on its volume (0.100 L) times its concentration, the original $PdCl_2$ solution contained 1.00×10^{-3} mol of Pd.

The difference between 1.00×10^{-3} mol and 0.8466×10^{-3} mol is the Pd removed from solution by the action of the CO*(g)*. It is 0.1534×10^{-3} mol. One mole of Pd is removed for every 1 mol of CO*(g)* passed into the solution. Therefore, 0.1534×10^{-3} mol of CO*(g)* reacted.

This chemical amount of CO*(g)* came from 1000 ft^3 of air. There are 35.31 ft^3 in a m^3 and 1000 L in a m^3. Combination of these facts shows that the 1000 ft^3 of air is 2.83×10^4 L. The best way to verify this is to use dimensional analysis. The concentration of CO*(g)* in the air is:

$$[CO(g)] = 0.1534 \times 10^{-3} \text{ mol} / 2.83 \times 10^4 \text{ L} = 5.42 \times 10^{-9} \text{ mol L}^{-1}$$

---

29. ***a)*** In the graphs, the function:

$$\Delta G = \Delta H^\circ - T\Delta S^\circ$$

should be plotted. T increases to the right and ΔG increases upward. The slopes of the lines are $-\Delta S^\circ$. The lines have *positive* slopes since ΔS° is negative for all the of reactions *with the exception* of the reaction for the conversion of C*(s)* to CO*(g)*, which has a positive ΔS° and therefore has a graph with a negative slope.

b) At all temperatures for which the $2\ C(s) + O_2(g) \rightarrow CO(g)$ line is *below* the line for the conversion of a metal to its oxide, C*(s)* will successfully reduce the metal oxide. In these cases, the C*(s)* gains oxygen with a more negative free energy change than the metal. Hence C*(s)* wrests oxygen away from the metal oxide.

c) Al_2O_3 cannot be reduced to aluminum metal by C*(s)* unless the temperature exceeds 2500 K.

---

31. Metals with the most positive reduction potentials (like gold, silver and copper) are sometimes found in their elemental states in nature. Metals with somewhat smaller reduction potentials (like Fe, Ni, V, Cr) can be reduced from their oxides with coke or carbon monoxide. The most easily oxidized metals are locked most tightly with oxygen in their ores and require electrolytic means of reduction. (Al, Mg, Na, K).

Chapter 18

COORDINATION COMPLEXES

Coordination complexes contain transition metal atoms (the *central metal*) bonded to a small number of surrounding ions or neutral molecules (the *ligands*). The bonding can be thought of in acid-base terms. The ligands have unshared electron pairs which they donate, acting as Lewis bases. The metal atoms have vacant *d* orbitals (as well as vacant *s* and *p* orbitals) and act as Lewis acids, accepting electron pairs. The resulting *coordinate* bonds are intermediate in character between ionic and covalent. The bonded ligands are in the *coordination sphere* of the central metal.

The properties of the ligands (various ions and small molecules) and of the central metal are both altered by coordination. Coordination complexes persist in aqueous solution. The solutions have characteristic conductivities, colors, magnetism, and reactivities which are not the simple sums of the conductivities, colors, magnetism and reactivities of the parts. **See problem 1.** For example, coordinated Cl^- ion is often *not* precipitated by Ag^+ *(aq)*. Instead it is held tightly in the coordination sphere. **See problem 25.**

The overall charge on a coordination complex is the algebraic sum of the charges of its coordinated parts. Ligands are often neutral (*e.g.* H_2O, NH_3, CO) or negatively charged (Cl^-, CN^-, NCO^-, etc.). Thus, $K_2[PtCl_6]$ contains the $[PtCl_6]^{2-}$ complex ion which consists of a Pt(IV) and four Cl^- ions. Coordination complexes are often written enclosed in brackets.

18.1 *STRUCTURE OF COORDINATION COMPLEXES*

The number of bonds formed by the central atom to its surrounding ligands is its *coordination number* (*CN*). The geometrical structure of coordination complexes can be predicted from their coordination numbers using VSEPR theory with some modifications.

- Coordination numbers of 4 and 6 are more common than any others. When CN is 2, complexes are linear. Other CN's (3, 5 and 7 or more) also occur.

- All six-coordinated complexes have octahedral structures.

- Four-coordinated complexes are square-planar or tetrahedral. Tetrahedral coordination is the general rule for CN 4 except the square planar geometry is predominant for complexes of Pd^{2+} and Pt^{2+}. These metals have a d^8 valence electron configuration.

In reading the structural formulas of coordination complexes, remember that the coordination involves the central metal. Lines drawn from ligand to ligand (as in Figure 18-1, text p. 618 and Figure 18-4, text p. 620) are *not* bonds but merely suggest the geometry of the coordination sphere. In text Figure 18-1, there are exactly 6 coordinate bonds and 18 N−H covalent bonds within the 6 ligands. There are no ligand to ligand bonds.

Chelation

Many ligands occupy only one position in the coordination sphere. When such *mondentate* ligands coordinate, the number of ligands and the coordination number are the same. If a ligand has more than one *donor site,* then it can attach to the central metal in more than one place, provided that it can span the distance between the two points. Ligands which have more than one donor are *polydentate* ligands. Thus, $H_2\ddot{N}CH_2CH_2\ddot{N}H_2$ (ethylenediamine, "en") is a *bidentate* ligand. The dots represent electron pairs ready for donation by the two nitrogen atoms. Coordination complexes in which a polydentate ligand is coordinated at two or more donor sites are *chelates.* Chelates are more stable than non-chelate complexes with the same kind of donor atom. **See problem 37.**

Isomerism

Isomers are substances having the same number and kinds of atoms but arranged differently. Coordination complexes display many types of isomerism. In *geometrical isomerism,* the ligands are arrayed in different relative positions about the central metal atom. Two ligands which are near to each other, on the same side of a metal atom, are *cis.* Two ligands which are far apart (on opposite sides of a metal atom) are *trans.* In square planar complexes, geometrical isomerism is possible when the formula is Ma_2b_2 or Ma_2bc where a, b, and c are different ligands. In octahedral complexes, any given site (O in the following figure) has one site that is *trans* to it (labeled T) and four equivalent sites that are *cis* to it (labeled C). The C sites are equivalent although they do look different as usually sketched:

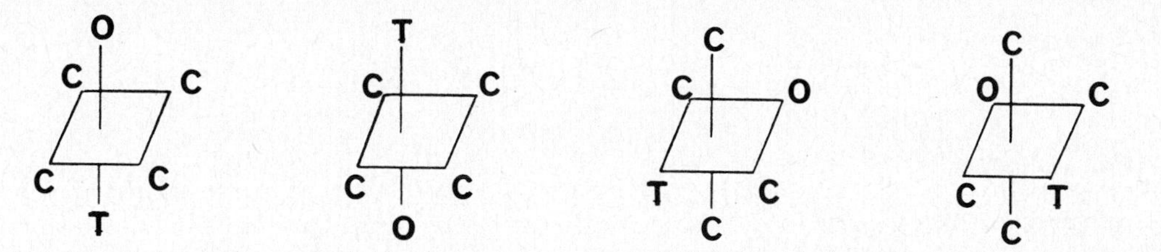

O = original site; C = *cis* position; T = *trans* position

In *optical isomerism* the difference is more subtle. The isomers differ by being mirror images of each other. Non-superimposable mirror image structures are *enantiomers.* Enantiomers have the same relationship to each other as the left and right hand. If the source of optical isomerism is the three-dimensional distribution of other, bonded atoms about a single atom, then that atom is a *chiral* center. Note that the Co(III)-EDTA complex drawn in Figure 18-6, text p. 621, has an enantiomer. The central Co atom is the chiral center.

Example. Draw the enantiomer of the Co(III)-EDTA structure in Figure 18-6, text p. 621.

Solution. Think of the chelate formed by the hexadentate ligand EDTA with Co(II) as having a backbone connecting the two N donors and four arms connecting the N donors to the four O donors. The mirror image of the structure in Figure 18-6 has one chelate arm leading from the front nitrogen atom to the *top* of the octahedron (instead of the bottom), and one chelate arm leading from the back nitrogen to the *bottom* of the octahedron (instead of the top). These two arms wrap around the Co(III) in opposite senses in the two enantiomers. The other two chelate arms keep the same relative arrangement. The wrapping of the chelate arms defines a twist or *chirality* about the metal center. The two enantiomers appear in further down this page.

In the tetrahedral geometry, optical isomerism with the metal as chiral center is possible only if there are four different ligands. In the square planar geometry, optical isomerism with the central metal as the chiral center is not possible.

A common task is to sketch the geometry of a complex, including all isomers, given its formula. Both optical and geometrical isomerism occur in **problem 3.** Attack such problems in the following way:

1. Start by determining the CN of the metal and the geometry of the complex.

2. Sketch a starting-point structure.

3. Test for geometrical isomerism by moving ligands around while systematically checking pair-wise relationships of all the ligands. Do the variant structures have the same *cis* and *trans* relationships as in the original sketch? If so, then the new structure is not a geometrical isomer of the original. If not, then it is.

4. Check for optical isomerism by drawing a vertical line next to the original sketch and using it as a mirror to create a mirror image of the original. Remember that atoms *near* the mirror line are still near it in the reflection on its other side. Atoms far from the mirror remain far from it in the reflection. This is illustrated for the Co-EDTA complex of text Figure 18-6 in the following sketches. The structure on the left is the starting structure:

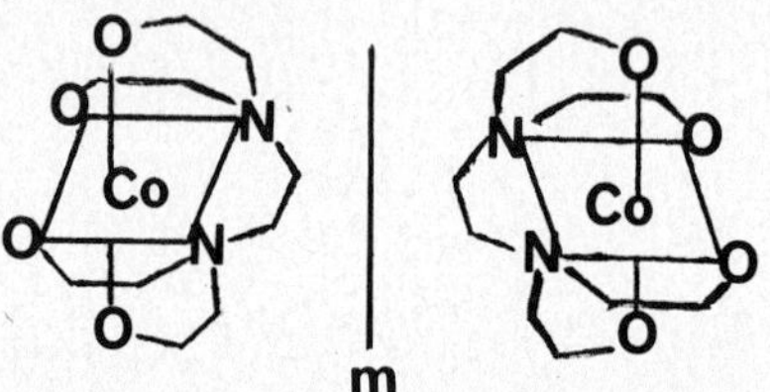

An effective alternative tactic is to turn the original drawing over and trace it through the back of the paper. The result is the mirror image of the original.

5. Check whether the new structure is superimposable on the original. Rotate it so as to put as many ligands as possible in the same positions as in the original sketch. This requires some practice in three-dimensional visualization. The best beginning approach is to perform the rotation by drawing several helper sketches each showing a portion of the total rotation. In the following figure, the structure on the right above is rotated 180° about the vertical axis so that it is as close as possible to being superimposable on the original structure:

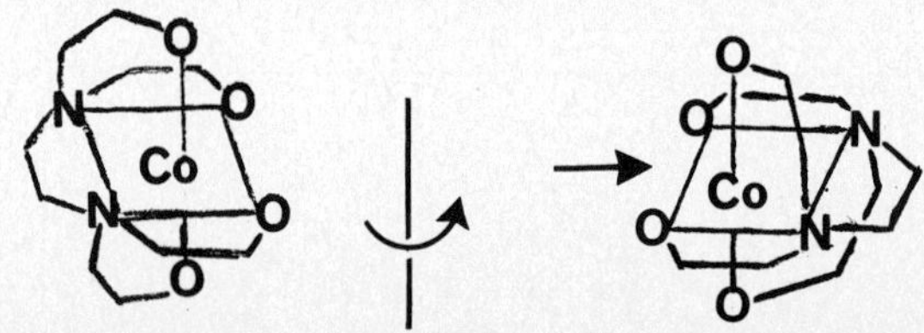

If the mirror image structure cannot be superimposed upon the original no matter how it is rotated, then the two are genuine enantiomers.

Nomenclature for Coordination Compounds

1. The cation is named before the anion.

2. Within the complex ion, the ligands are named first, followed by the metal ion. If more than one type of ligand is present, negatively charged ligands are listed first, followed by neutral ones. Within these categories ligands are listed in alphabetical order.

3. The names of anionic ligands end with the letter *o*, whereas neutral ligands are usually called by the names of the molecules. The exceptions are H_2O (aquo), CO (carbonyl), and NH_3 (ammine).

4. When several ligands of a particular kind are present, use the Greek prefixes di-, tri-, tetra-, penta-, and hexa-. Thus the ligands in $[Co(NH_3)_4Cl_2]^+$ are designated "dichlorotetraammine." If the ligand itself contains a Greek prefix, use the prefixes *bis* (2), *tris* (3), *tetrakis* (4) to indicate the number of ligands present. The ligand ethylenediamine already contains the term di; therefore "bis(ethylenediamine)" is used to indicate two ethylenediamine ligands.

5. If the complex is an anion, attach the suffix "-ate" to the name of the metal. Unfamiliar terms like "zincate", "cobaltate", "rhodate" result.

6. The oxidation number of the metal is written in Roman numerals following the name of the metal. Thus $Fe(CN)_6^{3-}$ is named hexacyanoferrate(III) ion.

Example. A coordination complex has the formula $[Co(NH_3)_5Cl]Cl_2$. Which atom is the central atom, what is the charge on the complex ion, what is the oxidation number of the central atom, and what is the name of the compound?

Solution. The transition metal *cobalt* is the central atom and the ligands are ammonia and the chloride ion. Since two chloride ions (charge of -1 each) are needed to balance its charge, the complex ion (in brackets) must have a charge of $+2$. Cobalt is in the $+3$ oxidation state because there are three -1 chlorides in the formula to be countered electrically, and the ammonia molecules are neutral. The name of the compound is chloropentaamminecobalt(III) chloride.

Familiarity with nomenclature is tested in two ways: by giving names and asking for formulas and by giving formulas and asking for names. **See problems 5 and 6.**

18.2 *BONDING IN COORDINATION COMPLEXES*

A successful bonding theory must explain: *a)* the *geometry* of coordination complexes; *b)* the often striking *colors* of the complexes; *c)* the variation in the *magnetism* of the complexes (paramagnetism, caused by unpaired electrons, *versus* diamagnetism, when all electrons are paired).

The text presents two bonding theories. They are *crystal field theory* and *ligand field theory.* The "field" in these names refers to the electrostatic influence exerted by the ligands upon the central metal.

Crystal Field Theory

In crystal field theory the bonding between metal and ligands is modelled as *ionic.* The ligands are approximated by point charges situated at the proper distance from and in the proper geometry around the central metal. The ligands exert a negative electrostatic field which perturbs the *d* orbitals of the central metal. If the symmetry of the ligands is *octahedral,* then the metal's 5 *nd* orbitals are split into two groups, the t_{2g} and e_g The t_{2g} contains *three* of the *d* orbitals (*t* for triple) and is lower in energy than the two orbitals of the e_g level. The degree of splitting is the *crystal field splitting* and is symbolized Δ_o where the subscript refers to the octahedral field.

Splitting occurs because the 5 *nd* orbitals have different spatial distributions. Consult **problem 27,** which tests understanding of the reasons for splitting among the *d* orbitals by asking about what happens to the *p* orbitals in fields of various symmetries. An octahedral field gives one kind of splitting pattern. If the symmetry of the crystal field is tetrahedral or square planar, instead of octahedral, then characteristic *different* patterns of splitting occur (see Figure 18-14, text p. 632). Figures depicting these patterns are *crystal field splitting diagrams.* The magnitudes of the crystal field splittings are symbolized Δ_t (subscript t for tetrahedral) and Δ_1, Δ_2 and Δ_3 (for square planar). A square planar field creates three intervals of splitting because it is inherently less symmetrical than the tetrahedral or octahedral fields. A completely asymmetric arrangement of ligands would split the 5 *d* orbitals to 5 different energies.

The magnitude of the splitting depends on the strength of the perturbing crystal field which in turn depends on the identity of the ligands. The magnitude of the splitting varies considerably from ligand to ligand. It is on the order a few hundred kilojoules per mole.

- *The crystal field splittings are on the general order of the strength of the chemical bond.*

The many colors exhibited by coordination complexes are interpretable in terms of this model. *Strong field* ligands split the *d* orbitals far apart. Their Δ is big. It requires *more* energy to excite a *d* electron from the low-lying set of *d* orbitals to an orbital of higher energy. Complexes with strong field ligands therefore absorb light more toward the blue end of the visible spectrum. Complexes with weak-field ligands absorb more toward the red (low frequency) end of the spectrum. The *spectrochemical series* is an empirical ordering of common ligands according to the strength of the field which they exert:

weak-field $I^- < Br^- < Cl^- < F^- < OH^- < H_2O < NH_3 < en < CO \simeq CN^-$ strong-field

To understand the visible spectra of coordination complexes, the colors of the spectrum should be memorized. From low-frequency to high-frequency they are:

Red Orange Yellow Green Blue Indigo Violet

The initials spell the mnemonic ROY G BIV. Figure 18-8b (text p. 625) bends this array of colors into a circle, the color wheel. Where violet and indigo overlap with red the color wheel has the color purple.

- When a compound absorbs light of a given color, it removes those frequencies from the spectrum. The color perceived is the given color's *complement*, the opposite color on the color wheel. This idea plays a major role in **problems 7, 9 and 15.**

The splitting of the metals d orbitals in the crystal field gives rise to two possible types of d-electron configurations. *Low-spin* configurations occur when Δ is large and the d electrons remain paired in the low-lying d orbitals. High-spin configurations occur when Δ is small and the d electrons remain unpaired because the energy it would require to pair them exceeds Δ, the energy required to push them up to occupy the higher set of d orbitals. The correspondence in terminology is:

strong-field...low-spin *versus* weak-field...high-spin

The magnetic properties of complexes depend on the number of unpaired electrons and are successfully predicted by crystal field theory.

Ligand Field Theory

In this theory the ligands are no longer regarded as point negative charges exerting a purely electrostatic field as they surround the central metal. Instead, the ligands are allowed to have orbitals. Molecular orbitals are constructed from the metal valence orbitals and the ligand orbitals. Orbital correlation diagrams which are similar to, but more elaborate than, the correlation diagrams of Chapter 13 result. Note that:

- The ligand field theory is intrinsically more realistic. It adds to crystal field theory an understanding of the *variations* in field strength of the ligands.

- The correlation diagrams of ligand field theory *include*, as a component, the crystal field splitting diagrams previously discussed. Hence, they do not contradict but include and expand upon the previous theory.

- The ligands may, if they have properly arrayed orbitals, receive electrons from as well as give electrons to the metals. This is π *back-bonding* and is associated with an increase in the field strength exerted by the ligand. Common ligands that can back-bond are CN^- and CO.

Organometallic Compounds

Organometallic compounds have bonds between metal and carbon atoms. The carbon monoxide (:C≡O:) molecule has a lone pair at both ends. It bonds through its carbon end to many metals. It donates σ electron probability density to the metal and accepts electron probability density back into its empty π^* (antibonding) orbitals.

Special stability occurs among organometallics when the central metal is surrounded by 18 valence electrons. **See problem 13.** Attaining 18 electrons corresponds to attaining a closed shell electron configuration for the metal. The rule of 18 is thus conceptually like the rule of 8 (the octet rule) and similarly subject to being broken.

18.3 *LIGAND EXCHANGE REACTIONS*

Ligand exchange is the substitution of one ligand for another. A complex whose ligands exchange slowly with other ions or molecules is *inert.* When exchange is rapid the complex is *labile.* The two terms refer to *kinetic stability.* Thermodynamic stability is a different matter. A complex that is thermodynamically unstable with respect to exchange of its ligands (with, for example a solvent) can still last for years in solution because of its kinetic inertness.

All problems involving complex-ion equilibria treat labile complexes. It would be pointless to try to apply equilibrium principles to solutions of inert complexes because the solutions take so long to come to equilibrium.

Complex-Ion Equilibria

A complex will lose ligands one by one just as molecules of polyprotic acids (such as H_3PO_4) in water solution lose their protons one by one. If a metal ion has four ligands surrounding it, then the metal-ligand complex may lose the four in a series of four consecutive reactions. Each of these is an equilibrium and has an equilibrium constant. The constants are K_{d1}, K_{d2}, etc. where the *d* stands for dissociation.

- *Complex-ion equilibria are fundamentally no different than any other equilibria in aqueous solution.*

The product, $K_{d1} \times K_{d2} \times K_{d3} \times K_{d4}$, is the *instability constant* of the complex ion. It is written K_{inst}. Taking the product of the *K*'s corresponds to *adding up* the four step-wise equilibria.

Consider the dissociation of the complex $Cd(NH_3)_4{}^{2+}$. The overall dissociation is:

$$Cd(NH_3)_4{}^{2+}(aq) \rightleftarrows Cd^{2+}(aq) + 4\ NH_3(aq)$$

K_{inst} is 3×10^{-7}. This value comes from Table 18-3, text p. 637. It is true at equilibrium that:

$$3 \times 10^{-7} = [Cd^{2+}]\,[NH_3]^4 \,/\, [Cd(NH_3)_4{}^{2+}]$$

It is *not* true that $[NH_3] = 4\,[Cd^{2+}]$. Much of the total cadmium concentration in the solution at equilibrium is tied up in the form of the $Cd(NH_3)_3^{2+}$ and the other intermediate complexes. The Cd^{2+} *(aq)* concentration is much less than ¼ the concentration of ammonia. Writing false relationships like the above is a major source of error in solving problems involving complex ion equilibria. The wisest policy is always to use the principles of electrical neutrality and material balance (p. 110 of this Guide) to generate needed relationships among concentrations.

There is another source of difficulty. In text Table 18-3, the successive K_d values for many complex ions are close together in magnitude. For example, for the $Cd(NH_3)_4^{2+}$ complex, K_{d1} is 0.16 and K_{d2} is 0.05. Calculations for complex ion dissociations often become tedious because it is necessary to treat two or more stages of the equilibrium simultaneously. This is rarely the case with the ionization of polyprotic acids (Chapter 6).

Fortunately, there is a common special case in practical problems. *If the ligand is present in excess and the K_d's are small, then:*

1. Most of the metal ion will be tied up in the complex with the highest coordination number.

2. The concentration of the free ligand will approximately equal the concentration of *excess* ligand that is present.

These are exactly the conditions that prevail in text Example 18-3. They also are the conditions in **problem 39.** When these conditions do *not* prevail, the calculations become complicated. If there is uncertainty about what assumptions might be valid for a given system, then set up the problem in complete terms, using the principles of electrical neutrality of the solution and material balance for the different species present. See p. 110 of this Guide. Then apply chemical and mathematical insight to find out what terms (if any) can be neglected. **See problem 6-41** for an example.

The formation of complex ions can exert a significant effect on the solubilities of salts. In such cases, the complex-ion equilibria run simultaneously with the K_{sp} equilibria that govern the solubility of the salt. It is useful to think of these reactions in terms of a competition among different species in binding to the metal ion. In **problem 17,** Ag^+ ion is distributed among three competitive forms, $AgCl$*(s)*, $AgCl_2^-$ *(aq)* and Ag^+ *(aq)*. Factors determining the outcome of the competition are the intrinsic stability of the different forms (as conveyed in the equilibrium constants for their formation or dissociation) and the relative concentrations of the Cl^- *(aq)* and Ag^+ *(aq)*. **Problem 21** can be analyzed in the same terms.

Hydrolysis and Amphoterism of Complex Ions

The aquo and hydroxo complexes of the metal ions are of particular interest because they furnish species that affect the pH of their solution directly. A metal ion that is six-coordinated with water has 12 protons that could in theory be donated if a strong enough base were present to accept them. In fact, a typical aquo complex of a metal ion is a weak acid. The values of K_{a1}, K_{a2}, etc. for its function as an acid are related to the dissociation constants of its *hydroxo* complexes containing, respectively, 1, 2, etc. OH^- ligands in place of coordinated water molecules. The relationship is:

$$K_d = K_w / K_a$$

This relationship is covered in detail in **problem 41b.** It is also the basis for **problem 19.**

18.4 *A MICROSCOPIC VIEW OF HYDRATION*

Solubilities of Ionic Solids

The free energy change of solution of ionic solids in water is the sum of enthalpy and entropy effects. The entropy change of solution of ionic solids can be either positive or negative. A negative entropy change of solvation means that dissolution of a solid leads to a situation of *lower* entropy. This result is contrary to common expectation based on considerations of randomness. It occurs because some ions *order* the structure of the solvent considerably when solvated.

The ΔH of solution is the difference between two large numbers, the ΔH of solvation of the free ions and the lattice energy. It too can be either positive or negative. The overall trends of ΔG of solution values parallel the trends in the ΔH values.

Hydrated Cations

Hydrated cations are categorized in two types.

- *First Type.* The cations have definite coordination numbers and are linked to the water molecules which surround them by bonds that have significant covalent character. The aquated cation is a distinct coordination complex and can (usually) be isolated in a solid by evaporation of the solvent or precipitation with some anion.

- *Second Type.* The cations have indefinite coordinate numbers and are only weakly complexed by water molecules in bonds with little covalent character. Evaporation of aqueous solutions gives precipitated solids in which the water is entirely lost.

DETAILED SOLUTIONS TO ODD-NUMBERED PROBLEMS

1. The four substances all dissolve in water to make 0.05 M solutions. Chemical equations for the dissolutions are:

$$Co(NH_3)_6Cl_3\,(s) \rightarrow 3Cl^-\,(aq) + [Co(NH_3)_6]^{3+}\,(aq)$$

$$K_2PtCl_6\,(s) \rightarrow 2\,K^+\,(aq) + [PtCl_6]^{2-}$$

$$KNO_3\,(s) \rightarrow K^+\,(aq) + NO_3^-\,(aq)$$

$$Cu(NH_3)_2Cl_2\,(s) \rightarrow Cu(NH_3)_2Cl_2\,(aq)$$

The more ions per mole of solute then the greater the conductivity of the solution. The order of increasing conductivity is therefore the same as the order in which the dissolution reactions have been written.

---

3. There are *three* isomeric $Fe(en)_2Cl_2^+$ complexes. They are the *trans*-dichlorobis(ethylenediamine)iron(III) ion and the two mirror image *cis*-dichlorobis(ethylenediamine)iron(III) ions. All involve octahedral coordination about the Fe(III). The en is a bidentate ligand, coordinating through its two $-NH_2$ group. It can span an edge of the Fe(III) octahedron but not opposite corners.

---

5. ***a)*** $Na_2[Zn(OH)_4]$ ***b)*** $[Co(H_2NCH_2CH_2NH_2)_2Cl_2]NO_3$
c) $[PtBr(H_2O)_3]Cl$ ***d)*** $[Pt(NO_2)_2]Br_2$
e) $Ag_4[Fe(CN)_6]$ ***f)*** $K_2[Co(NCS)_4]$

All answers involve the use of the rules of nomenclature of coordination compounds. Note that the subscript 4 on the silver in example **e)** occurs because there are 6 cyanide ions at -1 each and one iron at $+2$, and electrical neutrality is maintained with 4 Ag^+ ions.

---

7. Because a solution of $[Fe(CN)_6]^{3-}$ transmits red light it absorbs light in the green portion of the spectrum. See Figure 18-8b (text p. 625). This is light of *shorter* wavelength and *higher* frequency than red light. It means that Δ_0 is relatively large in $[Fe(CN)_6]^{3-}$ because the crystal field splitting is proportional to the frequency of the absorbed light. According to Figure 12-2 (text p. 403) or Figure 18-8a, absorption of green light corresponds to an electronic absorption at a wavelength of about 5.1×10^{-7} m. Using the relationship:

$$E = hc/\lambda$$

with values of Planck's constant and the speed of light in the proper units gives an energy of 3.9×10^{-19} J or 2.3×10^5 J mol^{-1}.

---

9. ***a)*** Since chromium(III) nitrate is blue in aqueous solution, it must absorb light at the red end (longer wavelength, lower frequency end) of the spectrum. Reference to Figure 18-8 suggests the compound absorbs at about 5.5×10^{-7} m. The experimentally observed transition is at 5.75×10^{-7} m.

b) The wavelength of maximum absorption will *decrease.* The CN^- group is higher up the spectrochemical series than H_2O. The octahedral splitting parameter Δ_0 will be larger. Hence *d* to *d* electronic transitions require a higher energy and involve *shorter* wavelength light.

---

11. ***a)*** In Mn^{2+} there are five $3d$ electrons. In a strong octahedral field 4 of them are paired and 1 is unpaired. In a weak field all 5 are unpaired.

b) Zn^{2+} has 10 $3d$ electrons. They are always paired irrespective of the strength of the octahedral crystal field.

c) Cr^{3+} has 3 $3d$ electrons. They always remained unpaired (in the t_{2g} level) no matter what the strength of the octahedral crystal field.

d) In Mn^{3+} there are 4 $3d$ electrons. In a strong octahedral field, the low-spin case, there are 2 unpaired electron spins. In a weak field there are 4 unpaired spins.

e) Fe^{2+} has 6 $3d$ electrons. All are paired in a strong octahedral field; 4 are unpaired in a weak field.

---

13. The central atom provides valence electrons according to its position in the periodic table. Ligands, with the exception of H, donate a pair of electrons. H can donate only a single electron. Thus, $Cr(CO)_4$ has 14 valence electrons about the central chromium, eight from the 4 CO's and 6 from the Cr itself.

Similarly, $Os(CO)_5$ has a total of 18 valence electrons about the metal, 8 from the metal and 10 from the ligands. $H_2Fe(CO)_4$ also has 18 valence electrons about the metal, counting 8 from the metal, 8 from the CO's and 2 from the H's.

$K_3[Fe(CN)_5CO]$ contains the $[Fe(CN)_5CO]^{3-}$ ion. Regard this complex ion as composed of Fe^{2+}, with 6 valence electrons, interacting with 6 ligands contributing a total of 12 valence electrons. The Fe atom sees 18 valence electrons.

In $HMn(CO)_5$ the Mn is surrounded by $1 + 7 + 10 = 18$ valence electrons. In $V(CO)_6$, vanadium sees 17 valence electrons. Regard $NaCo(CO)_4$ as the Na^+ salt of $Co(CO)_4^-$ ion. The electron count is then 9 (for Co) + 8 (for 4 CO's) + 1 = 18. $HTc(CO)_5$ is just like $HMn(CO)_5$. The technicium atom sees 18 valence electrons.

---

15. ***a)*** In the spectrochemical series F^- exerts a weaker field than H_2O. CN^- on the other hand exerts a stronger field. In aqueous $Fe(NO_3)_3$ the Fe^{3+} ion is coordinated to weak-field ligands: $Fe(H_2O)_6{}^{3+}$. The electronic absorptions involving *d-d* transitions on the Fe^{3+} are at low energy. They should be at even lower energies in $FeF_6{}^{3-}$ ion. That ion therefore should be very faintly colored, too.

b) In K_2HgI_4 the central Hg^{2+} is in a tetrahedral crystal field. The I^- ligand exerts only a weak field. More importantly, all of the *d* orbitals on the Hg^{2+} are filled. There are no *d-d* transitions. The solution of K_2HgI_4 should be colorless.

---

17. The problem really asks for a calculation of the solubility of AgCl in 1.0 M NaCl solution. Let *S* equal this solubility. For every mol L^{-1} of AgCl that dissolves either an Ag^+ or a $AgCl_2{}^-$ (dichloroargenate(I)) ion forms. Obviously, this assumes that all of the dissolved silver is present as one of two ions. In mathematical form:

$$S = [Ag^+] + [AgCl_2{}^-]$$

The solubility product constant equilibrium for AgCl assures that:

$$K_{sp} = 1.6 \times 10^{-10} = [Ag^+][Cl^-]$$

as long as any solid silver chloride is present in the system. The dissociation equilibrium of the $AgCl_2{}^-$ complex ion leads to the mass action expression:

$$6 \times 10^{-6} = [Ag^+][Cl^-]^2 / [AgCl_2{}^-]$$

Substitute these equations in the expression for *S:*

$$S = 1.6 \times 10^{-10} / [Cl^-] + 1.6 \times 10^{-10} [Cl^-] / 6 \times 10^{-6}$$

Assume that the concentration of Cl^- is so large at 1.00 M that it is not substantially reduced by reaction with Ag^+. Then $[Cl^-] \simeq 1$ and:

$$S = 1.6 \times 10^{-10} + 1.6 \times 10^{-10} / 6 \times 10^{-6} \simeq 3 \times 10^{-5} \text{ M}$$

The assumption that $[Cl^-]$ is only negligibly reduced is immediately vindicated by this low solubility. Only about 6×10^{-5} M of Cl^- is tied up in the complex. The solubility of AgCl in pure water is 1.3×10^{-5} M. Hence, AgCl*(s)* dissolves to a *greater* extent in 1.0 M NaCl than in pure water.

In 0.100 M NaCl the solubility becomes:

$$S = 1.6 \times 10^{-10} / [Cl^-] + 1.6 \times 10^{-10} [Cl^-] / 6 \times 10^{-6}$$

$$S = 1.6 \times 10^{-9} + 2.7 \times 10^{-6} = 3 \times 10^{-6} \text{ M}$$

This is *less* than the solubility of AgCl*(s)* in pure water.

---

19. The hard part of the problem is writing appropriate equilibria for the hydrolysis of the metal ions and selecting the correct constants from text Table 18-3. The rest is a routine weak acid ionization computation.

a) The Mn^{2+} from the dissolved $Mn(NO_3)_2$ is hydrated. This hydrated cation acts as an acid:

$$Mn^{2+}(aq) + 2\,H_2O(l) \rightleftarrows Mn(OH)^{+}(aq) + H_3O^{+}(aq)$$

The K_a for this reaction is K_w divided by K_{d1} for the $Mn(OH)^{+}(aq)$ ion. From text Table 18-3, K_{d1} is 1.3×10^{-4} so K_a is 7.7×10^{-11}. Set up a routine acid dissociation relationship for the 0.10 M solution and solve the resulting quadratic:

$$[H_3O^{+}]^2 + 7.7 \times 10^{-11}[H_3O^{+}] - 7.7 \times 10^{-12} = 0$$

$$[H_3O^{+}] = 2.77 \times 10^{-6}\ M \quad \textit{and} \quad pH = 5.6$$

b) The proper chemical equation is:

$$Cu^{2+}(aq) + 2\,H_2O(l) \rightleftarrows Cu(OH)^{+}(aq) + H_3O^{+}(aq)$$

The K_a for this reaction is K_w divided by K_{d4} for $Cu(OH)_4{}^{2-}(aq)$. K_a is 1.0×10^{-7}. Routine solution of the weak acid ionization gives $[H_3O^{+}] = 1.00 \times 10^{-4}$ M, pH = 4.0. A careful solver of this problem notes that K_{d3}, for the $Cu(OH)_3{}^{-}(aq)$ ion, is 2×10^{-7} (rather close to 1×10^{-7}), and verifies that the role of a second acid ionization is negligible in determining the pH.

c) The proper chemical equation is:

$$Cr^{3+}(aq) + 2H_2O(l) \rightleftarrows Cr(OH)^{2+}(aq) + H_3O^{+}(aq)$$

with $K_a = 1.25 \times 10^{-4}$. The pH comes out to 2.5.

---

21. The problem concerns the solubility of $Pb^{2+}(aq)$ in a solution adjusted to pH 13 by the addition of NaOH. The reaction:

$$Pb^{2+}(aq) + 2\,OH^{-}(aq) \rightleftarrows Pb(OH)_2(s)$$

acts to reduce the concentration of $Pb^{2+}(aq)$. The K for this reaction is the reciprocal of K_{sp} of $Pb(OH)_2(s)$. Its large value (K_{sp} is 4.2×10^{-15} so its reciprocal is 2.38×10^{14}) means that 1 M lead(II) ion would give a precipitate of $Pb(OH)_2(s)$ at pH 13. But there is a complication. The equilibrium:

$$Pb^{2+}(aq) + 3\,OH^{-} \rightleftarrows Pb(OH)_3{}^{-}(aq)$$

ties up Pb^{2+} ion in a *soluble* form and combats the tendency to precipitate $Pb(OH)_2(s)$. The K for this reaction is the reciprocal of K_{inst} of $Pb(OH)_3{}^{-}(aq)$ and is large. It is 3.33×10^{14}. Whether solid $Pb(OH)_2$ forms depends on the resolution of the competition between these reactions. Suppose $Pb(OH)_2(s)$ *does* form. Then, at pH 13, where $[OH^{-}] = 0.1$ M, $[Pb^{2+}]$ must fulfill the equation:

$$4.2 \times 10^{-15} = [Pb^{2+}][OH^{-}]^2$$

This means $[Pb^{2+}]$ would be 4.2×10^{-13} M. The mass-action expression for the second equilibrium is:

$$3.33 \times 10^{14} = [Pb(OH)_3^-] / [Pb^{2+}][OH^-]^3$$

If $[Pb^{2+}]$ were 4.2×10^{-13} M, then $[Pb(OH)_3^-]$ would be 0.139 M, assuming that there is solid $Pb(OH)_2$ at the bottom of the solution to maintain the K_{sp} equilibrium. With both equilibria in action there must be a total lead concentration of $0.139 + 4.2 \times 10^{-13} \sim 0.139$ M.

In the first case, the Pb^{2+} *(aq)* starts at 1 M. The critical concentration, 0.139 M, is less than 1 M, so $Pb(OH)_2$ *(s)* will precipitate.

If the Pb^{2+} starts at 0.0500 then there is *not* enough Pb^{2+} ion to furnish the critical 0.139 M, and there will be no precipitate of $Pb(OH)_2$ *(s)*. The K_{sp} equilibrium will *not* be in effect. Essentially all 0.05 M of the Pb^{2+} ion is tied up in the complex. The concentration of free Pb^{2+} *(aq)* is 1.5×10^{-13} M, computed using the K_{inst} equilibrium constant expression.

---

23. Assume that all of the compounds in both groups are dissolved in water. The problem requires some knowledge of what happens to the compounds in both columns when placed in water.

a) $[Fe(H_2O)_4Cl_2]Br$ dissolves in water as a 1:1 electrolyte. (a +1 positive ion and a −1 ion). Its molar conductivity will resemble that of the only 1:1 ionic compound among the list on the right, namely, NaCl.

b) $[Mn(H_2O)_6]Cl_3$ would give one +3 ion and three −1 ions in solution. It resembles $AlCl_3$. Note that the Al^{3+} would be aquated in water solution as $[Al(H_2O)_6]^{3+}$.

c) $[Zn(H_3O)_3(OH)]Cl$ resembles NaCl in giving one +1 and one −1 ion.

d) $[Fe(NH_3)_6]CO_3$ resembles $CaSO_4$ in giving a +2 and −2 ion in solution. Its behavior would be complicated by hydrolysis of the CO_3^{2-} *(aq)* ion to give HCO_3^- *(aq)*. $CaSO_4$ is not very soluble in water. Since the conductivities are to be compared on a molar basis, this does not matter.

e) $[Cr(NH_3)_3Br_3]$ is a non-electrolyte. In its conductivity it resembles HCN, which, as a weak acid, is scarcely ionized in water.

f) $K_3[Fe(CN)_6]$ gives 3 +1 ions and one −3 ion when dissolved in water. It resembles Na_3PO_4 in this respect. The conductivity behavior of the latter is complicated by the fact that the PO_4^{3-} *(aq)* ion hydrolyzes (to HPO_4^{2-} *(aq)*).

---

25. The three compounds are $[Cr(H_2O)_4Cl_2]Cl{\cdot}2H_2O$, $[Cr(H_2O)_5Cl]Cl_2{\cdot}H_2O$ and $[Cr(H_2O)_6]Cl_3$. The loosely bound waters of hydration (indicated after the dot in the formulas) are lost to the dehydrating agent. The other water is coordinated and is *not* lost. Based on the structures, 1 mol of Cl^- per mole of coordination compound is available to precipitate with Ag^+ *(aq)* in the first compound, 2 in the second and 3 in the third. This is the Cl^- that is not coordinated. The three compounds therefore release 1, 2 and 3 mol of AgCl per mol respectively, when treated with excess $AgNO_3$ solution. The molar weights of the three compounds are all the same: 266.45 g mol^{-1}. 100 g therefore comprises

0.375 mol in each case. 0.375 mol of AgCl (molar weight 143.32 g mol^{-1}) is 53.8 g. The first compound yields 53.8 g of AgCl; the second yields twice this, 108 g of AgCl; the third yields three times this, 161 g.

---

27. In a crystal field of octahedral symmetry all three *p* orbitals have the same energy. There is no splitting. Electrons in the *p* orbitals have their greatest probability density along a coordinate axis and differ among themselves only in which axis: p_x, p_y or p_z. These axes are indistinguishable in the octahedral case; each has one ligand on each end. Similarly, a tetrahedral field affects all three *p* orbitals equally.

A square planar field is derived from an octahedral field by removal of negative charge from one axis, say, the *z*. If so, the p_z orbital would be split to lower energy (since it does not encounter a negatively charged ligand) while the p_x and p_y orbitals would remain unsplit.

---

29. The problem concerns the standard reduction potentials of Mn^{3+} *(aq)*, Fe^{3+} *(aq)* and Co^{3+} *(aq)* to the corresponding +2 ions. All six of the ions (the three +3 and three +2) are hexa-coordinated by H_2O in acidic aqueous solution. Mn^{3+} is a d^4 species, Fe^{3+} is a d^5 species and Co^{3+} is a d^6 species. The hexaaquo complexes are all high-spin complexes because water is a weak field ligand. Their reduction products are high-spin d^5, d^6 and d^7 species respectively. The Fe^{3+} *(aq)* standard reduction potential is substantially less that the standard reduction potential of its neighbors which means it is *harder* to reduce to the +2 state than its neighbors.

Once these preliminaries are established the question can be answered. The Fe^{3+} *(aq)* ion has a high-spin d^5 electron configuration $(t_{2g})^3$ $(e_g)^2$, which is especially stable (the t_{2g} and e_g levels are both half-filled), and resists gaining another electron. The extra stability of the d^5 configuration which is produced *enhances* reduction of the Mn^{3+} *(aq)* ion and does not come into play in the Co^{3+} *(aq)* reduction.

---

31. Both $Fe(CN)_6{}^{3-}$ and $Fe(H_2O)_6{}^{3+}$ have a central Fe^{3+} ion surrounded by six ligands in octahedral symmetry. The central ion has in each case five *d* electrons. CN^- is a strong field ligand and H_2O is a weak field ligand. In the $Fe(CN)_6{}^{3+}$ ion, as many *d* electrons as possible pair. With five 3*d* electrons this means 1 unpaired electron is left over. In $Fe(H_2O)_6{}^{3+}$ the five Fe^{3+} 3*d* electrons remain entirely unpaired, one in each of the 3*d* orbitals. This high-spin complex has five unpaired electrons.

---

33. The shift reaction is:

$$H_2O(g) + CO(g) \rightarrow CO_2(g) + H_2(g)$$

The following series of reactions summarizes the action of the homogeneous catalyst in solution as described in the problem:

$$Fe(CO)_5(aq) + H_2O(g) \rightarrow Fe(CO)_4(COOH^-)(aq) + H^+(aq)$$

$$Fe(CO)_4(COOH^-)(aq) \rightarrow HFe(CO)_4^-(aq) + CO_2(g)$$

$$HFe(CO)_4^-(aq) + CO(g) \rightarrow Fe(CO)_5(aq) + H^-(aq)$$

The sum of these three reactions is:

$$H_2O(g) + CO(g) \rightarrow CO_2(g) + H^+(aq) + H^-(aq)$$

But $H^+(aq)$ and $H^-(aq)$ react instantly in solution to form $H_2(g)$. The net reaction is thus the shift reaction.

---

35. The dissociation of $Cu(NH_3)_4{}^{2+}(aq)$ in basic aqueous solution follows the equation:

$$Cu(NH_3)_4)^{2+}(aq) \rightleftarrows Cu^{2+}(aq) + 4\ NH_3(aq)$$

The ΔG_f°'s of the products and reactant are in Appendix D. Combining them in the usual way gives ΔG° (reaction) = 70.56 kJ mol^{-1}. This corresponds to K of 4.3×10^{-13} (using the equation $-RT \ln K = \Delta G^\circ$).

At pH = 0 the course of the reaction is different:

$$Cu(NH_3)_4{}^{2+}(aq) + 4\ H_3O^+(aq) \rightleftarrows Cu^{2+}(aq) + 4\ NH_4{}^+(aq) + 4\ H_2O(l)$$

The data in Appendix D also allow computation of ΔG° of this reaction. It is -140.68 kJ mol^{-1}. Hence, the K for this reaction is big, 4.4×10^{24}. The complex *does* dissociate under acidic conditions. The "driving force" in acid is the formation of $NH_4{}^+(aq)$ ions.

---

37. The problem requires the comparison of the two reactions:

$$[Cd(NH_2CH_3)_4]^{2+}(aq) \rightarrow Cd^{2+}(aq) + 4\ NH_2CH_3(aq)$$

$$[Cd(en)_2]^{2+}(aq) \rightarrow Cd^{2+}(aq) + 2\ en(aq)$$

where en stands for ethylenediamine $H_2NCH_2CH_2NH_2$. Judging from the number of particles formed among the products, ΔS_1° (where the subscript refers to the first reaction) is a larger positive number than ΔS_2°. The ΔH°'s of the two reactions are about the same, according to the problem. For the two reaction:

$$-RT \ln K_1 = \Delta H_1^\circ - T\Delta S_1^\circ \quad \textit{and} \quad -RT \ln K_2 = \Delta H_2^\circ - T\Delta S_2^\circ$$

Subtracting the first equation from the second gives:

$$RT \ln K_1 - RT \ln K_2 = T\Delta S_1^\circ - T\Delta S_2^\circ$$

$$R \ln (K_1 / K_2) = \Delta S_1^\circ - \Delta S_2^\circ$$

ΔS_1° exceeds ΔS_2° so K_1 exceeds K_2. This mathematical comparison proves that the first reaction lies farther to the right at equilibrium; the instability constant of $Cd(NH_2CH_3)_4^{2+}(aq)$ is larger than the instability constant of $Cd(en)_2^{2+}(aq)$. The chelate effect stabilizes the $Cd(en)_2^{2+}$ complex.

---

39. ***a)*** The aqueous HCl reacts with $CdS(s)$ to give both $CdCl_4^{2-}(aq)$ and $H_2S(aq)$:

$$CdS(s) + 2\,H_3O^+(aq) + 4\,Cl^-(aq) \rightleftarrows CdCl_4^{2-}(aq) + H_2S(aq) + 2\,H_2O(l)$$

The HCl solution acts on the $CdS(s)$ by tending to remove both sulfide (as H_2S) and cadmium (as the tetrachloro complex). The problem states that some $CdS(s)$ remains, so the above equation is an accurate description of the final equilibrium.

b) The equilibrium in part a) can be constructed as the *sum* of the reactions:

$$CdS(s) \rightleftarrows Cd^{2+}(aq) + S^{2-}(aq) \qquad K_{sp}$$

$$S^{2-}(aq) + 2\,H_3O^+(aq) \rightleftarrows H_2S(aq) + 2\,H_2O(l) \qquad K = 1 / K_{a1}K_{a2}$$

$$Cd^{2+}(aq) + 4\,Cl^-(aq) \rightleftarrows CdCl_4^{2-}(aq) \qquad K = 1 / K_{inst}$$

The problem gives the K's of the component reactions so the equilibrium constant for the equation in part a) is 7.68×10^{-7}, (equal to $K_{sp} / K_{a1}K_{a2}K_{inst}$).

c) The equilibrium constant expression is:

$$7.68 \times 10^{-7} = \frac{[CdCl_4^{2-}][H_2S]}{[H_3O^+]^2[Cl^-]^4}$$

Let S equal the solubility of the $CdS(s)$. If all of the cadmium in solution is in the form of $CdCl_4^{2-}(aq)$ and if all of the sulfur in solution is in the form of $H_2S(aq)$, then:

$$S = [CdCl_4^{2-}] = [H_2S]$$

Every Cd^{2+} that goes into solution consumes 4 Cl^- ions; every S^{2-} consumes 2 H_3O^+ ions. Thus, at equilibrium:

$$[H_3O^+] = 6 - 2\,S \quad \textit{and} \quad [Cl^-] = 6 - 4\,S$$

The 6 comes from the original concentration of $HCl(aq)$, which was 6 M. The equilibrium constant expression becomes:

$$7.68 \times 10^{-7} = S^2 / (6 - 2\,S)^2\,(6 - 4\,S)^4$$

To solve this equation for S use the method of successive approximations. Suppose at first S is negligible compared to 6. Then:

$$7.68 \times 10^{-7} \simeq S^2 / 46656$$

This gives $S = 0.189$ mol L^{-1}. But this answer is about 3 percent of 6 and not really negligible compared to it. Successive approximations starting from this point give improved values of S. $S = 0.146$ fits the equation acceptably. The solubility of the CdS(*s*) is 0.146 mol L^{-1}, or, to 2 significant figures, 0.15 M.

---

41. ***a)*** From text Example 18-5, K_{a1} for $Fe(H_2O)_6^{3+}$ is 7.7×10^{-3}. The problem quotes K_{a2} as 2.0×10^{-5}. This constant is about 400 times smaller than K_{a1} so it is probable that the second stage of the acid ionization does *not* exert an important effect on the pH. According to Example 18-5, the pH of the 0.100 M solution of $Fe(NO_3)_3$, based on the first ionization alone, is 1.62. The $[H_3O^+]$ is therefore 2.4×10^{-2} M. The equilibrium constant expression for the *second* ionization is:

$$K_{a2} = 2.0 \times 10^{-5} = [H_3O^+][Fe(H_2O)_4(OH)_2{}^+] / [Fe(H_2O)_5(OH)^{2+}]$$

Substitute the concentrations which come from the first-stage-only calculation into this expression. They are:

$$[H_3O^+] = [Fe(H_2O)_5(OH^{2+}] = 2.4 \times 10^{-2} \text{ M}$$

Hence, the Fe-containing product of the second stage has a concentration of only 2.0×10^{-5} M. This concentration is negligible compared to 0.024 M, the concentration of $Fe(H_2O)_5OH^{2+}$*(aq)*. Therefore the $[H_3O^+]$ that arises in the second stage is also negligible compared to 0.024 M.

b) The question requires writing the two dissociation equations:

$$Fe(OH)_2{}^+(aq) \rightleftarrows Fe(OH)^{2+}(aq) + OH^-(aq)$$

$$Fe(OH)^{2+}(aq) \rightleftarrows Fe^{3+}(aq) + OH^-(aq)$$

These equilibria have dissociation constants K_{d1} and K_{d2}. The mass action expression for the first equation is:

$$K_{d1} = [Fe(OH)^{2+}][OH^-] / [Fe(OH)_2{}^+]$$

Divide this equation *into* the K expression for the autoionziation of water, $K_w = [H_3O^+][OH^-]$. The result is:

$$K_w / K_{d1} = [H_3O^+][Fe(OH)_2{}^+] / [Fe(OH)^{2+}]$$

The right-hand side of this equation is identical to the mass-action expression for K_{a2} in part a), *except* that the associated H_2O molecules are shown explicitly in the formulas in the K_{a2} expression. The point is that $Fe(OH)_2{}^+$ is the *same* as $Fe(H_2O)_4(OH)_2{}^+$. Hence, $K_{a2} = K_w / K_{d1}$. Similarly, $K_{a1} = K_w / K_{d2}$. Numerical values for K_{a1} and K_{a2} are known. Substitution gives $K_{d1} = 5.0 \times 10^{-10}$ and $K_{d2} = 1.3 \times 10^{-12}$. The K_{inst} of $Fe(OH)_2{}^+$*(aq)* is the product of these numbers because the dissociation of $Fe(OH)_2{}^+$*(aq)* proceeds through $Fe(OH)^{2+}$*(aq)* to Fe^{3+}*(aq)*. It is 6.5×10^{-22}. The K_{d1} and K_{d2} answers correspond respectively to the K_{d2} and K_{d3} of Table 18-3. The offset in numbering occurs because a different starting complex of iron is involved. It is $Fe(OH)_2{}^+$ in the problem but $Fe(OH)_3$ in the table.

Chapter 19

THE CHEMISTRY OF SILICON AND RELATED ELEMENTS

19.1 *INORGANIC STRUCTURES*

Silicon and its neighbors in the periodic table differ sharply from the metals in their structures. The following table summarizes the differences:

	Metals	*Silicon et al.* *
Nature of Bonds	Non-directional, metallic	Directional, Covalent
Valence electrons	Delocalized, free to move	Localized in Bonds
CN of Atom	High (12, 8, etc.), reflects Packing	Low, Related to Group Number
Structure	f.c.c., b.c.c.	Networks, layers, cages... covalent bonding in three dimensions

*Others are B, C, Ge, P, S, Se, Te

Networks are extended three-dimensional structures which are, as far as individual atoms are concerned, infinite in extent. Strong directed bonds between atoms lend coherence to the three dimensional structure. When strong bonding is confined to two dimensions, then the structure contains *layers* or *infinite sheets.* In a solid, such layers are held together by weaker intermolecular forces.

Cage structures consist of small groups of atoms covalently (hence strongly) linked in discrete small groups. Weaker bonds between cages maintain the structure of the solid. **See problem 3.** *Examples*: Elemental boron forms icosahedral B_{12} cages in which each B is bonded to 5 others. An *icosahedron* is a regular solid that has 20 faces, 12 corners and 30 edges. White phosphorus consists of tetrahedral P_4 cages. A diatomic molecule could be thought of as a degenerate (two-membered) cage. Usually however, the term cage is reserved for structures of four or more atoms that enclose or partially enclose a region of space.

Chains and *rings* arise from atoms (or groups of atoms) that form only *two* covalent bonds. Polymeric chains can grow to enormous length. Silicon forms long-chain polymers in which Si and O atoms alternate. The ability of carbon to form long hydrocarbon chains is well-known.

The oxyacids of boron, phosphorus and sulfur can be dehydrated to give new oxyacids with E−O−E linkages (where E stands for the element). The simplest cases are dimers (two units sharing an oxygen). Longer chains also occur. The links in these polyacids are tetrahedral EO_4 or triangular EO_3 units sharing *corners.* The electrical charge per individual unit depends on the identity of the element E. **See**

problem 1. Remember that every time two EO_n units join at a corner, one O^{2-} is lost.

Study this and other descriptive material by consciously re-trying the concepts of previous chapters in the novel settings the new information provides.

Example. Write the Lewis structure of disulfur dinitride (Figure 19-7d, text p. 659).

Solution. The cyclic S_2N_2 molecule has 22 valence electrons. Giving each atom an octet and at the same time maintaining the cyclic structure with single bonds would require 24 valence electrons. Hence, there must be one double bond in the ring. The 4 possible positions of the double bond are equivalent, so the complete answer is a resonance combination of four Lewis structures. One of the four is shown.

The double-bonded S has a +1 formal charge and the double-bonded N has a 0 formal charge. The formal charges of the *other* S and N are 0 and −1 respectively. Counting all 4 resonance structures, the *average* formal charge on the S atoms is $+\frac{1}{2}$ and $-\frac{1}{2}$ on the N atoms.

```
 ..   ..
 S == N
 |    |
:N -- S:
 ..   ..
```

The average bond order in the molecule is $1\frac{1}{4}$. Compare to **problem 1.**

19.2 *SILICATES*

The orthosilicate anion, SiO_4^{4-} provides the structural basis of many minerals. The *ratio* of Si to O in this anion is 1:4. As several SiO_4^{4-} tetrahedra link together by sharing corners, the number of oxygen atoms per silicon *diminishes.*

Example. Determine the charge on the cyclic trisilicate ion pictured in Figure 19-8c, text p. 661.

Solution. Any trisilicate contains three Si atoms. There are three links between SiO_4 tetrahedra in this cyclic structure. This reduces the number of oxygen atoms from 12, the number in $(SiO_4)_3$, to 9. In silicates the oxidation number of Si is +4 and of O −2. Hence, the ion is $Si_3O_9^{6-}$. Another approach notes that silicon atoms have a zero formal charge in silicates, that oxygens with just one bond in a silicate always have a −1 formal charge, and that oxygens which are shared between silicons have a zero formal charge. There are three oxygens shared between the SiO_4 tetrahedra in Figure 19-8c, but six oxygens on the perimeter of the structure. The sum of the formal charges of all atoms is −6, hence the net charge on the ion is −6.

The Si:O ratio provides a criterion for the *type* of silicate structure. The criteria are summarized in Table 19-1 (text p. 661). There are two complications:

- Not all of the oxygen atoms in a mineral are part of the silicate structural system. Some may be present in water of crystallization, for example. These oxygens are not counted in determining the Si:O ratio. **See problem 5.**

- *Aluminum* can substitute for silicon in the silicate structures. If it does, it must be counted as an Si atom in predicting silicate structures. **See problem 7.** Note that aluminum in silicate minerals does not *always* replace silicon. Sometimes it is present merely to maintain electrical neutrality.

Zeolites

Zeolites are *aluminosilicates* with three-dimensional network structures which are open enough to allow the inclusion of small molecules and ions. Every time an aluminum(III) replaces a silicon(IV) in a silicate structure, the resulting aluminosilicate gains a negative charge. There must be cations nearby to maintain electrical neutrality. These cations reside in the openings in the zeolite framework.

Zeolites are used for ion-exchange. An equilibrium is established as one type of positive ion becomes lodged in the pores and tunnels of the aluminosilicate structure and another is released. In addition, small neutral molecules like water can be taken up and held in the internal cavities of zeolites. Finally zeolites can hold small molecules in favorable orientations for reaction, thus serving as catalysts.

19.3 *GEOCHEMISTRY*

This section covers many applications of chemical principles. Understanding earth science requires use of nearly all the principles covered in the first half of the text. For example, **see problems 5, 15, 23, 27,** which require determination of oxidation number (Chapter 2), a determination of stoichiometry (Chapter 1), equation-balancing, and an acid-base equilibrium calculation (Chapter 6), respectively.

The text introduces a new device for displaying chemical data. It is the *potential versus pH diagram* (Figure 9-13, text p. 670). The key points in understanding and using these diagrams are:

- Inconsistencies in plotting and scaling give confusing results. Always plot *reduction* potentials (*not* oxidation potentials) on the vertical axis. Remember that as pH increases (to the right on the horizontal scale) the concentration of H^+ *(aq)* becomes less, (not more).

- Half-reactions in which H^+ *(aq)* or OH^- *(aq)* do not occur explicitly have no dependence of potential upon pH.

- Breaks in potential versus pH lines occur when one half-reaction supersedes another. Thus, in **problem 9,** the species Co^{2+} *(aq)* is reduced in acid solution, but $Co(OH)_2$ *(s)* is reduced in basic solution. A bend in the potential versus pH curve marks the cross-over.

- The slope of the reduction potential versus pH line derives from the logarithm term in the Nernst equation. At 25° C, the slope of this line is some multiple or submultiple of 0.0592 V per pH unit.

- The potential versus pH diagrams identify *stability regions* for the elements in different oxidation states. They help to decide which minerals will form under certain conditions (oxidizing versus reducing and acidic versus basic).

Clapeyron Equation

The Clapeyron equation states the effect of pressure on a chemical equilibrium:

$$dP \,/\, dT = \Delta H \,/\, T\Delta V$$

This equation gives the *slopes* of phase coexistence lines in the phase diagrams discussed in Chapter 4.

In using the Clapeyron equation, remember that it is valid *only* at combinations of temperature and pressure for which the system is at equilibrium. A common error is to substitute a temperature which does *not* fit the equation:

$$\Delta G = 0 = \Delta H - T\Delta S$$

and expect the Calpeyron equation to work. **See problem 13d.**

As long as ΔG is 0 for a change taking place at constant temperature and pressure, the quotient ($\Delta H / T$) equals ΔS. An alternate version of the Clapeyron equation then is:

$$dP/ \,dT = \Delta S \,/\, \Delta V$$

This form of the equation is used in **problems 11 and 29.**

Difficulties arise with the Clapeyron equation because it concerns equilibrium at constant temperature and pressure, yet the term dP / dT clearly involves *changes* in both temperature and pressure. Imagine that a given system is brought to equilibrium under constant temperature and pressure in many separate experiments at many different (P,T) combinations. It would certainly be possible to plot these sets of values on a graph with P on the vertical axis and T on the horizontal axis. The term dP / dT is the slope of the resulting line, telling how the observed pressures and temperatures interrelate.

A second difficulty in applying the Clapeyron equation occurs because it has constantly been assumed in problems that the ΔS of a change does not depend strongly on the temperature. This is true, at least across limited ranges of temperature. Unfortunately, it encourages the thought that ΔS values are also only weakly dependent upon the *pressure.* If the change in the number of moles of gas, Δn_g, is non-zero in a process, then the ΔS of the process is *strongly* dependent upon pressure. **See problem 11.**

The units of dP / dT are Pa K^{-1} (pascal per kelvin) in the SI system. These will be the units of the answer if ΔS (or $\Delta H / T$) is in J K^{-1} and ΔV is in m^3. Volume changes however nearly always derive from densities, which rarely are quoted with m^3 as part of their unit. Do not forget to convert if ΔV is in liters or cubic centimeters or if ($\Delta H / T$) is in kJ K^{-1}. The unfamiliar Pa K^{-1} unit can readily be converted to atm K^{-1}, if desired. **See problem 13.**

19.4 *INORGANIC MATERIALS*

Major categories of inorganic materials include glasses, ceramics and cement.

Glasses are amorphous solids of variable composition. Soda lime glass has the approximate formula $Na_2O{\cdot}CaO{\cdot}6SiO_2$. Its structure consists of SiO_4 tetrahedra linked in a random three dimensional network. The cations (Na^+ and Ca^{2+}) occupy voids in the network and maintain electrical neutrality.

Ceramics start with aluminosilicate minerals that contain hydrated cations between the layers of an infinite sheet structure. These materials are clays. Firing an object formed of clay drives away the water and, among other reactions and phase changes, leads to the formation of mullite ($Al_6Si_2O_{13}$), which lends strength to the ceramic.

Portland cement is made from ground limestone mixed with aluminosilicates. Firing the mixture in a cement kiln causes many reactions; the major product is tricalcium silicate (Ca_3SiO_5), and the minor product is tricalcium aluminate ($Ca_3Al_2O_6$). This material, in the form of *clinkers* , is mixed with a small percentage of gypsum. Cement hardens when mixed with water because of the hydration of the tricalcium silicate.

DETAILED ANSWERS TO ODD-NUMBERED PROBLEMS

1. The structure of $Si_2O_7^{6-}$ is:

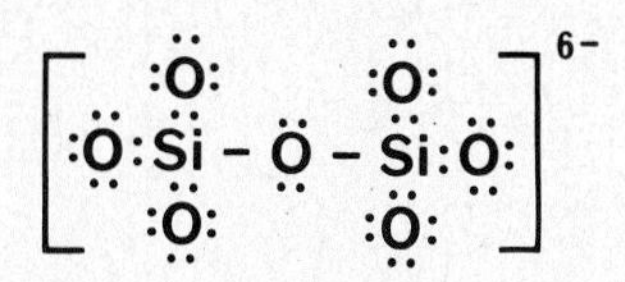

The 6 oxygen atoms on the perimeter of the structure each have three lone pairs and a single bond to an Si atom. All 6 perimeter oxygens have a formal charge of -1. The O atom between the Si atoms has a zero formal charge. The Si atoms also have a formal charge of zero. There are 56 valence electrons in the Lewis structure.

The $P_2O_7^{4-}$ and $S_2O_7^{2-}$ ions also have 56 valence electrons and have the *same* Lewis structure except for the replacement of the central atoms. Both P atoms in the $P_2O_7^{4-}$ version of this structure have a $+1$ formal charge. Both S atoms have a $+2$ formal charge in the $S_2O_7^{2-}$ structure.

If the octet rule is broken for the Si (or P or S) atoms in these structures, then many resonance Lewis structures are possible. Such structures would have one or more double bonds between the surrounding O atoms and the central atoms. They would increase the average bond order and would give a pattern of formal charge in which the central atoms have lower positive values.

---

3. In white phosphorus the intermolecular forces (among P_4 molecules) are *dispersion* interactions. In black phosphorus the structure is maintained by covalent bonds.

---

5. In each example, determine the Si:O ratio for the network. Ignore oxygen atoms found in, for example (OH) groups. Then use Table 19-1.

a) Tetrahedra. Ca, +2; Fe, +3; Si, +4; O, −2.

b) Infinite sheets. Na, +1; Zr, +2; Si, +4; O, −2.

c) Pairs of tetrahedra. Ca, +2; Zn, +2; Si; +4; O, −2.

d) Infinite sheets. Mg, +2; Si, +4; O, −2; H, +1.

---

7. The problem is exactly like the preceding except that Al atoms grouped in the formulas with the Si atoms are counted as Si atoms in determining the Si:O ratio.

a) Network. Li, +1; Si, +4; Al, +3; O, −2.

b) Infinite sheets. K, +1; Al, +3; Si, +4; O, −2; H, +1.

c) Closed rings or single chains. Al, +3; Mg, +2; Si, +4; O, −2.

---

9. The standard reduction potential for the half-reaction:

$$Co^{2+}(aq) + 2\,e^- \rightarrow Co(s)$$

is −0.28 V at pH 0. As the pH rises the potential remains constant until, after a brief intermediate regime, cobalt(II) hydroxide precipitates, and the half-reaction:

$$Co(OH)_2(s) + 2\,e^- \rightarrow Co(s) + 2\,OH^-(aq)$$

more accurately describes what is going on. The standard potential for this half reaction is −0.73 V. The Nernst equation for the new half-reaction is:

$$\xi = \xi^\circ - (0.0592 / 2) \log [OH^-]^2 = -0.73 - 0.0592 \log [OH^-]$$

The term −0.0592 log [OH⁻] equals +0.0592 pOH which is 0.0592 (14 − pH). Thus, with the understanding that the potential is in volts:

$$\xi = -0.73 + 0.0592\,(14 - pH) = 0.0988 - 0.0592\,pH$$

This equation tells how the half-cell potential varies with pH. A plot of ξ versus pH is a straight line with a slope of −0.0592. If extended back to pH 0, its intercept would be 0.0988 V. Of course the actual behavior at low pH is governed by the −0.28 V half-reaction.

---

11. ***a)*** The reaction of calcite and quartz to give wollastonite is:

$$CaCO_3(s) + SiO_2(s) \rightarrow CaSiO_3(s) + CO_2(g)$$

b) The ΔH° of the reaction is 89.41 kJ, using the data from Appendix D in the usual way. ΔS° is 160.81 J K^{-1}.

c) Assume that ΔH° and ΔS° are independent of temperature. Then:

$$\Delta G = \Delta H^\circ - T\Delta S^\circ$$

The temperature that makes ΔG equal 0 is 556 K. At this temperature the equilibrium constant for the above reaction equals 1.

d) Although ΔH for the process is approximately independent of the pressure, ΔS *does* depend on pressure. The $CO_2(g)$, if released at a higher P, has less entropy. Hence ΔS for this reaction at 500 atm is less than ΔS°. Compressing 1 mol of $CO_2(g)$ from 1 atm to 500 atm changes the entropy by $-R\ln(^{500}/_1)$ which is −51.67 J K^{-1}. This formula is discussed in Chapter 8 (p. 167 of this Guide). Therefore, replace ΔS° = 160.81 J K^{-1} in part b) above with a ΔS of (160.81 − 51.67) J K^{-1} or 109.14 J K^{-1}. Then set ΔG equal to zero and compute T. It is 819 K. At high pressure, it takes a *higher* temperature to make the equilibrium constant of the reaction equal 1. This fits with the prediction of leChatelier's principle. Higher pressure favors the reactants.

---

13. ***a)*** The two forms of calcium carbonate are calcite and aragonite. Represent the conversion from the first to the second as:

$$\text{calcite} \rightarrow \text{aragonite}$$

The ΔH° for this change is the *difference* between the ΔH°_f's of calcite and aragonite. Refer to Appendix D for the necessary enthalpies of formation. ΔH° is -0.21 kJ. The ΔS° is the difference in the two materials' absolute entropies at 298.15 K. Taking values from Appendix D, it is -4.2 J K^{-1}. The ΔG° of the reaction is 1.04 kJ, by a similar calculation with ΔG°_f's, using the same source for data.

b) The ΔG° for the calcite to aragonite conversion is positive. This means that *calcite* is favored over aragonite at 298.15 K and 1.00 atm.

c) Assume that ΔH° and ΔS° depend only weakly on the temperature. Then:

$$\Delta G = \Delta H^\circ - T\Delta S^\circ$$

When the two forms are equally favored, the equilibrium constant for the conversion betweeen them equals 1.0, and ΔG (which equals $-RT \ln K$) is 0. Setting ΔG equal to zero gives T = 50 K. Below 50 K *aragonite* is favored. At all temperatures above 50 K, calcite is favored (as long as the pressure is 1.00 atm).

d) Higher pressure favors the *more* dense form. The molar volume of aragonite is 34.16 cm^3 mol^{-1}. It is more dense than calcite, which occupies 36.94 cm^3 mol^{-1}. *Aragonite* is favored at high pressure.

e) The Clapeyron equation gives the slope of the P versus T coexistence line:

$$dP \,/\, dT = \Delta H \,/\, T\Delta V$$

For the transition of 1 mol of calcite to 1 mol of aragonite ΔH° is -210 J mol^{-1}. The ΔV for one mole of the transition is 34.16 cm^3 $-$36.94 cm^3 or -2.78 cm^3 which is 2.78×10^{-6} m^3 mol^{-1}. The coexistence line for the two solids passes through the point defined by the (P,T) combination: (P=1 atm, T=50 K). See part c) above. Assume that the ΔH and ΔV values are still good at this point on the P versus T graph. Then the slope of the line at this point is:

$$dP \,/ dT = -210 \text{ J mol}^{-1} \,/\, (50 \text{ K} \times 2.78 \times 10^{-6} \text{ m}^3 \text{ mol}^{-1})$$

$$dP \,/\, dT = 1.5 \times 10^6 \text{ Pa K}^{-1}$$

Converting all units to SI assures that the answer comes out in the SI units of pressure over temperature. The SI unit of pressure is the pascal (Pa). One pascal is 101,325 atm, so an equivalent answer is 15 atm K^{-1}.

The result tells the rate of change of pressure with temperature at (P=1 atm, T=50 K). As long as ΔH and ΔV are independent of T and P, the coexistence line is a straight line with this slope. Compare to **problem 4-31.**

---

15. The problem recalls principles from Chapter 1. Suppose someone gathers a 100 g representative sample of the earth's crust. It contains 8 g of Al, 28 g of Si and 47 g of O. If these figures are *divided* by the three elements' molar weights, then the sample contains 0.296 mol of Al, 1.00 mol of Si and 2.93 mol of O. The molar ratio of the three is about 1 to 3.4 to 10.

---

17. ***a)*** The equation for the production of silicon carbide is:

$$SiO_2(s) + 3\ C(s) \rightarrow SiC(s) + 2\ CO(g)$$

b) The standard enthalpy change per mole of SiC(*s*) is just the ΔH° of the above reaction expressed on a per mol basis. ΔH° is $2 \times (-110.52) - 65.3 - (-910.94)$ kJ. The values come from Appendix D. Completing the arithmetic gives $\Delta H^\circ = 624.6$ kJ. The answer is 624.6 kJ mol^{-1}.

c) Silicon carbide will be a poor conductor of electricity, high melting and very hard. Its other physical properties will also resemble those of diamond.

---

19. The valence electron configurations of P, S and Cl are $3s^2 3p^3$, $3s^2 3p^4$ and $3s^2 3p^5$ respectively. This suggests that they form 3, 2 and 1 covalent bonds as they attain the argon closed shell electron configuration. Their structural differences stem from this variation in bond-forming capacity in much the same way that different silicate structures stem from variations in Si to O ratio.

---

21. The crucial point is the ratio of Si to O *within the silicate network.* Other oxygen atoms (in water molecules or hydroxide ions, for example) do *not* count in determining this ratio. The fewer oxygen atoms there are per silicon atom then the more the Si atoms share O atoms. See text Table 19-1.

a) In $KCa_4Si_8O_{20}F \cdot 8H_2O$, the Si:O ratio is 1:2½. The silicate is an infinite sheet. The oxidation numbers are: K, +1; Ca, +2; Si, +4; O, −2; F, −1; H, +1.

b) In $Mg_3Si_2O_5(OH)_4$, the Si:O ratio is 1:2½. The silicate is an infinite sheet. The oxidation numbers are: Mg, +2; Si, +4; O, −2; H, +1.

c) In $CaMn_4Si_5O_{15}$, the key ratio is 5 to 15 or 1:3. The silicate consists of infinite single chains. Oxidation states: Ca, +2; Mn, +2; Si, +4; O, −2.

d) In $CaAl_2(Si_2Al_2)O_{10}(OH)_2$, aluminum atoms *substitute* for half the Si atoms. The key ratio is 1:2½. The structure is an infinite sheet. The oxidation numbers are: Ca, +2; Al, +3; Si, +4; O, −2; H, +1.

---

23. The balanced equation for the production of mullite and cristobalite from kaolinite is:

$$3\ Al_2Si_2O_7 \cdot 2\ H_2O(s) \rightarrow Al_6Si_2O_{13}(s) + 4\ SiO_2(s) + 2\ H_2O(g)$$

---

25. The volume of a sphere is:

$$V = {}^4/_3\, \pi\, r^3$$

where r is its radius. The volume of the top 1 km of the earth's crust is the volume of a sphere of radius 6370 km minus the volume of a sphere of radius (6370 − 1) km. It is:

$$V(\text{top of crust}) = {}^4/_3\, \pi\, [6370^3 - 6369^3]\ \text{km}^3$$

The answer is in km³. A km is 10^5 cm, so a km³ is 10^{15} cm³. The volume of the top kilometer of crust is 5.11×10^{23} cm³. Each cm³ weighs 2.8 g. Accordingly, the mass of the top 1 km of crust is 2.8 g cm⁻³ times the volume in cm³ or 1.43×10^{24} g. The text (Section 19-2) states that the crust is 8 percent Al. The mass of Al in the crust is therefore $0.08 \times 1.43 \times 10^{24}$ g which is 1.14×10^{23} g or 1.14×10^{20} kg.

---

27. ***a)*** The dissolution of $CaCO_3$ *(s)* in water involves the equilibria:

$$CaCO_3(s) \rightleftarrows Ca^{2+}(aq) + CO_3{}^{2-}(aq) \qquad K_{sp}$$

$$CO_3{}^{2-}(aq) + H_2O(l) \rightleftarrows HCO_3{}^-(aq) + OH^-(aq) \qquad K_{b1}$$

$$HCO_3{}^-(aq) + H_2O(l) \rightleftarrows H_2CO_3(aq) + H_2CO_3(aq) \qquad K_{b2}$$

The mass action expressions for the two acid-base equilibria are:

$$K_{b1} = [OH^-][HCO_3{}^-]/[CO_3{}^{2-}] \quad \textit{and} \quad K_{b2} = [OH^-][H_2CO_3]/[HCO_3{}^-]$$

K_{b1} is K_w divided by the K_{a2} for H_2CO_3 (text Table 6-2) and K_{b2} is K_w divided by K_{a1} for H_2CO_3. Hence, K_{b1} is 2.08×10^{-4}, and K_{b2} is 2.32×10^{-8}. Completing the list of constants is K_{sp}, from text Table 5-1, at 8.7×10^{-9}.

Since the river water is at pH 7, $[OH^-] = 1 \times 10^{-7}$ M. Substituting this value into the two K_b expressions gives:

$$[HCO_3{}^-] = (2.08 \times 10^3)[CO_3{}^{2-}] \quad \textit{and} \quad [H_2CO_3] = (0.482 \times 10^3)[CO_3{}^{2-}]$$

Let S equal the equilibrium solubility of the $CaCO_3$ *(s)*. Then S is equal to $[Ca^{2+}]$ and is also equal to the sum of the concentrations of the three carbon-containing species. This latter is a material balance condition for the carbonate. See p. 110 of this Guide. Expressed mathematically:

$$S = [Ca^{2+}] = [CO_3{}^{2-}] + [HCO_3{}^-] + [H_2CO_3]$$

Substituting the independent expressions for the bicarbonate and carbonic acid concentrations into this equation gives:

$$S = [CO_3{}^{2-}]\{1 + (2.08 \times 10^3) + (0.482 \times 10^3)\}$$

Finally, combining this equation with the K_{sp} expression for $CaCO_3(s)$ allows the calculation of S:

$$8.7 \times 10^{-9} = S^2 / 2.56 \times 10^3 \qquad ergo \qquad S = 4.72 \times 10^{-3} \text{ M}$$

One error is to assume that $HCO_3^-(aq)$ is the only significant carbon-containing species. This corresponds to omitting the first and third terms within the curly brackets above. Committing this error yields an S equal to 4.25×10^{-3} M, which is 10 percent less than the correct answer. A more gross error ignores the acid-base interaction of the carbonate ion with the water entirely. This corresponds to omitting the second and third terms within the curly brackets and gives a wrong S of 9.3×10^{-5} M.

The river's annual flow is 8.8×10^{12} L. This much water would, at equilibrium, dissolve 4.2×10^{10} mol of $CaCO_3$ (molar weight 100 g mol^{-1}) which is 4.2×10^6 metric tons.

---

29. ***a)*** The balanced equation is:

$$Mg_3Si_4O_{10}(OH)_2(s) + Mg_2SiO_4(s) \rightarrow 5\,MgSiO_3(s) + H_2O(g)$$

Trial-and-error balancing gives this answer. Another approach is to note that Mg_2SiO_4 *loses* O^{2-} ions on per silicon basis, and $Mg_3Si_4O_{10}(OH)_2$ *gains* O^{2-} ions on the same basis. Hence, the oxide ion in this reaction plays a role similar to that of the electron in oxidation-reduction reactions. Arranging the oxide-gain to equal the oxide-loss rapidly gives a balanced equation:

$$\text{Oxide loss: } Mg_2SiO_4(s) \rightarrow MgSiO_3(s) + O^{2-} + Mg^{2+}$$

$$\text{Oxide gain: } O^{2-} + Mg^{2+} + Mg_3Si_4O_{10}(OH)_2(s) \rightarrow 4\,MgSiO_3(s) + H_2O(g)$$

b) The answer requires the use of LeChatelier's principle. Increasing the total pressure increases the activity of $H_2O(g)$ and shifts the reaction to the left. The products are disfavored.

c) The slope of the coexistence curve is given by the Clapeyron equation:

$$dP / dT = \Delta H / T\Delta V = \Delta S / \Delta V$$

For the reaction as written above, ΔV is positive. The volume of the products includes the volume of 1 mol of gas, a large number, and the reactants include no gas. ΔS is also positive. The slope of the curve is, of course, positive.

Chapter 20

NITROGEN, PHOSPHORUS AND SULFUR

20.1 *NITROGEN CHEMISTRY*

The characteristic fact about $N_2(g)$ is its great stability and consequent lack of reactivity in most situations. Nitrogen molecules contain N to N triple bonds with a very large dissociation energy.

The most important compounds of nitrogen are ammonia, a base, and nitric acid. The text reviews the preparation (in the *Haber-Bosch* process) of ammonia with particular emphasis on the compromise that nature requires between favorable thermodynamics (at low temperatures) and favorable kinetics (at high temperature).

Hydrazine (H_2NNH_2), produced from $NH_3(aq)$ by oxidation with $ClO^-(aq)$ in the Raschig synthesis, is related to ammonia. Note that its formula is twice the formula of ammonia with H_2 subtracted. Loss of H_2 amounts to oxidation. Accordingly, the oxidation number of nitrogen in hydrazine is -2, *versus* -3 in ammonia. Ammonia is NH_3 and its conjugate acid, NH_4^+, is the ammonium ion. Similarly, H_2NNH_2 is hydrazine and its conjugate acid is the hydrazinium ion. In fact, hydrazine can gain *two* protons; both nitrogen atoms are basic. **See problem 21.**

Another compound related to ammonia is *hydroxylamine* (H_2NOH), which formally is NH_3 with an OH^- replacing an H or H_2NNH_2 with an OH^- replacing an NH_2. Both of these replacements are oxidations (because a negative group replaces a neutral group), and the oxidation number of N in hydroxylamine is -1, higher than it is in hydrazine.

The preparation of nitric acid illustrates several important principles of thermodynamics and kinetics in a practical context. HNO_3 has a variety of uses, for example, in fertilizer production.

Nitrogen forms 6 oxides, in which its oxidation number ranges from $+1$ to $+5$. Nitrogen oxide (NO) and nitrogen dioxide (NO_2) both have an odd number of valence electrons. Much of the chemistry of these gases can be rationalized on the basis of their tendency to react to form products with even numbers of valence electrons. Thus, $NO(g)$ rather easily loses an electron to form the nitrosyl ion (NO^+), and NO_2 dimerizes to N_2O_4. Both NO_2 and N_2O_4 are fairly strong oxidizing agents in aqueous solution. N_2O_5 is the acid anhydride of nitric acid.

The solution chemistry of nitrogen and its oxides is best studied by following the descriptive comments on the various reactions with constant reference to the reduction potential diagrams on text p. 692. These diagrams are applied in **problem 7.** Note the frequent necessity for distinctions between the thermodynamic stability of a compound and its kinetic stability. This sounds an important theme in descriptive chemistry. Thus, NO_2^- is thermodynamically unstable in basic aqueous solution with respect to disproportionation but such solutions can be kept indefinitely (text p.

692). Similarly, the oxidation of ammonia with oxygen is strongly favored thermodynamically at room temperature but does not proceed at any perceptible rate because no effective kinetic pathway is available. Addition of a Pt catalyst speeds the reaction. **See problem 25.**

20.2 *PHOSPHORUS CHEMISTRY*

Phosphorus has less tendency than nitrogen to form double and triple bonds in its compounds. Instead it tends to form a larger number of single bonds. Thus, it often violates the octet rule in its compounds.

Elemental phosphorus exists in several allotropic forms. White phosphorus consists of tetrahedral P_4 molecules. In red and black phosphorus the element is extensively bonded in network structures.

The important oxides of phosphorus are P_4O_6 and P_4O_{10}. They are both acidic oxides and react with excess water to give H_3PO_3 (phosphorous acid) and H_3PO_4 (phosphoric acid) respectively.

Solution Chemistry of Phosphorus

Phosphorus compounds in aqeuous solution are more difficult to reduce than the analogous nitrogen compounds and therefore easier to oxidize. The solution chemistry of phosphorus should be studied with reference to the reduction potential diagrams given on text p. 695 and in comparison to the similar diagrams for nitrogen (text p. 692). Phosphate ions link together to form polyanions. Tetrahedral PO_4^{3-} groups join at their corners to give chains that are analogous to the polysilicates. **See problem 19-1.**

Example. Phosphoric acid (H_3PO_4) condenses to give the dimer pyrophosphoric acid ($H_4P_2O_7$) and the trimer tripolyphosphoric acid ($H_5P_3O_{10}$). Predict the formula of the polyphosphoric acid with four phosphates linked in a straight chain. Predict the formula of the polyphosphoric acid with four phosphates linked in a ring.

Solution. Each upward step corresponds to adding H_3PO_4 and subtracting H_2O. This is equivalent to adding HPO_3. The next formula is $H_6P_4O_{13}$. Closing a tetraphosphate chain into a a ring would involve one more condensation step and the concomitant loss of one more H_2O. The cyclic tetraphosphoric acid would have the formula $H_4P_4O_{12}$. This example is closely related to **problem 30.**

The existence of condensed phosphates is very important in biochemistry. The crucial biological molecules adenosine diphosphate and adenosine triphosphate, ADP and ATP, are derivatives of pyrophosphoric acid and triphosphoric acid respectively, just as adenosine monophosphate, AMP, is a derivative of phosphoric acid. See Figure 22-6b, text p. 767.

20.3 *SULFUR CHEMISTRY*

Sulfur has less tendency to form multiple bonds in its compounds than oxygen, the element above it in Group VI of the periodic table, and more tendency to form a larger number of single bonds. In addition, its lower electronegativity means that it tends more to form covalent bonds than oxygen.

Sulfur forms a series of fluorides ranging from S_2F_2 to SF_6. All are reactive with the exception of SF_6 which is kinetically stable despite being thermodynamically unstable with respect, for example, to reaction with water. **See problem 17.**

Sulfuric acid is an industrial chemical of great importance. It is produced by the *contact process* which replaced the *lead-chamber process.* The details of the contact process present another example of the way compromise between thermodynamic and kinetic factors shapes the methods used in the production of a needed chemical.

The solution chemistry of sulfur involves sulfates, sulfites, and sulfides, all represented in the reduction potential diagrams for sulfur, text p. 699. In addition, the *thiosulfate* ion, in which sulfur is in a +2 oxidation state, exists in basic solution. The reduction potential diagram shows that the thiosulfate ion in acid tends thermodynamically to disproportionate to S and SO_2. In base, the reverse is true. S*(s)* and SO_3^{2-}*(aq)* react spontaneously to give $S_2O_3^{2-}$. Know how to use the reduction potential diagrams to verify that the standard potential for the disproportionation of thiosulfate *in acid* is +0.10 V and that the standard potential for the reverse of this reaction *in base* is +0.16 V.

DETAILED ANSWERS TO ODD-NUMBERED PROBLEMS

1. ***a)*** The equations representing the two decomposition routes of hydrazine are:

$$NH_2NH_2(l) \rightarrow N_2(g) + 2\,H_2(g)$$

$$NH_2NH_2(l) \rightarrow {}^4/_3\,NH_3(g) + {}^1/_3\,N_2(g)$$

b) The thermodynamically favored pathway at constant temperature and pressure is the one with the more negative ΔG. Compare the above two reactions, both of which involve 1 mol of hydrazine, at 298.15 K and 1 atm. The ΔG° of the first is the negative of ΔG°_f of $NH_2NH_2(l)$. It is -149.24 kJ. The ΔG° of the second reaction is ${}^4/_3\,\Delta G^\circ(NH_3) - \Delta G^\circ(NH_2NH_2)$. This is $({}^4/_3)(-16.48) - 149.24 = -171.21$ kJ. The *second* pathway is thermodynamically favored.

Another approach is to compute the free energy difference between the two sets of products. This is the ΔG° for the reaction:

$$N_2(g) + 2\,H_2(g) \rightarrow {}^4/_3\,NH_3(g) + {}^1/_3\,N_2(g)$$

This reaction has a ΔG° of -21.97 kJ. The ammonia-containing product mix is favored.

---

3. The two pathways for the decomposition of $NH_4NO_3(s)$ are:

$$NH_4NO_3(s) \rightarrow N_2O(g) + 2\,H_2O(g)$$

$$NH_4NO_3(s) \rightarrow N_2(g) + \tfrac{1}{2}\,O_2(g) + 2\,H_2O(g)$$

Using the data in Appendix D, ΔH° of the first reaction is $2(-241.82) + 82.05 - (-365.56) = -36.03$ kJ. ΔH° for the second reaction is $2(-241.82) - (-365.56) = -118.08$ kJ. In both reactions one mol of ammonium nitrate is consumed. The molar weight of NH_4NO_3 is 80.04 g mol^{-1}. Thus, 1 g of NH_4NO_3 is 0.01249 mol (this number is the reciprocal of 80.04). The ΔH°'s *per gram* of the two reactions are -0.4501 kJ g^{-1} and -1.475 kJ g^{-1}.

The products of the reactions as written have the volumes of 3 and 3½ mol of ideal gas respectively. Assuming the product gases are ideal ($PV = nRT$) at 298.15 K and 1 atm, these volumes are 73.40 L and 85.62 L. If only 1 g (instead of 1 mol) of $NH_4NO_3(s)$ reacts, the volumes are 0.01249 times these values: 0.9167 L and 1.069 L. The volume of the reactant, 1 g of solid NH_4NO_3, is less than 1 mL. If it is neglected, then ΔV is 0.917 g^{-1} for the first reaction and 1.07 L g^{-1} for the second.

---

5. The decomposition of TNT is:

$$2\,C_6H_2(NO_2)_3CH_3(s) \rightarrow 3\,N_2(g) + 7\,CO(g) + 5\,H_2O(g) + 7\,C(s)$$

Mixing $NH_4NO_3(s)$ with TNT provides a source of oxygen for a further reaction to consume the 7 mol of $C(s)$ on the right. The ammonium nitrate decomposes:

$$NH_4NO_3(s) \rightarrow N_2(g) + \tfrac{1}{2}\,O_2(g) + 2\,H_2O(g)$$

The O_2 combines with the $C(s)$ to form $CO(g)$ or $CO_2(g)$. More heat is produced and the explosion is more powerful. For instance, conversion of all the product $C(s)$ to $CO(g)$ would produce an additional 386.8 kJ of heat per mole of TNT.

---

7. ***a)*** The reduction diagram for the nitrogen oxides in base is given at the end of Section 20-1. It is:

$$NO_3^- \xrightarrow{0.01\ V} NO_2^- \xrightarrow{-0.46\ V} NO \xrightarrow{0.76\ V} N_2O \xrightarrow{0.94\ V} N_2 \xrightarrow{-0.73\ V} NH_3$$

From left to right nitrogen occurs in the +5, +3, +2, +1, 0 and −3 oxidation states. Reduction potentials for couples not directly connected are *not* simply sums of the potentials of the steps connecting them. Instead the reduction potentials are *weighted* by the number of electrons transferred. Thus, for NO_3^-——NO, the potential is [2(0.01) − 0.46] / 3 = −0.15 V. The −0.46 V step transfers 1 electron as does the 0.76 V step. The overall reduction transfers 2 electrons. For the NO_2^-——N_2O couple, the reduction potential is (−0.46 + 0.76) / 2 = 0.15 V. This is the standard reduction potential for the half-reaction:

$$2\,e^- + {}^3/_2\,H_2O(l) + NO_2^-(aq) \rightarrow \tfrac{1}{2}\,N_2O(g) + 3\,OH^-(aq)$$

b) The disproportionation of the NO_2^- ion to N_2O and NO_3^- in basic solution is represented by the equation:

$$2\,NO_2^-(aq) + \tfrac{1}{2}\,H_2O(l) \rightarrow \tfrac{1}{2}\,N_2O(g) + NO_3^-(aq) + OH^-(aq)$$

This equation is the first of the following half-equations subtracted from the second:

$$2\,e^- + NO_3^-(aq) + H_2O(l) \rightarrow NO_2^-(aq) + 2\,OH^-(aq)$$

$$2\,e^- + {}^3/_2\,H_2O(l) + NO_2^-(g) \rightarrow \tfrac{1}{2}\,N_2O(g) + 3\,OH^-(aq)$$

$\Delta\mathcal{E}^\circ$ for the disproportionation is therefore $\mathcal{E}^\circ_2 - \mathcal{E}^\circ_1$, or 0.15 − 0.01 = 0.14 V. The positive answer means that $NO_2^-(aq)$ is thermodynamically unstable in base relative to disproportionation to $NO_3^-(aq)$ and $N_2O(g)$.

---

9. NF_3 will be pyramidal and polar, according to VSEPR theory. The X−N−X angle should *increase* as X becomes larger in the sequence F, Cl, Br.

---

11. On the basis of the data given in the problem, NH_2OH should be *less* basic than NH_3. In NH_2OH the central nitrogen atom is bonded to O. Oxygen is more electronegative than H as a substituent on the N and draws over electron density. The result is that the lone pair on the N is held more closely (less effectively donated) in NH_2OH than in NH_3.

---

13. The same equilibrium, $P_4(g) \rightleftarrows 2\ P_2(g)$, between tetraphosphorus and diphosphorus is the basis for **problem 5-11.** See p. 88 of this Guide. Let P_{di} equal the partial pressure of P_2 and P_{tet} the partial pressure of P_4. Then, at an equilibrium total pressure of 1.00 atm:

$$P_{di} + P_{tet} = 1.00$$

$$K = P_{di}^2 / P_{tet}$$

The problem asks for the computation of the mole fraction of diphosphorus at several temperatures. This mole fraction, f, is equal to P_{di} divided by the total pressure. Therefore f is just P_{di} without its units because the total pressure is 1.00 atm. (The gases are assumed to be ideal.) Eliminate P_{tet} between the two equations:

$$P_{di}^2 + K P_{di} - K = 0$$

$$f^2 + Kf - K = 0$$

Getting values for f requires values of K. The ΔH° and ΔS° values for the equilibrium come from the data in Appendix D:

$$\Delta H^\circ = 2(144.3) - 58.91 = 229.69 \text{ kJ}$$
$$\Delta S^\circ = 2(218.02) - 279.87 = 156.17 \text{ J K}^{-1}$$

As long as these values are independent of temperature, K at any T is given by:

$$-RT \ln K = \Delta H^\circ - T\Delta S^\circ$$

The problem requires values of K at three different temperatures. They are computed in the following table:

t(° C)	T (K)	ΔH° (kJ)	$-T\Delta S^\circ$ (kJ)	K
800	1073.15	229.69	−167.594	9.50×10^{-4}
1000	1273.15	229.69	−198.828	5.42×10^{-2}
1200	1473.15	229.69	−230.062	1.03

The equilibrium constant for this reaction at 1200° C is not 0.612 (as given in problem 5-11) but 1.03. This of course affects the point of neither problem.

The values of K are now in turn substituted into the quadratic equation derived above. Solving it for f gives:

At 800° C $f = 0.0325$
At 1000° C $f = 0.204$
At 1200° C $f = 0.623$

---

15. Sulfuryl chloride, SO_2Cl_2, has a central S atom with a steric number of 4. The geometry about the S will be distorted tetrahedral.

---

17. From the statement of the problem, the ΔH° for the reaction:

$$SF_4(g) \rightarrow S(s) + 2\,F_2(g)$$

is +774.9 kJ. If this chemical equation is added to:

$$2\,F_2(g) \rightarrow 4\,F(g) \qquad \Delta H^\circ = 4 \times 78.99 \text{ kJ}$$

and also added to:

$$S(s) \rightarrow S(g) \qquad \Delta H^\circ = 278.80 \text{ kJ}$$

the result is:

$$SF_4(g) \rightarrow S(g) + 4\,F(g) \qquad \Delta H^\circ = 1369.66 \text{ kJ}$$

The average enthalpy change for breaking a mole of S−F bonds in $SF_4(g)$ is one-fourth of this ΔH° or 342.41 kJ mol^{-1}. A similar calculation gives the average enthalpy of a mole of S−F bonds in $SF_6(g)$. From Appendix D, the ΔH_f° of $SF_6(g)$ is −1209 kJ mol^{-1}. Hence:

$$\text{Bond enthalpy} = {}^1/_6\,[+1209 + (6 \times 78.99) + 1278.80] = 326.96 \text{ kJ mol}^{-1}$$

The bond enthalpy in $SF_4(g)$ is larger than in $SF_6(g)$. Despite the fact that the bonds in $SF_4(g)$ are thermodynamically more stable than in $SF_6(g)$, $SF_4(g)$ is much more reactive than $SF_6(g)$. $SF_6(g)$ is kinetically stable (see text p. 696).

---

19. ***a)*** The steps in the Frank-Caro process are:

$$CaCO_3(s) \rightarrow CaO(s) + CO_2(g)$$

$$CaO(s) + 3\,C(s) \rightarrow CaC_2(s) + CO(g)$$

$$CaC_2(s) + N_2(g) \rightarrow CaCN_2(s) + C(s)$$

$$CaCN_2(s) + 3\,H_2O(l) \rightarrow CaCO_3(s) + 2\,NH_3(g)$$

b) The standard enthalpy change for the second step is 464.8 kJ mol^{-1}. This is computed using the enthalpies of formation given in Appendix D.

---

21. The problem concerns a 0.100 M aqueous solution of hydrazine (NH_2NH_2) with pH 0. At this pH $[H_3O^+]$ is 1.0 M and $[OH^-]$ is 1.0×10^{-14} M. Hydrazine is a base in water:

$$N_2H_4(aq) + H_2O(l) \rightleftarrows N_2H_5^+(aq) + OH^-(aq)$$
$$N_2H_5^+(aq) + H_2O(l) \rightleftarrows N_2H_6^{2+}(aq) + OH^-(aq)$$

for which:

$$K_{b1} = 8.5 \times 10^{-7} = [N_2H_5^+][OH^-] / [N_2H_4]$$

$$K_{b2} = 8.9 \times 10^{-16} = [N_2H_6^{2+}][OH^-] / [N_2H_5^+]$$

Because $[OH^-] = 10^{-14}$ M, the ratio $[N_2H_5^+] / [N_2H_4]$ is 8.5×10^7, and the ratio $[N_2H_6^{2+}] / [N_2H_5^+]$ is 8.9×10^{-2}. The huge size of the first ratio means that effectively no N_2H_4 *(aq)* is in the solution at equilibrium. It is virtually all converted to the acid forms by the large concentration of H_3O^+ *(aq)* present at pH 0. Let x equal $[N_2H_6^{2+}]$. Then $[N_2H_5^+]$ is $0.100 - x$. The equation:

$$8.9 \times 10^{-2} = x / (0.100 - x)$$

holds, based on the K_{b2} expression. Solving the quadratic equation for x gives $[N_2H_6^{+2}] = 8.17 \times 10^{-3}$ M. Hence $[N_2H_5^+] = 9.18 \times 10^{-2}$ M.

---

23. The problem refers to the disproportionation of $NO(g)$ to $N_2O(g)$ and $NO_2(g)$:

$$3\ NO(g) \rightarrow N_2O(g) + NO_2(g)$$

The ΔG° for this reaction is $51.29 + 104.18 - 3\ (86.55) = -104.18$ kJ. The negative ΔG° means that the products of this reaction are favored (at 298.15 K and 1 atm) over the reactants. The equilibrium constant ($\ln K = -\Delta G^\circ / RT$) is large ($1.83 \times 10^{18}$), which says the same thing.

---

25. The Pt catalyzes the oxidation of $NH_3(g)$ to $NO(g)$:

$$2\ NH_3(g) + {}^5/_2\ O_2(g) \rightarrow 2\ NO(g) + 3\ H_2O(g)$$

The $NO(g)$ then immediately reacts with oxygen from the air to give the brown $NO_2(g)$:

$$2\ NO(g) + O_2(g) \rightarrow 2\ NO_2(g)$$

---

27. The N-nitroso group absorbs visible light and gives its compounds a characteristic yellow color. This means it absorbs purple, the color complement of yellow. See text Figure 18-8b. Purple is a combination of violet and red. The N-nitroso group therefore absorbs both in the violet (around 420 nm) and in the red (around 700 nm). See text Figure 18-8a.

---

29. Use VSEPR theory. In PCl_4^+ ion the central P atom has a SN of 4. The ion should be tetrahedral. In PCl_6^+ the central P atom has an SN of 6 and the ion should be octahedral.

---

31. **a)** The reduction potential for the half reaction $Ag^+(aq) + e^- \rightarrow Ag(s)$ is +0.800 V (see Appendix E). In its standard state (1 M), $Ag^+(aq)$ will oxidize any species (in *its* standard state) having a standard reduction potential less than +0.800 V. This follows because the $\Delta\xi^\circ$ for such a process is:

$$\Delta\xi^\circ = 0.800 - \xi^\circ\,(\text{anode})$$

and a positive $\Delta\mathcal{E}^\circ$ means a spontaneous process.

Refer to the reduction potential diagrams for phosphorus (text page 695). In basic solution, Ag^+ *(aq)* would precipitate as $AgOH$*(s)* or $Ag_2O\ xH_2O$*(s)*. Since the problem states that the solution contains silver ions, we can restrict discussion to the reduction potentials in acidic solution. All of the listed reduction potentials are less than +0.800 V. Therefore, as long as there is plenty of silver ion, the P_4*(s)* should be oxidized all the way to H_3PO_4*(aq)*. The balanced oxidation half reaction is:

$$P_4(s) + 16\ H_2O(l) \rightarrow 4\ H_3PO_4(aq) + 20\ H^+(aq) + 20\ e^-$$

The standard reduction potential for this half-equation is computed by combining the standard potentials available in the diagram and the number of electrons each transfers. Of course, a preliminary step is to determine the oxidation state of P at the different points in the diagram:

$$\mathcal{E}^\circ = {}^1/_5\ [2(-0.28) + 2(-0.50) + (-0.51)] = -0.414\ V$$

The overall reaction is:

$$20\ Ag^+(aq) + P_4(s) + 16\ H_2O(l) \rightarrow 4\ H_3PO_4(aq) + 20\ Ag(s) + 20\ H^+(aq)$$

This reaction has $\Delta\mathcal{E}^\circ = 0.800 - (-0.414) = 1.214$ V.

b) Zn^{2+}*(aq)* is a much poorer oxidizing agent then Ag^+*(aq)*. Its standard reduction potential is −0.763 V. The standard reduction potentials for P_4*(s)* in acidic media are all algebraically greater than this value. P_4*(s)* is not oxidized by Zn^{2+}*(aq)* in acid solution. The potentials for P_4*(s)* in *base* are not considered because Zn^{2+}*(aq)* does not exist in base. Instead there is $Zn(OH)_4{}^{2-}$*(aq)* (see Chapter 18).

---

33. Convert the 1000 g of S to moles of S by dividing by 32.064 g mol^{-1}, the molar weight of sulfur. It takes 1 mol of O_2*(g)* to react with each mole of sulfur so (1000 / 32.064) mol is the chemical amount of O_2*(g)* required. At STP, O_2*(g)* occupies 22.4 L mol^{-1} so 698.6 L of O_2*(g)* at STP is required. Air contains only 21 percent O_2 by volume. Therefore *more* than 698.6 L of air is needed. The answer is 698.6 L divided by 0.21 or 3.3×10^3 L of air.

---

35. The molecules FNO and NSF both have central atoms with a steric number of 3. VSEPR theory predicts a bond angle of approximately 120° in each.

The sulfur becomes the central atom in NSF because it is *less* electronegative than O. The structure tends to be more ionic with a build-up of positive charge on the S. The situation is also covered in the solution to problem 45 in Chapter 13 (p. 290 of this Guide).

---

37. ***a)*** The balanced equation for the production of superphosphate fertilizer must represent the production of phosphoric acid from fluoroapatite in the wet acid process followed by reaction of the phosphoric acid with more fluoroapalite. Equations for these two steps are given on text p. 693 and 704 respectively. Multiply the first through by 7 and the second through by 3. Then add them. The H_3PO_4 cancels out. The sum, with smallest whole-number coefficients is:

$$2\ Ca_5(PO_4)_3F(s) + 7\ H_2SO_4(aq) + 17\ H_2O(l) \rightarrow 7\ CaSO_4 \cdot 2H_2O(s) + 3\ Ca(H_2PO_4)_2 \cdot H_2O + 2\ HF(g)$$

The process produces 7 mol of by-product gypsum for every 3 mol of superphosphate.

b) The molar weight of $Ca(H_2PO_4)_2$ is 0.23405 kg mol^{-1}. Since the fertilizer plant produces 1×10^5 kg of this material per day, it produces (10^5 kg / 0.234 kg mol^{-1}) or 4.273×10^5 mol of superphosphate each day. Hence, it also makes $^7/_3 \times 4.273 \times 10^5$ mol of gypsum each day. This is 9.970×10^5 mol gypsum per day or (at 0.1722 kg mol^{-1}) 1.72×10^5 kg of gypsum. This weight is equivalent to 172 metric tons of gypsum per day.

If the formula of superphosphate is taken as $Ca(H_2PO_4)_2 \cdot H_2O$ (molar weight 0.25106 kg mol^{-1}) instead of $Ca(H_2PO_4)_2$ (molar weight 0.23405 kg mol^{-1}), then the answer is 160 metric tons of by-product gypsum each day. For this rough calculation the difference is of little significance.

Assuming that the plant operates 365 days a year, it produces between 5.8×10^4 and 6.3×10^4 metric tons of gypsum in a year.

c) Express the mass of the gypsum in grams. It is 6.3×10^{10} g because a metric ton is 10^6 g. Using the density of 2.32 g cm^{-3}, the volume of the by-product gypsum is (6.3×10^{10} g / 2.32 g cm^{-3}) which is 2.7×10^{10} cm^3 or 2.7×10^4 m^3. This is a small mountain of gypsum. It would cover a 100 meter $\times$ 100 meter field to a depth of 2.7 meters.

Chapter 21

THE HALOGEN FAMILY AND THE NOBLE GASES

21.1 *THE HALOGENS*

The first part of the chapter concerns the properties, reactions and compounds of the elements of Group VII, the halogens. These elements exhibit many regular trends in their properties in accord with the predictions of the periodic law. All are reactive non-metals which serve as oxidizing agents. Their oxidizing strength decreases regularly going down the group.

The chemistry of fluorine is discussed under a separate heading because of the extreme difference between it and the other halogens. These differences include:

- Fluorine exhibits only the 0 and -1 oxidation states unlike the other halogens, all of which can be oxidized to positive oxidation states ($+1$, $+3$, $+5$, and $+7$).

- Only fluorine of all the halogens has been shown to form compounds with any of the noble gases (Group VIII).

- The dissociation energy of the F_2 molecule is abnormally low.

- The oxidizing ability of fluorine is extremely high. This is a consequence of its high electronegativity, its low dissociation energy, and the strong bonds it forms with other elements.

The solution chemistry of the halogens can in large part be summarized in a set of reduction potential diagrams. (See below.) Many of the potentials come from Appendix E.

Reduction Potential Diagrams of the Halogens

Acidic Solution

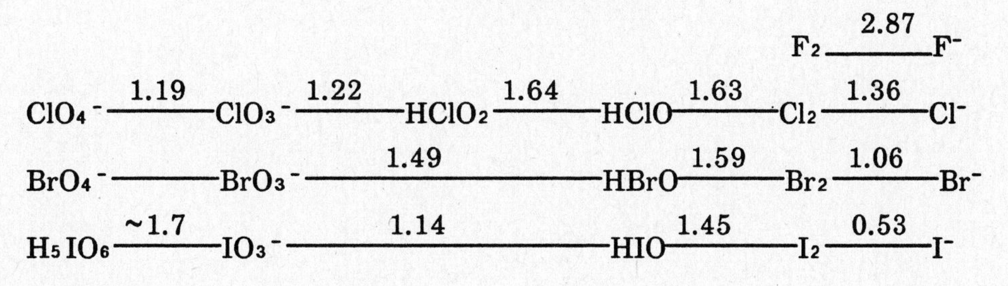

Basic Solution

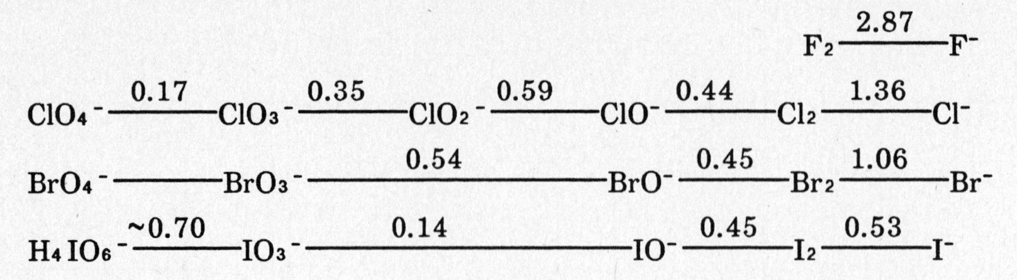

Observe how much these diagrams can convey. They show, for example:

- The high reduction potential of $IO_4^-(aq)$ in acid solution. The periodate ion can oxidize $Mn^{2+}(aq)$ to $MnO_4^-(aq)$ because the reduction potential of the MnO_4^- to Mn^{2+} couple is only about 1.5 V and is exceeded by 1.7 V.

- That F_2 oxidizes Cl^-. In addition the diagrams show that each elemental halogen will oxidize all halides (X^- ions) in compounds of the halogens beneath it in the periodic table. **See problem 25c.**

- That BrO_4^- is likely to be a powerful oxidizing agent, despite the absence of its reduction potential from the diagram.

- That the chemistry of periodic acid involves complications not present in perbromic and perchloric acids. The formulas of the major compound of iodine(VII) both in acid and base show the expansion of the iodine coordination sphere to include two additional water molecules.

The use of the diagrams continues in **problem 11.**

Interhalogens and Polyhalides

The interhalogen compounds involve two different halogen elements and have the general formula XY_n where Y is the *less* electronegative halogen. The subscript n is 1, 3, 5, or 7. Twenty-four formulas fulfill these conditions; 14 compounds have actually been prepared. The VSEPR model provides correct predictions of the geometry of the interhalogens. **See problem 1.** All of the compounds are oxidizing agents.

Polyhalides are charged (ionic) species containing three or more halogen atoms. Examples are the triiodide ion (I_3^-) and the tetrafluorobromate ion (BrF_4^-). In polyhalides all of the halogens can be the same, or there can be two or three different halogens.

21.2 *THE NOBLE GASES*

The noble gases are in Group 0 (VIII) of the periodic table. Their principal uses come from their physical properties. The unreactive character of these elements prevents their entry into chemical compounds with most other elements. Fluorine and oxygen form several compounds with Xe. Note that the ΔH_f°'s of all of the xenon fluorides are negative. The compounds are thermodynamically stable with respect to their elements.

Problem 28 amounts to the construction of a reduction potential diagram for Xe in aqueous acid. An obvious extension would be to set up the reduction potential diagram for Xe in aqueous base. The Xe(VII) in the hydrogen xenate ion, $HXeO_4^-$ disproportionates in basic solution to Xe*(g)* and the perxenate ion, XeO_6^{4-}, in which xenon is in the +8 oxidation state. This fact would be part of such a diagram. The unstable gas XeO_4 arises from the dehydration of a perxenate salt. It is the acid anhydride of perxenic acid.

- Problems in Chapter 21 all involve the application of previously studied principles to the particular cases of the halogens and noble gases.

DETAILED SOLUTIONS TO ODD-NUMBERED PROBLEMS

1. The shapes of all of these compounds can be predicted using the valence shell electron pair repulsion (VSEPR) theory. See p. &VESPRef. in this Guide.

a) OF_2 has a central O atom with a steric number of 3. This includes the two F atoms and one lone pair of electrons. The molecule is bent.

b) BF_3 has a central B atom with a steric number of 3. The molecule is trigonal planar.

c) In BrF_3 the central Br has an SN of 5. The molecule has a distorted T shape.

d) In BrF_5 the central Br is surrounded by five F atoms and one lone pair for an SN of 6. The molecule has a square pyramidal shape.

e) In IF_7 the central I is surrounded by seven F atoms. The arrangement that minimizes replusions among 7 electron pairs distributed about a center is a pentagonal bipyramid.

---

3. The problem is a reversal on the more common task of determining an equilibrium constant from ΔG data. It gives an equilibrium constant (in slightly disguised form) and asks for an estimate of ΔG°. The reaction is:

$$Br_2(l) \rightleftarrows Br_2(aq)$$

At 25° C (298.15 K), 33.6 g of bromine dissolves in one liter of water. Once the solution is saturated, the above equilibrium exists. The concentration of $Br_2(aq)$ at the point of saturation is 0.210 M (taking the molar weight of Br_2 as 159.82 g mol^{-1}). The equilibrium constant expression for the reaction is:

$$K = [Br_2](aq)$$

so K is 0.210. The equilibrium constant and ΔG° are related:

$$\Delta G^\circ = -RT \ln K$$

Substitution of R = 8.314 J $mol^{-1}K^{-1}$ and T = 298.15 K gives ΔG° equal to 3.87×10^3 J for the reaction as written, the reaction giving 1 mol of $Br_2(aq)$. The value tabulated in Appendix D as ΔG_f° of $Br_2(aq)$ is 3.93×10^3 J. Note that *some* Br_2 does dissolve despite the fact that ΔG° is positive.

---

5. The ΔG° of the reaction:

$$2\ Br^-(aq) + Cl_2(g) \rightleftarrows Br_2(g) + 2\ Cl^-(aq)$$

is:

$$\Delta G^\circ = 2(-131.23) + 3.14 - 2(-103.96)\ kJ = -51.4\ kJ$$

where the data come from Appendix D. ΔG is related to K by the equation:

$$\Delta G^\circ = -RT \ln K$$

Substituting R = 8.314 J $mol^{-1}K^{-1}$ and T = 298.15 K gives $K = 1.0 \times 10^9$. The K is large enough to make the extraction of Br_2 *(g)* from sea water by oxidation with Cl_2 *(g)* thermodynamically feasible. The same conclusion can be reached by getting K from the standard voltage of the reaction.

---

7. ***a)*** The balanced equation is:

$$Br^- (aq) + H_3PO_4 (aq) \rightarrow H_2PO_4^- (aq) + HBr(g)$$

Other versions are possible (involving NaBr as a reactant and giving PO_4^{3-} or Na_3PO_4 or other forms of phosphate as a product). This ionic equation is best.

b) The 100 mL of solution contains 0.050 M × 0.100 L = 0.0050 mol of Br^- *(aq)*. According to the balanced equation, 0.0050 mol of HBr*(g)* will form. At standard temperature and pressure (STP), HBr*(g)*, if ideal, occupies 22.4 L mol^{-1}. Hence, this portion of HBr*(g)* occupies 1.1×10^{-1} L.

---

9. Use the ΔG_f° data in Appendix D to compute ΔG° of this reaction. The set-up is:

$$\Delta G^\circ = \underset{\text{for } OCl^-(aq)}{2(-36.8)} + \underset{\text{for } H_2O(l)}{-237.18} - \underset{\text{for } OH^-(aq)}{2(-157.24)} - \underset{\text{for } Cl_2O(g)}{2(97.9)} \text{ kJ}$$

The answer is −192.1 kJ. Then substitute this value into the equation $\Delta G^\circ = -RT \ln K$. Use T = 298.15 K and R = 8.314 J $mol^{-1}K^{-1}$. Be sure to convert from the units of ΔG° from kJ to J. The ln K is 77.49, and K is 4.5×10^{33}.

---

11. ***a)*** The standard reduction potential diagram for iodine in strong base shows that I_2 *(s)* disproportionates spontaneously to IO^- *(aq)* and I^- *(aq)*:

$$I_2 (s) + 2\ OH^- (aq) \rightarrow I^- (aq) + IO^- (aq) + H_2O(l)$$

But IO^- *(aq)* itself disproportionates spontaneously to IO_3^- *(aq)* and I^- *(aq)*:

$$3\ IO^- (aq) \rightarrow IO_3^- (aq) + 2\ I^- (aq)$$

The first statement is true because 0.535 V, the standard reduction potential for I_2 *(s)* to I^- *(aq)*, exceeds 0.45 V, the reduction potential for IO^- *(aq)* to I_2 *(aq)*. The second follows from similar comparison between the IO_3^- *(aq)* to IO^- *(aq)* and IO^- *(aq)* to I^- *(aq)* reduction potentials. The final balanced reaction for the disproportionation is:

$$6\ OH^- (aq) + 3\ I_2 (aq) \rightarrow IO_3^- (aq) + 5\ I^- (aq) + 3\ H_2O(l)$$

The $\Delta\mathcal{E}^\circ$ for this reaction is (0.49 − 0.13) = 0.35 V. It is spontaneous as written.

b) The half-reaction:

$$IO_3^-(aq) + 3\ H_2O(l) + 6\ e^- \rightarrow I^-(aq) + 6\ OH^-(aq)$$

is a combination of the half-cell reductions:

$$IO_3^- \text{—} IO^- \qquad \textit{and} \qquad IO^- \text{—} I^-$$

The potentials for these are 0.14 V and 0.49 V respectively. The first transfers 4 electrons and the second transfers 2 electrons for an overall transfer of 6 e^-. The reduction potential is $(4 \times 0.14 + 2 \times 0.49) / 6 = 0.26$ V.

---

13. ***a)*** The reaction for the reduction of $XeF_6(g)$ with $H_2(g)$:

$$XeF_6(g) + 3\ H_2(g) \rightarrow Xe(g) + 6\ HF(g)$$

has a ΔH° equal to the sum of the ΔH_f°'s of the products minus the sum of the ΔH_f°'s of the reactants:

$$\Delta H^\circ = \Delta H_f^\circ(Xe) + 6\,\Delta H_f^\circ(HF) - 3\,\Delta H_f^\circ(H_2) - \Delta H_f^\circ(XeF_6)$$

Since ΔH_f° of an element in its standard state is zero:

$$-1282 \text{ kJ} = 6\,\Delta H_f^\circ(HF) - \Delta H_f^\circ(XeF_6)$$

The standard enthalpy of formation of $HF(g)$ is -271.1 kJ mol^{-1}. (See Appendix D). Inserting this value and solving give the ΔH_f° for $XeF_6(g)$. It is -345 kJ mol^{-1}.

b) From part a), the reaction:

$$XeF_6(g) \rightarrow Xe(g) + 3\ F_2(g)$$

has ΔH° equal to $+345$ kJ. According to Appendix D, the reaction:

$$\tfrac{1}{2}\ F_2(g) \rightarrow F(g)$$

has ΔH equal to 78.99 kJ. Multiplying the second equation by 6 and adding it to the first give:

$$XeF_6(g) \rightarrow Xe(g) + 6\ F(g) \qquad \Delta H^\circ = 818.94 \text{ kJ}$$

This is the energy required to dissociate 1 mol of $XeF_6(g)$. The average bond enthalpy of the Xe$-$F bond is $^1/_6$ this or 136 kJ mol^{-1}.

---

15. The SF_5^- ion has 42 valence electrons. The central S atom is surrounded by 5 F atoms and has a steric number of 5. This predicts a trigonal bipyramidal geometry about the S. See text Chapter 13.

---

17. The balanced reaction:

$$4\,HF(aq) + SiO_2(g) \rightarrow SiF_4(g) + 2\,H_2O(l)$$

shows that 4 mol of HF*(aq)* dissolves away 1 mol of quartz rock. The rock contains traces of gold. (1 part in 1.0×10^{-5} by weight). Suppose enough rock is dissolved to recover 1 troy ounce (31.3 g) of Au. Clearly, 31.3×10^5 g of rock needs to be dissolved away. Assume the rock is pure quartz. Quartz is $SiO_2(s)$ and has a molar weight of 60.08 g mol^{-1}. 31.3×10^5 g of quartz is $(31.3 \times 10^5 / 60.08)$ mol or 5.21×10^4 mol of quartz. The process consumes four times this chemical amount of HF so 2.08×10^5 mol of HF is used up to recover the troy ounce of gold. The molar weight of HF is 20.0 g mol^{-1}. Hence, 4.16×10^6 g of HF is consumed.

The HF is delivered in the form of an aqueous solution of density 1.17 g cm^{-3}. 3.56×10^6 cm^3 of such solution contains the needed 4.16×10^6 g of HF. This volume of HF*(aq)* is 3.56×10^3 L which, at $0.25 per liter, would cost $889. But a troy ounce of gold is worth only $350. The method is a losing proposition. It breaks even on *materials* when the ore is 2.54 times richer (2.54 is the ratio of $889 to $350). For real feasibility the cost of labor and equipment would have to be covered, too.

---

19. Inspection of Appendix E discloses listings of several half-reactions involving chlorine-containing species in acid solution. The species are $ClO_4^-(aq)$, $ClO_3^-(aq)$, $ClO_2^-(aq)$, $ClO^-(aq)$, $Cl_2(g)$ and $Cl^-(aq)$. The oxidation numbers of the Cl atom in six species are 7, 5, 3, 1, 0, −1 respectively. The layout of the reduction potential diagram therefore is:

$$ClO_4^- \text{———} ClO_3^- \text{———} ClO_2^- \text{———} ClO^- \text{———} Cl_2 \text{———} Cl^-$$

(Acid Solution)

Several of the reduction potentials needed to complete this diagram come immediately from Appendix E:

$$ClO_4^- \overset{1.19\text{ V}}{\text{———}} ClO_3^- \text{———} ClO_2^- \overset{1.64\text{ V}}{\text{———}} ClO \overset{1.63\text{ V}}{\text{———}} Cl_2 \overset{1.358\text{ V}}{\text{———}} Cl^-$$

Appendix E also gives the potential for the ClO_3^- to Cl_2 reduction (1.47 V). This reduction involves gaining 5 electrons. It achieves the same product as the ClO_3^- to ClO_2^- reduction (2 electrons) followed by ClO_2^- to ClO^- (2 electrons) and ClO^- to Cl_2 (1 electron). Let x equal the standard reduction potential that is missing in the above diagram, the potential of the ClO_3^- to ClO_2^- step. Then:

$$5(1.47) = 2\,x + 2(1.64) + (1.63)$$

and x is 1.22 V. Therefore, in acid solution:

$$ClO_4^- \overset{1.19\text{ V}}{\text{———}} ClO_3^- \overset{1.22\text{ V}}{\text{———}} ClO_2^- \overset{1.64\text{ V}}{\text{———}} ClO^- \overset{1.63\text{ V}}{\text{———}} Cl_2 \overset{1.358\text{ V}}{\text{———}} Cl^-$$

---

21. The reaction is the disproportionation of $KClO_3(s)$:

$$4\ KClO_3(s) \rightarrow 3\ KClO_4(s) + KCl(s)$$

The $\Delta G°$ is -133.43 kJ mol^{-1}. This answer comes *via* the standard method of computing $\Delta H°$ and $\Delta S°$ and combining them to give $\Delta G°$. **See problem 8-13** for example. This reaction is *spontaneous.*

For comparison, $\Delta\mathcal{E}°$ for this disproportionation taking place in acid solution is 0.26 V. This is computed using the data in the previous problem. Thus, in acid aqueous solution, the similar disproportionation:

$$4\ ClO_3^-(aq) \rightarrow 3\ ClO_4^-(aq) + Cl^-(aq)$$

is also spontaneous under standard conditions. Its $\Delta G°$ is -150.5 kJ mol^{-1} (from $-nF\Delta\mathcal{E}°$, with n = 6). The difference between the two $\Delta G°$ values stems from the difference between the solid state and the aqueous medium.

---

23. Use the definition of the Pauling electronegativity:

$$\Delta^2 = 0.0104\ [\Delta E_d(AB) - (\Delta E_d(A_2) \times \Delta E_d(B_2))^{\frac{1}{2}}]$$

together with the electronegativities in Figure 13-6, text p. 461. Take the ΔH values in Table 13-1 as an acceptable approximation for the dissociation energies (ΔE_d) of the halogens. This short-cut is discussed in **problem 13-11.** See p. 283 in this Guide. The six enthalpies of dissociation, in kJ mol^{-1}, are: ClF, 261; BrF, 275; IF, 322; BrCl 220; ICl, 216; BrI 179.

---

25. ***a)*** The trend among the hydrogen halides for acid strength is: HI > HBr > HCl.

Therefore HAt should be an even stronger acid than HI.

b) The question concerns the reduction potential for the half-reaction:

$$At_2(s) + 2\ e^- \rightarrow 2\ At^-(aq)$$

Zn*(s)* reduces $At_2(s)$. As it does it forms $Zn^{2+}(aq)$. The standard reduction potential for the $Zn^{2+}(aq)$–$Zn(s)$ couple is -0.763 V. Therefore, $\mathcal{E}°$ for the $At_2(s)$ to $At^-(aq)$ reduction exceeds -0.763 V. On the other hand, At_2 does not react with $Fe^{2+}(aq)$. The standard reduction potential for reduction of $Fe^{3+}(aq)$ to $Fe^{2+}(aq)$ is 0.770 V. Therefore $\mathcal{E}°$ for $At_2(s)$ is less than 0.770 V.

Furthermore, on the basis of the trends in Group VII, the standard reduction potential for $At_2(s)$ should also be less than 0.535 V, the value for $I_2(s)$.

c) The $Cl_2(g)$ will easily oxidize the $At^-(aq)$:

$$2\ At^-(aq) + Cl_2(g) \rightarrow 2\ Cl^-(g) + At_2(s)$$

d) Solid At_2 should have an even greater tendency to disproportionate in aqueous base than $I_2(s)$. Thus:

$$6\ At_2(s) + 12\ OH^-(aq) \rightarrow 2\ AtO_3^-(aq) + 10\ At^-(aq) + 6\ H_2O(l)$$

---

27. The reaction can be viewed as the sum of the donation of a fluorine ion by XeF_2 and the acceptance of a fluoride ion by AsF_5:

$$2\,XeF_2 \rightarrow Xe_2F_3^{+} + F^{-} \qquad \textit{and} \qquad AsF_5 + F^{-} \rightarrow AsF_6^{-}$$

The F^- is a bearer of an electron pair. Thus the AsF_5 is an electron-pair acceptor (a Lewis acid) and the XeF_2 molecules are electron pair donors (Lewis bases).

---

29. The ΔH° of the hydrolysis reaction:

$$XeF_6(g) + H_2O(l) \rightarrow XeOF_4(l) + 2\,HF(g)$$

is calculated by the usual combination of ΔH°_f values:

$$\Delta H^\circ_{hydrolysis} = \Delta H^\circ_{f\,XeOF_4} + 2\,\Delta H^\circ_{f\,HF} - \Delta H^\circ_{f\,XeF_6} - \Delta H^\circ_{H_2O(l)}$$

The first and third ΔH°_f's are given in the problem. From Appendix D, the ΔH°_f of $HF(g)$ is -271.1 kJ mol^{-1}, and the ΔH° of liquid water is -285.83 kJ mol^{-1}. Therefore:

$$\Delta H^\circ\,(\text{hydrolysis}) = 148\text{ kJ} + 2\,(-\,271.1\text{ kJ}) - (-298\text{ kJ}) - (-285.83\text{ kJ})$$

$$\Delta H(\text{hydrolysis}) = +189.63\text{ kJ}$$

The standard enthalpy change for hydrolysis of $XeF_6(g)$ is, to the correct number of decimal places, 190 kJ mol^{-1}.

Chapter 22

ORGANIC CHEMISTRY AND BIOCHEMISTRY

Organic chemistry is the study of the reactions and properties of compounds containing carbon. Carbon has an intermediate electronegativity and readily forms strong covalent bonds. The bonds it forms to H and to other C's resist reaction with air and water. The text first considers *hydrocarbons*, which contain only carbon and hydrogen, and then takes up derivatives of hydrocarbons in which other atoms attach as a *functional group* to the hydrocarbon skeleton.

22.1 *HYDROCARBONS*

Hydrocarbons fall into four categories. The first three, *alkanes*, *alkenes*, and *alkynes*, are the *aliphatic* hydrocarbons. *Aromatic* hydrocarbons (*arenes*) are the fourth. Carbon's unique propensity for forming stable chains allows the existence of straight chain, branched chain and *cyclic*, or ring structures. Aliphatic hydrocarbons lacking any rings are *acyclic*. Other aliphatic hydrocarbons are cyclic. All aromatic hydrocarbons contain at least one ring.

Saturated Acyclic Hydrocarbons

Alkanes are called *saturated* hydrocarbons because all the carbon atoms in the molecule are combined with as many hydrogen atoms as possible. All C to C (and C to H) bonds are *single* bonds. The generic formula for acyclic alkanes in $C_n H_{2n+2}$.

The carbon atoms in acyclic alkanes are bound together in chains which are branched or unbranched. The names of the straight chain alkanes having from 1 to 15 C atoms are listed in textbook Table 22-2.

- Memorize the name of at least the first ten straight-chain alkanes.

Note that after $n=4$ the names employ familiar stems followed by the suffix "-ane." The compounds are a *homologous* series.

There is free rotation about C$-$C single bonds and therefore many *conformations* for acyclic alkanes as various segments of the chain rotate into proximity with each other. Different conformations of a given carbon chain look different but are the same compound as long as they have the same sequence of atoms.

Compounds that have the same chemical formula but *different* atom-to-atom bonding sequences are structural *isomers*. No amount of free rotation can convert one structural isomer into another. Instead, bonds must be broken and re-formed. The standard (IUPAC) system of naming casts light on the nature of isomers by it way it deals with the problem of naming them. Follow these steps:

1. Identify the longest continuous chain of C atoms. Chain length can be concealed by writing structures in zig-zags on the page. Do not be deceived. What counts is *sequence.* Follow the chain around corners.

2. Number the atoms in the chain, starting from the end nearer any branches. The aim is to give the lowest possible number to the positions of side-groups. *Substituent groups* (or side-groups) are branches from the main chain. Alkyl side-groups are named by removing the suffix "-ane" from the name of the alkane and adding "-yl."

4. Write the name of the compound starting with a number for each side group to tell where it is attached, followed by the name of each side-group and finally by the name for the main chain.

5. Side-groups may be named either in alphabetical order or in order of increasing complexity. The presence of identical side-groups at different locations on the chain is indicated by the appropriate multiplying prefix. **See problem 1.**

The above procedure does not deal with every possible naming situation in the alkanes, but it does work with most common compounds.

Example: Name the following alkane:

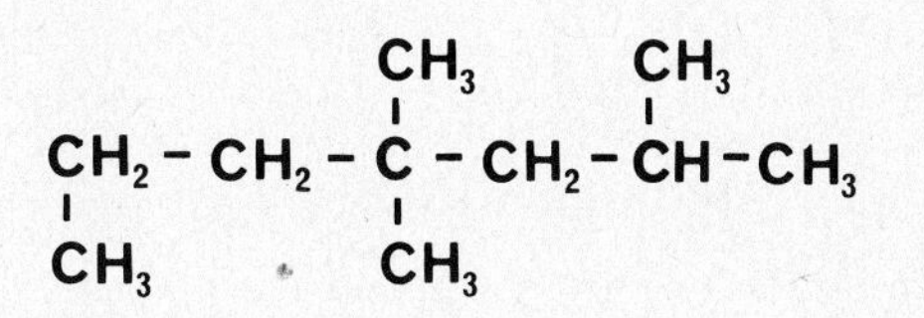

Solution. The name is 2,4,4-trimethylheptane. One common wrong answer is "1,3,3,5-tetramethylhexane." The methyl group which is apparently a side-group on the left-most carbon in the main line of the structure is really part of the chain. Once the longest chain is identified, there is the question of which direction to use for numbering. Numbering this chain from the right to the left, instead of the reverse, gives *smaller* prefix numerals in the name. That is, "2,4,4-" is preferable to "4,4,6-".

Unsaturated Acyclic Hydrocarbons

Alkenes contain at least one carbon-carbon double bond. They are *unsaturated* because they can take up H_2 to form alkanes. Acyclic alkenes have the generic formula C_nH_{2n}. If there are two double bonds in an acyclic hydrocarbon, it is a *diene* and has the generic formula C_nH_{2n-2}. Acyclic trienes have the formula C_nH_{2n-4}, etc. There is no free rotation about the C to C double bond. This allows the existence of *cis* and *trans* isomers. See text page 738. *Cis* and *trans* isomers have their atoms connected in the same sequence but lying in different geometrical relationships.

Alkenes are named similarly to alkanes. Always pick the longest chain *that includes the double bond.* The position of the double bond is signified in the name by the number of the carbon atom which is followed by the bond in the structural formula. Number the atoms of the chain so that this number is as small as possible. Then comes the root name of the parent hydrocarbon followed by the ending "-ene" to indicate the alkene. The trivial names "ethylene" and propylene" for ethene and propene respectively must be memorized.

Alkynes contain C to C triple bonds. The simplest alkyne is ethyne (acetylene). Again, the chain selected for naming must include the triple bond. In naming these compounds, the position of the triple bond is indicated by a number, and the ending "-yne" is attached. When there is no ambiguity the number is omitted. For example, 1-propyne and 2-propyne are the same compound, propyne. The same policy of simplification applies to alkenes (above). There is only one possible propene, so numbers are not needed in its name.

Alicyclic Hydrocarbons

These hydrocarbons all contain at least one ring. They are named by using the prefix "cyclo'" in conjunction with the appropriate root name for the number of carbons found in the ring. Thus, propane is C_3H_8, an alkane or saturated hydrocarbon, but, *cyclo*propane is C_3H_6. It has the same formula as prop*ene.* Formation of a ring from an acyclic aliphatic hydrocarbon reduces the number of H atoms by 2. Do not forget to include cyclic structures when listing the possible isomers of a hydrocarbon which is, by its formula, unsaturated. **See problem 5d.**

Aromatic Hydrocarbons

Benzene, C_6H_6, is the archtypal aromatic hydrocarbon. It is an unsaturated six carbon ring with a system of delocalized π electrons. See text p. 484. Each C to C bond is effectively a 1 ½ bond. The benzene ring is particularly stable thanks to the *resonance stabilization energy.* **See problem 13-55,** page 291 of this Guide, for a computation of benzene's resonance stabilization energy.

Substituent groups may be attached to the benzene ring. It is customary to omit the hydrogen atoms when drawing the structure of the benzene ring. Avoid the related errors of not counting these H's in molecular formulas and forgetting to subtract them from molecular formulas when they are replaced by substituent groups. Thus, $C_6H_5-CH_3$ is methylbenzene. One of benzene's six hydrogen atoms leaves and the methyl group replaces it.

The $-C_6H_5$ group is the *phenyl* group. If the $-C_6H_5$ group is viewed as a side-group in $C_6H_5-CH_3$, the compound is named phenylmethane. The generic name for aromatic side-groups is *aryl.*

If two side-groups are attached to the benzene ring, *three* different isomers are possible. Relative positions are distinguished by numbers. **See problems 5c and 23.** The 1,2-, 1,3-, and 1,4- di-substitution patterns are named *ortho, meta and para,* respectively.

22.2 *FUNCTIONAL GROUPS*

Hydrocarbon derivatives are compounds that contain elements such as oxygen, nitrogen, phosphorus, sulfur and the halogens in addition to carbon and hydrogen. Derivatives occur with specific combinations of non-hydrocarbon atoms acting as *functional groups.* Common functional groups are given in text Table 22-4 (text p. 744).

Many functional groups can be regarded as derived from a simple inorganic molecule by substitution of an alkyl or aryl group. The general symbol for an alkyl group is R and for an aryl group Ar.

- *Alkyl halides* and *aryl halides* have similar generic formulas R−X and Ar−X. Formally, they are derivatives of the hydrohalic acids H−Cl, H−Br and H−I.

- *Alcohols* and *phenols,* R−OH and Ar−OH, can be regarded as derived from water (H−OH) by replacing one H with an alkyl or aryl side group. In alcohols, the −OH group must be attached to a carbon atom that is *saturated,* one with four single bonds.

- *Ethers* all contain the R−O−R′ functional group where R and R′ are the same or different alkyl or aryl group. They are like water (H−O−H) with *both* H's replaced. Neither of the carbon atoms attached to the ether O may itself be double-bonded to an O atom, because then the molecule would be an ester (see below).

- *Amines* can be regarded as derivatives of ammonia, NH_3, in which one, two or three H atoms are replaced by alkyl or aryl groups. They are subclassified: RNH_2 is a primary amine, R_2NH is a secondary amine, and R_3N is a tertiary amine.

- *Aldehydes* and *ketones* contain a carbonyl group: $\rangle C=O$. The aldehydes have at least one H bonded to the carbonyl. The other group bonded to the carbonyl carbon is −R (an alkyl group) or −Ar (an aryl group). The simplest aldehyde is $H_2C=O$ (formaldehyde). The ketones have alkyl or aryl groups replacing both H atoms in $H_2C=0$.

- *Carboxylic acids* have the formula R−COOH or Ar−COOH. They can be viewed as deriving from HO−COOH (carbonic acid) by the formal replacement of an −OH group with an −R. They act as acids, forming *salts* upon neutralization with bases.

- *Esters* derive from carboxylic acids when the acid reacts with an alcohol:

$$R-COOH(acid) + HO-R'(alcohol) \rightarrow R-COOR'(ester) + H_2O$$

Reaction Types

In concurrence with its listing of organic functional groups, the text sketches the *kinds of reactions* that the different functions undergo. These fall into four classes:

• *Elimination,* • *Substitution,* • *Addition,* • *Oxidation-Reduction*

Addition reactions are common for the alkenes and alkynes. H_2 (or Br_2 or HX or another small molecule) adds across the double bond (or triple bond) to reduce it a single bond (or double bond). **See problem 27.**

Alcohols undergo both *substitution* and *elimination* reactions. In elimination, the HOH molecule is split out of an alcohol, giving an alkene. Many different groups substitute for the −OH group of the alcohol.

The carbonyl group in aldehydes and ketones is the site for their *addition* reactions. Addition of H−H across this bond forms alcohols. This is formally an oxidation-reduction reaction. The C=O bond gains electrons from the H−H. Addition of the *Grignard reagent*, R−MgX, across the carbonyl double bond in aldehydes and ketones is a particularly useful synthetic reaction since it lengthens carbon chains. **See problem 11d.** The carbonyl group can also be oxidized to form carboxylic acids. Aldehydes are much easier to oxidize than ketones, which is the reason they have a distinct name. Finally, aldehydes *disproportionate* (in base) to an alcohol plus a carboxylic acid. This, the Cannizzaro reaction, echos the disproportionation of many inorganic species from intermediate oxidation states. **See problem 11b.**

Carboxylic acids can be *reduced* to aldehydes and even further, to alcohols, by suitable reducing agents. They react with alcohols to form esters. In this reaction the −OR group of the alcohol substitutes for the −OH group of the carboxylic acid. Water is the by-product. **See problem 13d.**

22.3 *SUBSTITUTION REACTIONS IN THE AROMATIC RING*

The characteristic reaction of aromatic ring systems is *not* addition to the apparent double bonds. Because of the stability of the π system in arenes, substitution reactions are favored over addition reactions. Addition would disrupt the stable π electron system. The *Friedel-Crafts* reaction achieves the substitution of alkyl groups into an aromatic ring. In the reaction, an arene is mixed with an alkyl chloride and $AlCl_3$. The Lewis acid $AlCl_3$ abstracts a Cl^- from the alkyl chloride. The resulting alkyl carbonium ion, R^+, attacks the aromatic ring and displaces an H^+. Note that the name "carbonium ion" designates a *positive* ion (-onium) with the charge mainly located on a carbon atom.

22.4 *SYNTHESIS OF ORGANIC COMPOUNDS*

The text outlines several organic syntheses to illustrate how *strategies* for making desired structures are formulated. Designing an organic synthesis requires:

- Familiarity with the types of organic reactions reviewed in the previous sections. They are: addition, elimination, substitution and oxidation-reduction.

- Knowledge of the *nomenclature* of the compounds since the names, not the structures, of the starting materials and proposed product are often given.

- Familiarity with the oxidation-reduction relationships among functional groups. For example:

most oxidized			*most reduced*
carboxylic acid	aldehyde	alcohol	alkane
RCOOH	RCOH	RCH_2OH	RCH_3

In this series, each member is reduced by the gain of 2 electrons relative to its immediate neighbor to the left. Thus, a synthesis of a desired carboxylic acid might succeed *via* oxidation of the corresponding aldehyde. **See problem 11c.**

Another useful series is:

most oxidized		*most reduced*
alkyne	alkene	alkane

In this series each member is also reduced by the gain of 2 electrons relative to its neighbor to the left. The compounds however include no oxygen. Once a hydrocarbon is reduced all the way to an alkane, continued efforts to force in hydrogen would break up the C–C bonds and cause CH_4 to form.

Proposing a synthesis draws on all skills covered in the chapter and makes an ideal question to test knowledge of organic chemistry. Often, proposed schemes of synthesis might be possible in the sense that at least some product would arise if the reactions were actually performed. A synthesis that gives poor yields of impure product is, however, not a decent synthesis and not an acceptable answer to a question that requires proposal of a synthesis.

22.5 *SYNTHESIS OF POLYMERS*

Polymers are molecules built up by the linking together, in long chains, sheets, or three-dimensional networks, of many identical structural units called *monomers*. Silicate minerals (Chapter 19) are polymers. Man-made polymers are usually based on organic starting materials.

The acid or base catalyzed polymerization of phenol-formaldehyde mixtures to give the rigidly cross-linked polymer *Bakelite* is a *condensation* polymerization. The reaction splits out one molecule of water for every link between monomer units. Other condensation polymerizations are the synthesis of Nylon, a polyamide, and Dacron, a polyester. In both of these cases, monomers with functional groups on *two* ends link into long chains. In Nylon, a dicarboxylic acid reacts with a diamine. In Dacron, a dicarboxylic acid reacts with a diol (di-alcohol).

The polymerization of alkenes such as ethylene (ethene) is an *addition polymerization*. In this kind of polymerization, no small molecules are split out as by-products. The double bonds which exist in the monomer vanish in the polymer. **See problem 17.** Polymerization can be initiated by catalysts (often peroxides) or by radiation, high temperature or high pressure.

22.6 *BIOCHEMISTRY*

The text takes an historical approach. It identifies four groups of chemicals found under prebiotic conditions and involved in the evolution of life:

- *Amino acids.* These molecules contain an amine ($-NH_2$) and carboxylic acid group ($-COOH$) attached to the same C atom. Amino acids are amphoteric. **See problem 6-51** (p.136 of this Guide). They are also, with the exception of glycine, optically active (available in mirror image forms; see p.349 in this Guide). Twenty different amino acids commonly occur in proteins.

- *Monosaccharides.* Carbohydrates have the generic formula $C_nH_{2n}O_n$. The name reflects the fact that their formulas are the formulas of carbon that has been hydrated ($nC + nH_2O$). In more organic terms, monosaccharides are polyhydroxyaldehydes and ketones. There is one O atom per carbon atom, mostly as directly attached $-OH$ groups. The following 5-carbon pentahydroxyketone is an example of a carbohydrate:

$$\begin{array}{c} \quad\quad\quad\;\; OH \;\;\; OH \;\;\; O \\ \quad\quad\quad\;\; | \quad\;\;\; | \quad\;\; \| \\ HO-CH_2-CH-CH-C-CH_2-OH \end{array}$$

It is also a monosaccharide or simple *sugar.* Further, it is a *pentose* since it has 5 carbon atoms. There are also trioses, tetroses and hexoses. The "-ose" ending signifies a saccharide. Monosaccharides with long enough carbon chains often exist as five membered or six-membered rings. Suppose the first C atom on the left of the above *open-form* structure curls back and links to the carbonyl O atom. A $C-O-C$ linkage forms, and the cyclic-form sugar *ribose* (Figure 22-5, text p. 766) forms.

- *Nucleotides.* These molecules are more complex. They consist of three parts: a base, a cyclic sugar and phosphoric acid. The base portion of a nucleotide is a complex nitrogeneous molecule. Certain bases are so much involved in biochemistry that their structures should be memorized. They are thymine, adenine, cytosine and guanine. The sugar part of nucleotides is ribose or deoxyribose. Phosphoric acid is H_3PO_4.

- *ADP and ATP.* These sets of initials stand for adenosine diphosphate and adenosine triphosphate. Adenosine monophosphate (AMP) is the nucleotide that includes adenine, ribose and the phosphate group. Addition of a second and third phosphate group to AMP gives first ADP and then ATP, which are important *high-energy* compounds.

Anaerobic Fermentation

Glycolysis, the breakdown of glucose to lactic acid in the absence of oxygen, has a negative ΔG. It fuels the conversion of ADP^{3-} to ATP^{4-}, a process with a positive ΔG. The ATP^{4-} molecule then serves as an excellent source of free energy to drive non-spontaneous reactions which cells must perform to survive. The -4 ion is the principal form of the tetraprotic acid ATP at equilibrium in aqueous solution with pH values near 7. The reason for the -4 is clear when ATP is written as a derivative of triphosphoric acid ($H_5P_3O_{10}$). The adenosine replaces one H so the formula of

ATP in these terms is ($H_4P_3O_{10}$ – adenosine). The four remaining H's are typical phosphoric acid protons. NAD^+ (nicotinamide adenine dinucleotide) catalyzes glycolysis. During the first step of glycolysis, NAD^+ is reduced to NADH but then reoxidized in a subsequent step. Study glycolysis by writing out balanced chemical equations for the transformations in Figure 22-7 on text p. 769.

Note that Figure 22-8 on text p. 770 gives the structures of *four* compounds. They are NAD^+, NADH, $NADP^+$ and NADPH. Only the first two are involved in glycolysis. All four compounds are derivatives of diphosphoric acid (pyrophosphoric acid). The structures of all four of the compounds are important because *photosynthesis* involves the latter two.

Photosynthesis produces supplies of NADPH, the reduced form of $NADP^+$, as well as the high energy compound ATP. *Anaerobic* photosynthesis uses the energy of light to produce ATP and NADPH, consuming H_2S and other reducing agents as hydrogen sources. *Aerobic* photosynthesis uses the energy of light in another photosystem to produce the same products but consuming water as the reducing agent and liberating O_2 *(g)*. The other products, ATP and NADPH, provide energy for biosynthesis, in the *Calvin cycle*, of *polysaccharides* from carbon dioxide and water.

Respiration and Metabolism

Anaerobic metabolism (glycolysis) produces only two molecules of ATP per molecule of glucose. Aerobic metabolism is more efficient and produces as many as 38 molecules of ATP per molecule of glucose. The *citric acid cycle* (text p. 774) summarizes the central chemistry of aerobic metabolism. The cycle is preceded by the breakdown of the glucose molecule to two pyruvate anions. The citric acid cycle then produces ATP and the reduced forms of *two* enzymes, NADH and $FADH_2$, as it oxidizes the pyruvate the rest of the way to CO_2. FAD stands for flavine adenine dinucleotide. In a final, post-cycle step, the reduced forms of the enzymes are oxidized with oxygen to regenerate the original enzymes (NAD^+ and FAD) and produce additional supplies of ATP.

Proteins and Enzymes

Proteins are long-chain condensation polymers of amino acids. The amino group on one amino acid links to the acid group on the next in the *peptide linkage*. Every time a link is forged, a molecule of water is split out. Creation of a peptide linkage is non-spontaneous at room conditions.

Molecules with relatively few peptide links are called polypeptides. Really long-chain molecules are proteins. No matter how many links there are in such a chain the two ends are distinguishable. One is the amino end, and one is the acid end. **See problem 31.** There is essentially an infinite number of possible different protein molecules. Not only is the length of the polymer chain variable, but also different amino acids may be linked in different orders. Twenty amino acids are commonly found in proteins. A choice of 20 different monomers at every step in the lengthening of the chain gives an astronomical number of possible molecules very quickly. **See problem 29.**

The structure of proteins involves more than simply the order of the amino acid residues going down the chain. *Fibrous proteins* have regular three-dimensional structures. The polymeric chain may coil into a spiral from which the side-groups of the amino acids extend outward in a helical pattern. The coiling is governed by the formation of hydrogen bonds within the strand. Alternatively, neighboring polypeptide chains may hydrogen-bond together in sheet-like structures. *Globular proteins* have chains that are folded irregularly into a more or less compact globular shape. The folding causes amino acids which are widely separated in the sequence along the chain to lie adjacent to each other in the globule. This kind of folding is vital in the functioning of hemoglobin, the oxygen-carrying molecule in the blood. The presence of coordinated iron is of course also essential. **See problem 35.**

The same folding occurs in the structures of *enzymes*. Enzymes are globular proteins that catalyze particular reactions in the cell. The exact folding creates active sites at which steric and electronic factors combine to hold substrate molecules in such as way as to enhance the rates of specific reactions.

Nucleic Acids

Nucleic acids preserve and transmit information about the amino acid sequence in proteins. They are *nucleic* because they occur in the nuclei of cells. They are *acids* because they include $-(HO)OPO_2-$ groups which are proton donors. Nucleic acids are polymers. The backbone of the polymer chain consists of alternating phosphate and cyclic pentose sugars. In *ribonucleic acid* (RNA), the sugar is ribose, and in *deoxyribonucleic acid* (DNA) it is the closely related sugar deoxyribose. See Figure 22-6b, text p. 767. Each sugar has as a side-group a nitrogeneous base. In DNA, the bases are cytosine (C) guanine (G), adenine (A) and thymine (T). Expressed in somewhat different terms, a nucleic acid molecule is built by linking four different nucleotide units together in a long chain. Recall that a nucleotide consists of a base plus a sugar plus a phosphate.

The order of the bases along the polymeric chain encodes the information which the nucleic acid maintains or transmits. A nucleic acid molecule only 10 nucleotide units long has 4^{10} = 1,048,580 possible isomers based on the different sequences of the nucleotides. The concept is the same as the theme of **problem 29.** The difference is that the subject now is nucleotides linking together, not amino acids.

In DNA, two polymeric strands intertwine in a double helix. The cystosine side-group on one strand links through hydrogen bonds to a guanine side-group on the other. Each adenine links through hydrogen bonds to a thymine. The base-pairings are quite specific:

C...G
A...T

When the DNA double helix is unwound in the present of a supply of nucleotides and a suitable means for their delivery, each single strand serves as a template for the creation of a new complementary strand. The result is two new DNA molecules identical in their base sequences to each other and to their progenitor.

DETAILED ANSWERS TO ODD-NUMBERED PROBLEMS

1. *a)* 2-methyl-3-ethylhexane *b)* 3-methyl-1-butene *c)* 2-pentene
d) 2,5-hexadiene

---

3. Answers require: **1.** drawing a hydrocarbon chain with the proper number of C atoms. Memorize the names for C_1 through C_{10} straight-chain alkanes; **2.** inserting hydrocarbon side-groups (substituents, radicals) at proper positions (according to the numbers) starting from either end of the chain; **3.** inserting H atoms so that every C atom has 4 covalent bonds. See text p. A.81 for actual diagrams.

---

5. *a)* The pentane chain is 5 carbons long. If either of the methyl groups in dimethylpentane were at the 1 or 5 position of this chain, it would become a hexane chain. These possibilities are, therefore, not considered. The possible substitutions at atoms 2 through 4 are *four* in number: 2,2-; 2,3-; 2,4-; 3,3-. Note that 3,4 is the *same* as 2,3- and that 4,4- is the *same* as 2,2-. The point is that the two ends of the alkane chain are interchangable.

b) The cyclopentene ring contains 5 carbon atoms in a ring. Between any two of them is a double bond. Let these two C atoms be numbered 1 and 2. Then the methyl group can be at three unique locations, the 1, 3 and 4 C atoms. The isomers are: 1-methylcyclopentene, 3-methylcyclopentene and 4-methylcyclopentene. Note that 2-methylcyclopentene is the *same* as 1-methylcyclopentene.

c) Trimethylbenzene has three isomers. They are 1,2,3-, 1,2,4- and 1,3,5-,trimethylbenzene. All of these compounds have trivial names: hemimellitene, pseudocumene, and mesitylene, respectively. The merit of systematic nomenclature is apparent.

d) The hydrocarbon C_4H_8 has the general formula of an alkene. If the four carbons are set in a straight chain, there are two possible compounds, 1-butene and 2-butene. If the chain is branched, then 2-methylpropene results. Cyclic structures must be considered. Cyclobutane, with a four-membered ring, and methylcyclopropane, with a three-membered ring, are both C_4H_8 hydrocarbons. Thus, there are *five* possible structures. In addition, 2-butene has two geometrical isomers. They are the *cis* and the *trans*. See text p. 738. The problem asks for structural isomers, isomers which are interconverted by the breaking and making of chemical bonds. The two 2-butenes are geometrical isomers, interconverted by rotation about the C=C bond, and are not counted among the answers.

---

7. *a)* There are three *families* of isomers among the tri-substituted benzenes, the 1,2,3-, 1,2,4- and 1,3,5- families. **See problem 5c.** The numbers tell the position of substitution of the side-groups on the benzene ring. Take the families one at a time. With three different substituents, there is just *one* 1,3,5- compound. The symmetry of the 1,3,5- pattern makes this compound stay the same when the positions of the substituents are interchanged.

There are *three* 1,2,3- compounds. The trick is that the 2-position is unique and can accommodate each one of the 3 different substituents to give a different compound. Exchanging the two outer substituents is the same as turning the molecule over and does not lead to new isomers.

In the 1,2,4- pattern, all three of the positions for substitution have unique surroundings. There are $3 \times 2 \times 1 = 6$ different isomers.

The compounds are best named as derivatives of *toluene* (methylbenzene). In toluene, the 1- position is always occupied by the methyl group. The 10 names:

1,3,5- family

3-chloro-5-nitrotoluene

1,2,3- family

2-chloro-3-nitrotoluene
3-chloro-2-nitrotoluene
2-chloro-6-nitrotoluene

1,2,4- family

2-chloro-4-nitrotoluene
2-chloro-5-nitrotoluene
3-chloro-4-nitrotoluene
4-chloro-2-nitrotoluene
4-chloro-3-nitrotoluene
5-chloro-2-nitrotoluene

b) There are *three* isomers of bromoanaline: *ortho*-bromoanaline, *meta*-bromoanaline and *para*-bromoanaline, or 2-bromoanaline, 3-bromoanaline and 4-bromoanaline.

c) There are *three* isomers of dichloropropionic acid. 2,3-dichloropropionic acid, 2,3-dichloropropionic acid and 3,3-dichloropropionic acid. The 1-carbon is the carboxylic acid carbon and cannot be further substituted upon.

---

9. **a)** The standard enthalpy of formation of $C_3H_6(g)$ is the ΔH° of the reaction:

$$3\ C(s) + 3\ H_2(g) \rightarrow C_3H_6(g)$$

The reaction is equivalent to the atomization of 3 mol of $C(s)$ and 3 mol of $H_2(g)$ followed by the formation of 3 C−C and 6 C−H bonds. It is the sum of the steps:

$$3\ C(s) \rightarrow 3\ C(g)$$
$$3\ H_2(g) \rightarrow 6\ H(g)$$
$$3\ C(g) + 6\ H(g) \rightarrow C_3H_6(g)$$

The difference in bond enthalpies combined over the three steps, is:

$$\Delta(\text{B.E.}) = 6(413) + 3(348) - 3(716.7) - 6(218.0)\ \text{kJ}$$

The ΔH°_f of $C_3H_6(g)$ is the *negative* of the difference in bond enthalpies; it is -63.9 kJ mol^{-1}.

b) From Hess's law, the ΔH° of the combustion of $C_3H_6(g)$:

$$C_3H_6(g) + {}^9/_2\ O_2(g) \rightarrow 3\ CO_2(g) + 3\ H_2O(g)$$

equals the difference between the enthalpy of formation of the products and the reactants. It is −1959 kJ mol⁻¹. Therefore:

$$-1959 \text{ kJ} = 3(-393.51) + 3(-241.82) - {}^{9}/_{2}(0) - \Delta H^\circ_f(C_3H_6) \text{ kJ}$$

$$\Delta H^\circ_f = 53.01 \text{ kJ mol}^{-1}$$

c) The calorimetric ΔH°_f of C_3H_6 *(g)* is considerably higher that the ΔH°_f based on bond enthalpies. It requires more energy to make C_3H_6 *(g)* than would be expected strictly on the basis of the formation of normal single bonds. Formation of the triangular ring of C atoms forces C−C−C bond angles of 60°, much smaller than the usual tetrahedral angle, 109.5°. The extra energy is the *strain energy* of cyclopropane. It is about 117 kJ mol^{-1}.

---

11. **a)** This is an *esterification.* It resembles in form an acid-base neutralization with the alcohol playing the part of the base:

$$CH_3CH_2CH_2CH_2OH + CH_3COOH \rightarrow CH_3COOCH_2CH_2CH_2CH_3 + H_2O$$

The organic product is the ester *n*-butyl acetate.

b) This is the Cannizzaro reaction:

$$2\ CH_3C(CH_3)_2COH + H_2O \rightarrow CH_3C(CH_3)_2COOH + CH_3C(CH_3)_2CH_2OH$$

The products are 2,2-dimethylpropionic acid and 2,2-dimethylpropanol. The reaction is profitably regarded as a redox system and easily balanced along those lines. The organic reactant disproportionates.

c) The reaction between benzaldehyde and dichromate ion in acid solution is an oxidation-reduction. The oxidation half reaction is:

$$C_6H_5COH(aq) + H_2O(l) \rightarrow C_6H_5COOH(s) + 2\ H^+(aq) + 2\ e^-$$

The reduction half reaction:

$$14\ H^+(aq) + Cr_2O_7^{2-}(aq) + 6\ e^- \rightarrow 2\ Cr^{3+}(aq) + 7\ H_2O(l)$$

Combination of the two gives:

$$3\ C_6H_5COH(aq) + 8\ H^+(aq) + Cr_2O_7^{2-}(aq) \rightarrow 3\ C_6H_5COOH + 4\ H_2O + 2\ Cr^{3+}(aq)$$

d) The ethyl group of the Grignard reagent attacks the carbonyl group of the ketone:

$$CH_3-CO-C_2H_5 + C_2H_5MgBr \rightarrow CH_3-C(C_2H_5)_2OMgBr$$

Treatment with water yields 3-methyl-3-hydroxypentane and Mg(OH)Br. The name 2-ethyl-2-hydroxybutane for this product is inferior.

---

13. ***a)*** $CH_2{=}CH-CH{=}CH_2 + 2\ Br_2 \rightarrow CH_2Br-CHBr-CHBr-CH_2Br$

The product is 1,2,3,4-tetrabromobutane. Br_2 adds across the double bonds.

b) The reaction is the *dehydration* of methanol:

$$2\ CH_3OH \rightarrow CH_3-O-CH_3 + H_2O$$

The organic product is dimethyl ether.

c) The reaction is the hydrolysis of the ester:

$$CH_3-COOC_2H_5 + KOH \rightarrow CH_3-COO^-K^+ + C_2H_5OH$$

The products are potassium acetate and ethanol.

d) The reaction is an esterification:

$$CH_3-CH_2COOH + HO-CH_2-CH_3 \rightarrow CH_3-CH_2-COOCH_2-CH_3 + H_2O$$

The product is ethyl propionate.

---

15. Treat benzene with CH_3Br in the presence of $AlBr_3$. The Lewis acid $AlBr_3$ accepts a Br^- from CH_3Br to create $AlBr_4^-$ and CH_3^+ The methyl carbonium ion then attacks the aromatic ring.

---

17. The "trivinyl chloride" fragment is:

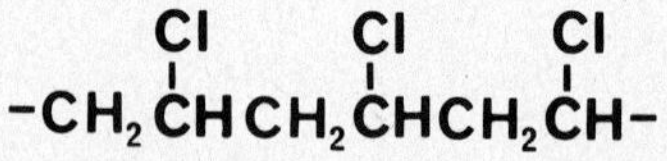

```
     Cl       Cl       Cl
     |        |        |
-CH2 CH  CH2  CH  CH2  CH-
```

$-CH_2CH(Cl)CH_2CH(Cl)CH_2CH(Cl)-$

The double bonds in the monomer disappear as each carbon atom links to a neighboring monomer.

---

19. The problem requires no application of principles, but, instead, reading and comprehension of the descriptive material on biochemistry. Full answers appear on text p. A.84.

---

21. The amino acid phenylalanine has the formula $C_6H_5CH_2CH(NH_2)COOH$, or $C_9H_{11}NO_2$. As it polymerizes to the polypeptide, one molecule of water is lost in the formation of each peptide bond. If n phenylalanine molecules link, the formula is $(C_9H_9NO)_n$ or, more exactly, $(C_9H_9NO)_n \cdot H_2O$. The extra H_2O is present because the number of bonds needed to hold n amino acids together in a linear chain is $n-1$.

The molar weight of C_9H_9NO is 147 g mol^{-1}. To make a polypeptide with a molar weight of 17500 g mol^{-1} would take 17500 / 147 or 119 phenylalanines. The molar weight of the extra H_2O is negligible in this calculation.

---

23. ***a)*** The problem presents five candidate structures of benzene. Structures *i*, *ii* and *iv* would form only one C_6H_5Cl upon reaction with Cl_2 *(g)*. The other candidate structures would undergo substitution in at least two distinct locations and give mixtures of isomeric C_6H_5Cl's.

b) Only structure *iv* gives exactly three isomers of formula $C_6H_4Cl_2$. They are the 1,2-, 1,3-, and 1,4- disubstituted benzenes. Structure *i* would give *two* 1,2-dichlorobenzenes. One would have a double bond between the adjacent Cl atoms and the other would not. There would be *two* 1,3- products along with one 1,4- dichlorobenzene for a total of five dichlorobenzenes. Structure *ii* would give only two $C_6H_4Cl_2$'s. Structure *iii* would give five $C_6H_4Cl_2$'s: 1,1-, 1,2-, 1,5-, 1,6-, and 2,5-.
Finally, structure *v* has no fewer than nine possible products of dichloro substitution. These include two pairs of *cis-trans* isomers. To draw them, first copy structure *v* so that its mirror plane of symmetry is emphasized. (See adjoining figure). Then replace 2 H atoms with 2 Cl atoms in a systematic way.

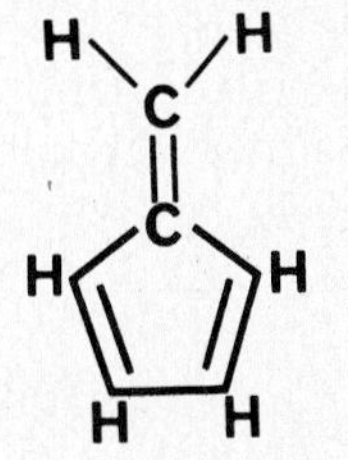

---

25. The unsaturated hydrocarbon must have a straight-chain skeleton of 6 carbon atoms because it gives *normal* hexane, or *n*-hexane, when reduced with hydrogen gas. When oxidized it is split in a four carbon acid (butanoic acid) and a two carbon acid (acetic acid). The double bond is therefore in the 2 position: $CH_3-CH_2-CH_2-CH=CH-CH_3$. The compound is 2-hexene. This compound has *cis* and *trans* isomers. There is no way to conclude from the available data which isomer is present. The balanced equations are:

$$C_6H_{12}(l) + H_2(g) \rightarrow C_6H_{14}(l)$$

$$27\ H^+(aq) + 9\ MnO_4^-(aq) + 5\ C_6H_{12}(l) \rightarrow$$
$$9\ Mn^{2+})(aq) + 5\ C_4H_8O_2(aq) + 5\ C_2H_3O_2(aq) + 16\ H_2O(l)$$

---

27. The unknown compound has empirical formula $C_{10}H_{10}$ and molecular weight of 130. The molecular formula of the unknown is therefore also $C_{10}H_{10}$ since 10 C (with atomic weight 12) and 10 H (with atomic weight 1.0) add up to 130. The compound adds 1 mol of both Br_2 and H_2, suggesting that it contains one double bond:

$$C_{10}H_{10} + Br_2 \rightarrow C_{10}H_{10}Br_2$$

$$C_{10}H_{10} + H_2 \rightarrow C_{10}H_{12}$$

The compound gives phthalic acid, 1,2-benzenedicarboxcylic acid, when oxidized with $KMnO_4$. Knowing the formulas and structure of phthalic acid are essential to solving this problem. The structure is given on text p. 755. The name on that structure is *o*-phthalic acid. When the prefix is omitted from the name the *ortho* isomer of benzenedicarboxylic acid is always intended. See, for example, the tables of organic compounds in the *Handbook of Chemistry and Physics*. The molecular formula of phthalic acid is $C_8H_6O_4$. The equation for the oxidation is:

$$12\ H^{+}(aq) + 4\ MnO_4^{-}(aq) + C_{10}H_{10}(l) \rightarrow 4\ Mn^{2+}(aq) + 8\ H_2O(l) + C_8H_6O_4(s) + 2\ CO_2(g)$$

The original compound is named 1,4-dihydronaphthalene. In the accompanying structure of this compound every C atom (at the intersections of the lines) has one hydrogen atom attached, except the top and bottom carbon atoms (positions 1 and 4) in the right-hand ring have two hydrogen atoms each and the two C's where the rings join no H atoms.

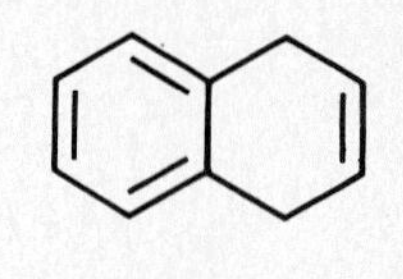

---

29. There are 20 amino acids in text Table 22-5. Any of the 20 could be in the first position of a 10 unit polypeptide. Also, any of the 20 could be in the second position, etc. The two ends of the polypeptide chain are different which means that polypeptides that have the same order of components in opposite directions are distinguishable. Hence there are 20^{10} or 1.024×10^{13} possible polypeptides.

If the chain is lengthened to 100 units, there are 20^{100} or 1.27×10^{130} possible polypeptides.

---

31. The polypeptide contains six amino acids: 2 Ala, 2 His, 1 Lys and 1 Trp. It is a hexapeptide. Hydrolysis breaks the peptide bonds linking the amino acids. Partial hydrolysis cleaves somes bond and leaves others untouched. It operates at random, and in this case four different dipeptide sequences are isolated. They are:

Ala−Trp His−Ala Lys−Ala Trp−His.

These sequences are *not* the same as their reversals, thus, Ala−Trp differs from Trp−Ala. Number the six positions in the hexapeptide:

(amino end) 1 2 3 4 5 6 (acid end)

The lysine is known to be at one end. Suppose it is in position 1. Then, reading from the order of the dipeptide sequences, Ala is at 2, Trp is at 3, His is at 4, Ala is at 5 and His is at 6, the other end. That is:

Lys−Ala−Trp−His−Ala−His

This sequence provides all four of the observed dipeptide fragments. It also predicts an Ala−His dipeptide fragment, which is not reported. However, its absence could be due to experimental problems of detection. It cannot be concluded that it was not formed.

Now, keep Lys at position 1, put a His at position 6 and work from there down. The result is a second answer:

Lys−Ala−His−Ala−Trp−His

which, when broken at all possible points to give dipeptides, predicts the observed four products and the unobserved fragment Ala–His as well.

If His is placed at the low end and Lys at the high end, the sequence:

His – Ala – Trp – His – Ala – Lys

results. This sequence would *not* however provide any Lys–Ala dipeptide and must be excluded.

---

33. The muscles resort to *glycolysis*, the anaerobic break-down of glucose, as a source of ATP^{4-} (and its free energy) during periods of violent muscular exertion. There is an insufficiency of oxygen to allow more efficient aerobic metabolism of glucose. The product of anaerobic metabolism is lactic acid, which accumulates in the muscles.

---

35. Hemoglobin has a molecular weight of 65000 g mol^{-1}. Consider 1 mol of hemoglobin. It weighs 65000 g and contains 0.00344×65000 g or 223.6 g of Fe. The molar weight of iron is 55.85 g mol^{-1} so the 223.6 g of Fe is 4.00 mol. The hemoglobin molecule contains 4 atoms of Fe.